Masters of All They Surveyed

Masters of All They Surveyed

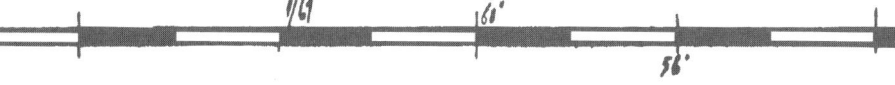

Exploration, Geography, and
a British El Dorado

D. Graham Burnett

THE UNIVERSITY OF CHICAGO PRESS
CHICAGO AND LONDON

D. Graham Burnett is assistant professor in the Honors College and a member of the history of science program at the University of Oklahoma.

The University of Chicago Press, Chicago 60637
The University of Chicago Press, Ltd., London
© 2000 by The University of Chicago
All rights reserved. Published 2000
Printed in the United States of America
09 08 07 06 05 04 03 02 01 00   1 2 3 4 5

ISBN: 0-226-08120-6 (cloth)

Library of Congress Cataloging-in-Publication Data

Burnett, D. Graham.
    Masters of all they surveyed : exploration, geography, and a British El Dorado /
D. Graham Burnett.
      p.   cm.
    Includes bibliographical references and index.
    ISBN 0-226-08120-6 (cloth : alk. paper)
    1. Cartography—Guyana—History—19th century.   2. Traverses (Surveying).
3. Guyana—Description and travel.   4. Guyana—History—1803–1966.
5. Schomburgk, Robert H. (Robert Hermann), Sir, 1804–1865—Journeys—Guyana.
I. Title.

GA718.7 .B87 2000
526'.09'034—dc21
                                                                                99-098199

Frontispiece. A panoramic vignette of Roraima and its neighboring summits, from the *Twelve Views*. Courtesy of the Yale Center for British Art, Paul Mellon Collection.

♾ The paper used in this publication meets the minimum requirements of the American National Standard for Information Sciences—Permanence of Paper for Printed Library Materials, ANSI Z39.48-1992.

Just as none of us is outside or beyond geography, none of us is completely free from the struggle over geography. That struggle is complex and interesting because it is not only about soldiers and cannons but also about ideas, about forms, about images and imaginings.

Edward Said,
Culture and Imperialism

CONTENTS

Preface		xi
1	Geography, Exploration, Colonial Territory: An Introduction	1
	POINT OF DEPARTURE 1	
	THE LAY OF THE SCHOLARLY LAND 5	
	A SKETCH OF THE ROUTE 13	
	THE DESTINATION: A LITTLE BACKGROUND ON BRITISH GUIANA 17	
2	MYTHS AND MAPS: MAKING EXPLORERS AND EMPIRES	25
	MYTHICAL EMPIRE AND INTERIOR EXPLORATION 1: FINDING EL DORADO 25	
	MYTHICAL EMPIRE AND INTERIOR EXPLORATION 2: REACHING EL DORADO 33	
	REMAPPING AND THE MAKING OF EXPLORERS 37	
	REMAPPING AND THE MAKING OF TERRITORY 46	
	THE OTHER MYTHICAL EMPIRE: MAHANARVA 54	
	EPILOGUE: THE AMBIVALENCE OF MYTH AND METALEPSIS 60	
3	Traversing Terra Incognita: Getting There and Making Maps	67
	ELUSIVE INCOGNITAE 67	
	LOCAL REORIENTATION AND DISORIENTATION 71	
	MAPPING TERRA INCOGNITA: THE TRAVERSE AND ITS WEAKNESSES 84	
	FIXED POINTS 1: THE HUMBOLDTIAN TRADITION 92	
	FIXED POINTS 2: DISCIPLINE AND NAVIGATION 99	
	FIXED POINTS 3: LANDMARK, IMAGE, ENCLOSURE 109	
4	Marks on the Land: Landmarks, Aesthetics, and the Image of the Colony	119
	VIEWS IN CONTEXT: WRITING THE JOURNEY 119	
	VIEWS AND STRATEGIES OF IMPERIAL VISIBILITY 126	
	PICTORIAL CONVENTIONS AND LANDMARK DEPICTIONS 131	
	SCHOMBURGK, BENTLEY, AND THE AESTHETIC SATURATION OF SITES 133	
	AESTHETIC CIRCUITS 145	

From Landmark to Icon 148
Epilogue: Widespread Views 161

5 Marks of Passage: Landmarks and the Practice of
   Geographical Exploration   167
   Re-viewed 167
   Tracks of Passage 169
   Landmarks and Methexis 174
   Amerindian Negotiations 181
   Icarian Anxieties: The Fall as Landmark 189
   Pictorial Conventions and Exploratory Subversions 194
   Epilogue: Place and Erasure 197

6 Boundaries: The Beginnings of the Ends   199
   Disputed Boundaries/Disputed Man 199
   Delimitation, Demarcation, "Deconstruction"? 202
   Consummatio, Landmarks, and the "Natural" Boundary 208
   Traversing the Boundary: The Surpassing Problem 214
   Narrative Part 1: Northwest 216
   Narrative Part 2: Southwest 228
   Narrative Part 3: East 242
   Epilogue: The Wake of the Boundary 253

7 Conclusion: History, Geography, Nation   255
   Up to This Point 255
   And Over the Line 257
   Epigram 265

   Bibliography   267
   Index   285

ILLUSTRATIONS

FIGURES

1	Seal of the colony	20
2	Horstman's sketch map	32
3	Schomburgk's Ralegh map	38
4	Van Heuvel's map of Parima	42
5	Hondius's map of Guiana	43
6	Jansson's map of Guiana	44
7	Van Heuvel's map of Parima (detail)	45
8	British Library Add. MSS 17940a	63
9	Hilhouse's general chart	70
10	Schomburgk's first proposal map	72
11	Schomburgk's second proposal map	73
12	Alexander's map (RGS)	75
13	Hilhouse's general chart (detail 1)	76
14	Hilhouse's general chart (detail 2)	77
15	Arrowsmith's map (RGS)	85
16	Hancock's sketch map	89
17	King William IV Cataract	115
18	Robert H. Schomburgk	121
19	*Victoria regia* house at Chatsworth	151
20	*Victoria regia* on show	153
21	Downing's *Victoria regia*	155
22	Brett's copies from *Twelve Views*	157
23	Brett's Roraima from *Twelve Views*	158
24	Tallis atlas map (details)	162
25	Hilhouse's map of the Mazaruni	173
26	"Palace of the Maranika-mama"	187
27	Wonotobo Falls	197
28	Abandoned Brazilian chapel	203
29	Schomburgk's northwest map	218
30	Victoria Point	219
31	The range of the boundary	227

| 32 | Sketch map of the Takutu | 235 |
| 33 | "All Guyana Is Ours to Defend" | 259 |

PLATES (FOLLOWING PAGE 112)

1 View of Pirara
2 *Victoria regia*
3 *Twelve Views,* frontispiece
4 *Twelve Views,* key map
5 Comuti or Taquiari Rock
6 Ataraipu, or Devil's Rock
7 Puré-Piapa
8 Roraima
9 Esmeralda
10 Fort São Gabriel
11 *Twelve Views,* key map (detail)
12 British Guiana stamps
13 Kaieteur Falls
14 Tallis atlas map
15 Bentley's Christmas Cataracts
16 Kundanama junction
17 Christmas Cataracts (detail)
18 Surveying scene near the Takutu

PREFACE

We send text into the world at our peril. Once the words have left our hands it falls to others to assess their significance, to construe their meaning. This must be on any author's mind at that anxious moment when he releases his grip on the pages, but no one knew it better than explorers, who drafted their narratives, compiled their reports, drew their maps, and slipped the whole lot into the mailbag with fingers crossed: their fortune lay in the hands of distant readers. Some of those readers, as it turned out, were less distant than others. I am reminded of the richest moment in researching this book, a marginal find too speculative and evanescent for scholarly significance, but that for a moment conjured up for me all the mysterious power of geographical exploration and cross-cultural encounter.

Early in 1842 an explorer in northeastern South America—a member of a boundary-surveying expedition working its way through the rough jungle tableland near the modern intersection of Brazil, Venezuela, and Guyana—complained in his journal about the sudden updrafts of warm air that would race through the canyons. These had the unfortunate effect of interrupting his work by sweeping away whatever papers he had out on his makeshift camp desk. These sheets, he wrote, would "spiral high up into the air, whence, to the very great chagrin of myself, but, on the other hand, to the intense enjoyment of the Indians, they would be carried on the breeze in all directions, often many miles away." From time to time the advance parties of the survey found that these missives of their own presence had preceded them: a page would appear up ahead on their route, flapping in a tree; they watched one sheet get sucked up the mountain face of Roraima, disappearing several thousand feet overhead.

This sibylline scenario has a charm all its own, suggesting nothing so much as a scattering aeolian breath puffed into the imperial archive, to the delight of colonial subjects. But texts, even airborne leaves, have consequences. Imagine my dark and queasy surprise as I read on and learned that fragments of these floating signatures turned up later in potent amulets among the apocalyptic Amerindians of Ibrima Yeng, where a messianic cult led by Awacaipu (one of the former porters of the boundary expedition) culminated in the mass suicide of several hundred indigenous people. In this circuitous way, it seems, colonial cartography claimed its first casualties. The story itself is a mere fragment, but drifting out of the nether reaches of a footnote, it became for me a kind of amulet in writing this book. In its opaque surface I thought I caught glimpses of the whole story I sought to tell: the implications of science displaced; maps and their consequences; the multiple meanings of exploration; the charged ambivalence of cultural exchange; the contingencies of a recalcitrant environment.

Masters of All They Surveyed explains how European explorers turned areas they called terra incognita into bounded colonial territories. In this sense it is an attempt to tell the story of how one went from one view of a foreign place (a view of the shore from the pitching deck of a ship or the prow of a canoe; a view of water, mud, and thick green) to another (a top-down image provided by a map, a map that defined a colonial possession). Another way to put it is to say something like this: if you are curious about how a soft-footed European gentleman, predisposed to seasickness (and unable to swim), made his way through thick jungles, arid savannas, and forbidding mountain ranges for the better part of a decade and emerged with a map that could be waved about among diplomats and military men in the metropolis, then this is the book for you. There were more such men than one might expect, and what they did had profound effects on the history of many parts of the world. What they did changed the way we look at the globe.

The great nineteenth-century oceanographer Charles Wyville Thomson wrote that in the early phases of deepwater exploration some researchers believed everything thrown into the sea sank to a particular level and then hung there, in suspension, in a subaquatic layer of similar articles: descending, one would have to pass through watery belts of jetsam—all lost shoes, all pocket change, all bones. I sometimes think of libraries this way: everything gets thrown in and sinks into textual layers, each stratum defined by call numbers and disciplinary isolines. Pitching this book overboard, I wonder where it will come to rest. As a text that examines the history of mapmaking and geography, *Masters of All They Surveyed* is a history of science; since the science in question operated at the colonial periphery, the history of empire is a central preoccupation. Having dragged my net though anthropology, art history, cultural geography, postcolonial studies, and scholarship on travel literature, it would be a great pleasure if this book could bob back through these regions as well. We take much more from the sea than we return.

Explorers fail places. I am more persuaded of this than ever. On preparing to come home, one senses one has left too much of the important stuff behind. Historians fail places too, and I cannot shake the fear that I have failed Guyana, the country that serves as one focus of this book. Too much remains in notes, in drafts, in the material that was hewn away in telling this story. An instance: When I first began this research, I stumbled on an early account of a strange upriver happening in the region—every once in a great while, a giant manta ray (disoriented? sick?) would make its way up the Demerara River and, reaching the upriver falls, begin to breach. The image of one of those distressingly sinister black rays, twelve feet across, leaping out of a creek in the jungle gave me a sense that Guyana would be like no place on earth. In this I was not wrong, but I was right for many of the wrong reasons. I hope those who know the place better than I, those who call it their home, will look with patience on this work, written by a visitor all too conscious of the fibs and slips employed by those who try to catch places on paper: in the hopes of finding "one good place" (our own El Dorado), we project coherence; in fear, we attend to our own borders, closing much out, taking in with fastidious care. This book is about these facts, but it surely reflects them as well.

I have incurred many debts in the course of this project, which began as doctoral research at the University of Cambridge. Can one thank a book? Perhaps not, but if I could I would single out Greg Dening's inspiring *Islands and Beaches.* When I was seriously considering abandoning the whole business, a kind spirit pointed me to that text, which changed my sense of what was possible in a scholarly book. I have never met Dening, but if this text ever finds its way into his hands, let me thank him here. Jim Secord, my adviser, continued to offer me his gentle guidance even after I left England. I am grateful. I would not have had the privilege of working at Cambridge without the support of a British Marshall Scholarship in 1993–95, and I would not have been awarded the Marshall without the training and support of the Princeton professors who where my recommenders. Particular thanks go to Tony Grafton, Gerry Geison, Mike Mahoney, and Gyan Prakash. After three lectures in Mike Mahoney's scientific revolution class, a freshman showed up during his office hour and said he wanted to be a historian of science. Graciously, Professor Mahoney took him at his word. Later, at Cambridge, I benefited from work with Simon Schaffer, Michael Bravo, and Stephan Collini, among others. Richard Drayton steered me into this material (which Francis Herbert and the other RGS archivists and librarians aided me in discovering), gave my first effort its most substantial critique, and kindly helped me establish myself in New Haven. The Yale community welcomed me.

The Program in Agrarian Studies, Jim Scott, and Kay Mansfield were superb resources, as were the other fellows in the program. Frank Turner and Emília da Costa offered me helpful criticism. The History of Medicine Section and Larry Holmes were patient with my comings and goings, afforded me a departmental affiliation, and supported my teaching. The librarians at Sterling, Mudd, Kline, and the map collection, as well as the staff of the Yale Center for British Art, assisted me a great deal. My students, particularly those in "Terra Incognita: Maps, Explorers, and Encounters, 1410–1900," stimulated and sharpened my thinking, and work on *The Mapmaker* with Tom Kelly and Josh Hanson was an added pleasure, giving this research a fresh theater. Thanks also go to the Society of Fellows in the Humanities, at Columbia University, and the Center for Scholars and Writers, at the New York Public Library.

My time in Guyana deeply influenced this work. Numerous funding sources made that trip possible, including Trinity College's Wyse Fund for Social Anthropology, the American Friends of Cambridge University, the Smuts and Sir Bartle Frere funds, and a number of other bursaries. Raleigh International subsidized my stay in return for lectures. For this (fruitful) relationship I would like to thank Bijan Nabavi, Jonathan Cook, Ian Purves, Andy Roe, and Stephen Smith. Without the help of Terry Roopnaraine (and his father, Rupert), Guyana would have been a much rockier landing. Their introductions and contacts shaped my encounter with Georgetown and the interior alike. Thanks also go to Bobby Fernandez, Desrey Fox, Nurse Sago, Diane McTurk, the Lancasters, Nigel Westmass, Amai Williams, and the staffs at the Caribbean Research Library and the National Archives. Sister Mary Noel Menezes is one of the most inspiring people I have ever met. Not only did she ensure that I had unlimited access to the historical resources of Guyana, she reviewed my project with her characteristic energy, then brought out sheaves of her own relevant materials, notes, and books—and gave me everything. Her generosity and spiritual intensity made Guyana more than a scholarly experience.

As the project came together, a number of scholars and friends offered me their wisdom and assistance, including Danielle Allen (who lost a week of her life on this text), Dorinda Outram (who offered Dening at a critical moment and rescued ill-formed drafts), Ray Craib, Chris Mooney, Matthew Edney, Jim Schulz, Aaron Hirsh, David Woodward, the readers on the Nebenzahl Prize Committee, and Penelope Kaiserlian at the University of Chicago Press. I am deeply grateful for their attentions, as I am for the generous and learned readings of Peter Rivière, Neil Whitehead, Richard Grove, and an anonymous referee for the Press. These readers (and others) saved me from many gaffes. For those that remain I must take full credit.

A small circle of people made this work possible by making its author possible. My mother and my father take pride of place in that category, but I wonder what would have become of me without my sister, Maria. To these, my family, I owe a deep debt of gratitude. To my beloved Christina, who braved Anegada on a metaleptic honeymoon, I can say only "Cuánto te habrá dolido . . ."

CHAPTER ONE

Geography, Exploration, Colonial Territory:

An Introduction

> The surveyor straightens from his theodolite.
> "Spirit-level," he scrawls, and instantly
> the ciphers staggering down their columns
> are soldier ants, their panic radiating in the shadow
> of a new god arriving over Aztec anthills.
>
> The sun has sucked his brain pith-dry.
> His vision whirls with dervishes, he is dust.
> Like an archaic photographer, hooded in shade,
> he crouches, screwing a continent to his eye.
>
> Derek Walcott, *Guyana*

A POINT OF DEPARTURE

At the end of the nineteenth century, when disputes over the boundary between Venezuela and British Guiana looked likely to draw the United States into conflict with Great Britain, President Grover Cleveland delivered a strongly worded call for congressional appropriations. He wanted money for a full investigation of claims to the disputed territory.[1] Things were getting ugly. Hastily cobbled together, a quintet of American scholars and diplomats received the charge "to determine . . . the true divisional line between the Republic of Venezuela and British Guiana."[2] Given United States willingness to stand on the recommendation of the committee—and given British resistance to sacrificing the gold-rich territory of its only colonial possession on the South American continent—the path of an obscure boundary through nearly uninhabited terrain was drawing two world powers toward war.[3] The assignment to find the

---

1. Braveboy-Wagner, *Venezuela-Guyana Boundary Dispute,* 104. For other summaries, see Simpson and Fox, *International Arbitration;* Prescott, *Political Frontiers and Boundaries;* and Rout, *Which Way Out?*

2. *Special Message of President Cleveland to the Congress,* 17 December 1895. Cited in *Official History,* 375.

3. De Villiers, "Foundation and Development of British Guiana," 22. See also Cleveland's closing remarks in his address to Congress on 17 December 1895: "It will in my opinion be the duty of the United States to resist by every means in its power." *Official History,* 379.

boundary, then, had about it all the gravitas of maintaining world peace, not to mention the patriotic fervor of an increasingly truculent United States imperialism. The well-heeled of Washington, however, spotted a distinctly comical side to the affair. Ribbing the committee members for the unseemly tropical spat they had been charged to resolve, the gentlemen of an Ivy League club in the capital presented each of the five with an outlandishly oversized package of quinine and a bottle of good sour mash Kentucky bourbon at a dinner held in their honor.[4]

This book is, in a way, about what made that gag seem funny.[5] The Yale Club, Washington, D.C., 1896: George Lincoln Burr, chair of the committee, unwraps apothecary pots on the podium. Guffaws. What is everyone laughing about? The joke, of course, was that the U.S. Boundary Committee, charged with locating the "true divisional line" running across the jungle between the Essequibo and Orinoco Rivers, had no intention of looking for that line anywhere near the equator. The line was nowhere to be "found" in the forests and swamps of the Orinoco delta. The boundary was not on the land, but in the brittle pages of documents: in dusty dispatches to the Dutch West India Company at The Hague; in the archives of the Spanish Capuchin missionaries at the Vatican; in the marginalia on manuscript reports of an anglicized Prussian geographer that were carefully filed in the Colonial Office, London; in the map library of the Royal Geographical Society.[6] Setting out to find the boundary meant equipping for a steamer trip across the Atlantic and lengthy stays in the great cities of Europe. It meant letters of introduction, not pith helmets; a keen eye for watermarks, not anacondas; cufflinks, not quinine.

And yet the historical maps that became important pieces of evidence in the disputes, as well as the map on which the arbitration tribunal drew the boundary after all the evidence was in, were products of geographical exploration on the ground. The joke played on this, drawing on the tacit understanding that the messy work of slogging through the swamps was done; earlier exploratory passages had made it possible for the gentlemen-scholars to sort out the issue of the boundary in libraries. This was obvious to everyone. Their shared laughter reflected the broad understanding that colonial territory came into being as a result of the passage of certain individuals—explorers and surveyors—who made distant land possessable by means of a set of powerful linked texts. Diplomats explored those texts, not jungles, to see who had a right to what. There can hardly be a more concise expression of the collective magnitude (and

4. Burr, "Search for the Venezuela-Guiana Boundary," 470. For a brief discussion of Burr's role in the dispute, see Wright, *Human Nature in Geography*, 183–84.
5. On reading jokes, see Darnton, "Workers Revolt," 81.
6. Abbreviated from here on as RGS.

success) of this project of possession than the simple statistic that serves as a point of departure for Edward Said's *Culture and Imperialism:* by the end of the nineteenth century, Europeans and their overseas descendants claimed sovereignty over some 80 percent of the land on earth.[7] This book seeks to explain just what those territorial claims meant and how they arose.

The world is big and books are small. In going after a big question like this one, some limits to scope are in order. In these chapters, then, I examine the "geographical construction" of one of these colonial territories, British Guiana, with an emphasis on the nineteenth century.[8] During this period scientific explorations and cartographic surveys turned a region that Europeans called terra incognita into a mapped and bounded colony. I show how geographical exploration of the region proceeded and how explorers produced key texts—maps, views, narratives—that constituted the foundation of a nineteenth-century British image of their unique colony in South America.[9]

But who cares? Or to put it differently, Who might care? Some readers may have come to this book wanting to learn about Guyana, that fabulously curious Caribbean nation that lies on the northern coast of South America. To them let me extend the warmest welcome: I hope you will not be disappointed. I hope also, however, that you will not be alone, because this book has been written not by a proper Caribbean historian, but by a historian of science with a particular interest in geography, exploration, and cross-cultural encounter. My main aim in researching and writing this book has been to understand the place of maps, explorers, and geographical knowledge in the history of imperialism. In this effort I am by no means alone, and I situate myself with respect to this growing literature in more detail in the section that follows. For now, suffice it to say that I am particularly interested in how certain explorers—lightly equipped, underfinanced, more or less solitary—struck out into difficult environments, spent months and even years wandering in the interiors of continents far from home, and came back with maps, maps that ended up on negotiating tables in London, Paris, and Berlin.

---

7. Said, *Culture and Imperialism,* 6.

8. I observe the following conventions with respect to the spellings of the region: "Guiana" is used to denote the region under Dutch, French, or British colonial authority; for the geographical area between the Orinoco and the Amazon, I follow Whitehead ("Carib Ethnic Soldiering," 379 n. 2) in referring to "Guayana"; "Guyana" indicates the modern South American nation. Maintaining consistency across citations is impossible, since nineteenth-century authors used a variety of spellings. I have left all quotations in the original. For more on orthographic entanglements for historians of the region, see Costa, *Crowns of Glory,* 295, and Whitehead, *Lords of the Tiger Spirit,* 200 n. 1.

9. For the particular importance of the map in the formation of such conceptions, see Taylor, "Official Geography and the Creation of 'Canada,'" 1–3, and Edney, *Mapping an Empire,* 2.

Because this kind of exploring work (called a "traverse survey") was the ubiquitous form of geographical investigation in the colonies during the nineteenth century, a better understanding of it will prove significant to those interested in regions beyond Guyana, beyond South America. In the best of all possible worlds, bits of this book will prove interesting to a broad range of scholars: historians of science will, I hope, see the enterprise as firmly situated in that field; to the imperial historian and the historian of cartography the relevance ought to be relatively direct. There are others: a large portion of the text is devoted to the close reading of landscape imagery in the colonial context, material decidedly art historical in approach; cultural geographers may think of this as a nonhostile reconnoiter in their domain; ethnohistorians and anthropologists will see that I have struggled with the problem of incorporating Amerindian realities into the narrative of European territorial expansion. In addition, much of this book is concerned with travel literature and narratives of exploration, literary genres with their own critical communities. Finally, I would be happy indeed if my culminating emphasis on landmarks, landmarking, and boundary making caught the attention not only of those considering boundary formation in the colonial and postcolonial context generally, but even of an architect or environmental historian.[10] This wish list of readers is fantasy, perhaps, but I have visited all these disciplines and hope not to have returned empty-handed.

Putting disciplines and scholars aside for a moment, I should add that there is a story here too, and something of a main character as well. Robert Herman Schomburgk, the explorer who proves so important in the geographical construction of the colony, will shortly be the subject of a full biography (by Oxford anthropologist Peter Rivière).[11] The present book is by no means biographical, but readers who push through to the end will have seen a fair bit of this most interesting man—will have made a traverse with him, both geographical and (I hazard) spiritual. For my part, I have lived with this material—with Schomburgk, with El Dorado, with the maps—too long to know for certain if a thinking person, intrigued by exploration but not a specialist or a scholar, could pick up this text and read it, savoring the ironies of colonial hubris, the charged moments of solitude and risk, the ambiguous passing incidents that suggest the bottomless complexity of the encounter between people who cannot understand one another. I can say only this: if that never happens, I will consider the project to have failed in a very significant way.

---

10. I use the term "postcolonial" guardedly. See McClintock, "Angel of Progress."
11. A work in progress (which I have not consulted), announced in Rivière, *Absent-Minded Imperialism,* 28 n. 6.

THE LAY OF THE SCHOLARLY LAND

Before turning to the stuff of this account, however, I need to review briefly how this book fits with respect to some other recent work on imperialism, science, geography, exploration, and mapping. Most studies of British imperialism have focused on the political and economic structures that sustained the empire or on the ideologies that gave it shape and meaning. If an early narrow definition of empire led historians of diplomacy and politics to focus on the evolution of the formal empire—that "area painted red on the map"—then subsequent closer attention to economics led them to see the cartographic boundaries of empire as something of a historical red herring.[12] Maps were not the story. Rather than limit their inquiry by nineteenth-century boundaries, John Gallagher, Ronald Robinson, and others drew attention to the extensive "informal empire" of economic dependents and debtor nations that might not have been red on the map but were *in the red* in the minds of metropolitan industrialists and bankers. Whether such an informal empire existed, and whether the empire drew its primary support from local elites (at the periphery) or metropolitan figures (at the center)—these are some of the broader questions that have driven academic debates among historians of imperial Britain over the past forty years.[13]

The relative significance of center and periphery became a key point of disagreement among historians of science and technology who turned to the relation between science and empire.[14] Although initial contributions of historians of science to imperial history were primarily instrumental histories drawing attention to the technologies that made empire possible, the cross-pollination of science studies and imperial history has given rise to a jungle of lush and hardy hybrids.[15] Nearly every scientific discipline, from physics to ethnography, botany to astronomy, has been

---

12. Cain and Hopkins, *British Imperialism*, 6.
13. For a recent review article see Wolfe, "History and Imperialism."
14. Basalla, "Spread of Western Science"; Gizycki, "Center and Periphery"; MacLeod, "On Visiting the Moving Metropolis"; and MacLeod, "Passages in Imperial Science."
15. Relevant literature includes Headrick, *Tentacles of Progress;* Headrick, *Tools of Empire;* Adas, *Machines as the Measure of Men;* Sangwan, *Science, Technology and Colonisation;* and Sangwan, "Technology and Imperialism in the Indian Context." See also Reingold and Rothenberg, *Scientific Colonialism;* MacKenzie, *Imperialism and the Natural World;* Kuklick and Kohler, *Science in the Field;* Petitjean, Jami, and Moulin, *Science and Empires;* Bell, Butlin, and Heffernan, *Geography and Imperialism;* Osborne, *Nature, the Exotic, and the Science of French Colonialism;* Brockway, *Science and Colonial Expansion;* MacLeod and Lewis, *Disease, Medicine and Empire;* Baber, *Science of Empire;* Grove, *Green Imperialism;* and Drayton, "Imperial Science and a Scientific Empire." Review essays include Drayton, "Science and European Empires," and Schaffer, "Visions of Empire." Some specific studies include Nicholson, "Alexander Humboldt, Humboldtian Science"; Stafford, "Geological Surveys"; Schaffer, "Empires of Physics"; and Browne, "Science of Empire."

situated in its imperial context, a project often accompanied by the discovery that substantial gaps in the evolution of such disciplines were being filled in the process. Early ideas about the "diffusion" of science from metropolis to periphery came to be revised. Initially expecting to find anthropology or plant collecting being used as hegemonic tools by imperial agents, historians of science and empire have taken long strides toward showing that such disciplines were in many cases as much (or more) the *result* of the imperial relationship as its cause. Ideas, instruments, whole disciplines appear to have been carried back to Europe in the mental and physical baggage of European travelers as often as they were borne dutifully the other way.

Since Said focused scholarly attention on the process by which colonial discourses produced the colonial "other" as an object of thought, the map has gradually found its way back into the study of empire. "Imaginative geographies" and the "projections" of alien space made possible by cartographic technology (and the associated geographical sciences) have provided an exemplary arena for exploring how the representational productions of empire produced a diorama of the world, a stage for dramatic imperial gestures.[16] Those "areas painted red" gained a new lease on life as objects of scholarly interest as aesthetic, cultural, and discursive approaches to the study of empire emerged: by ordering chaotic spaces maps created imperial places; by making distant places visible they satisfied the scopic and gnostic drives of a conquering people; by abetting territorial control in practical ways they made colonies into large-scale Benthamite panopticons; by providing a textual base map they enabled European nations to inscribe their ambitions on inaccessible places; by making places portable they conformed to (and even exemplified) the Latourian notion of the immutable mobile. Their collation, storage, and maintenance made them particularly good instances of the texts that composed the imperial archive.[17]

This recent interdisciplinary and "spatial" approach to the history of empire has taken much of its momentum from a burst of innovative studies of cartography that began in the mid-1980s with the work of Brian Harley, Denis Wood, and others who attempted to pull the traditionally staid field of cartographic history into the tumultuous arena of the post-

16. Wolfe, "History and Imperialism," 409. See Said, *Orientalism,* 50 and passim; Cook, "Reconstruction of the World"; Huggan, "Decolonizing the Map"; Harley, "Victims of a Map"; Harley, "Maps, Knowledge, and Power"; Stone, "Imperialism, Colonialism and Cartography"; and Edney, "Patronage of Science and the Creation of Imperial Space."

17. "Donnant à voire, la carte donne à posséder." Lestringant, "Fictions de l'espace brésilien," 207. See also Latour, "Les 'vues de l'esprit,'" 21. For a discussion of the empire as textual archive, see Richards, *Imperial Archive.* For a detailed account of an atlas that served as an "imperial archive," see Herbert's study of the *London Atlas of Universal Geography:* Herbert, "'London Atlas,'" 116.

structuralist revolution. Tracing burnish marks and plate changes went out of vogue, and in came a flurry of articles reexamining maps as "texts," with scholars like John Pickles, John Fels, and Barbara Belyea bringing the techniques of Barthes, Derrida, and Foucault to bear on cartographic artifacts from Europe and elsewhere.[18] Harley was one of the earliest voices in this project, and his influential 1989 paper "Deconstructing the Map"[19] typified map historians' new questioning of a broad range of cartographic assumptions: not least that maps were "straightforward, value-free, accurate transcripts of the earth's features."[20] Harley's commitment to the project bore fruit in his collaboration with David Woodward to produce the first volumes of a multivolume reference work, *The History of Cartography*. The new program in map history achieved its most accessible and popular articulation to date in Denis Wood's *The Power of Maps*, published in conjunction with the author's curatorship of the eponymous exhibition at the Cooper-Hewitt Museum in New York City in 1993. The book and the exhibition, both dynamic and well received,[21] could be considered the elaboration of Wood's italicized assertion at the beginning of an ideologically charged article published in *Cartographica* the same year: "The truth is . . . *maps are weapons.*"[22]

This trend offered those interested in empire a particularly useful tool.[23] As Harley suggested in a brief section on maps and empire in his 1988 article "Maps, Knowledge and Power," maps carried a particularly significant burden in the business of empire making, where they not only served "as much as guns and warships" as the "weapons of imperialism" but also legitimized, symbolized, and mythologized. Maps had been obvious symbols of empire to imperialists themselves, but it took the development of a theoretical apparatus before maps could be folded back into the critical history of empire. This new critical attention to maps also provided a new tack for solving a problem that had dogged historians of empire since the abandonment of the Whiggish histories of the 1960s: What was to be done with explorers? In the hagiographic treatments of early imperial his-

18. The most extensive theoretical reappraisal of the history of cartography is in French: Jacob, *L'empire des cartes.*
19. Harley produced several other contemporaneous pieces that outlined an alternative approach to cartography, emphasizing the map as a Geertzian "thick text" demanding context-sensitive interpretation. See Harley, "Historical Geography and the Cartographic Illusion," and Harley, "Texts and Contexts in the Interpretation of Early Maps."
20. Belyea, "Images of Power," 3.
21. With the exception of Belyea's polemical review "Images of Power," published with Wood's rejoinder in *Cartographica.*
22. Wood, "How Maps Work," 66.
23. Recent monographs include Mignolo, *Darker Side of the Renaissance;* Mundy, *Mapping of New Spain;* Edney, *Mapping an Empire;* Ryan, *Cartographic Eye.* Baber (*Science of Empire,* chap. 5) discusses British mapping in India, though his work is considerably less detailed than Edney's.

tories, European explorers were accorded foundational roles, and much was made of their place as intrepid vanguards of imperial fervor. Even the more nuanced works of scholars like R. A. Skelton or William Goetzmann retained an awestruck tone, though they provided invaluable encyclopedic résumés of who went where.[24] In light of increasingly social-scientific and economically driven accounts of European imperial hegemony, explorers fell through the historical cracks, and though their writings continued to receive some treatment by literary scholars, their stock as historical figures dropped precipitously.[25] Work on geography and empire has recently remedied this to some degree, but the work of literary scholars like Paul Carter and Mary Louise Pratt has best focused historians' attention on the problem of placing explorers back in a revised history of empire.[26]

By linking explorers with the production of maps and geographical knowledge, several recent works have attempted spatial histories of specific regions of empire. These studies share a concern with diversity and the specificity of individual colonial situations, rejecting the broad generalizations and anodyne formulations that characterize the least effective examples of both an older synthetic historiography and some of the most modish colonial discourse theories.[27] Paul Carter's important work on Australia has been joined by more recent contributions by Simon Ryan.[28] Thongchai Winichakul has explored mapping and geography in relation to state formation in Southeast Asia in *Siam Mapped*, and most recently Matthew Edney's *Mapping an Empire* has looked at the place of surveyors and cartography in the geographical construction of British India. These works share several characteristics: each explores what could be called the cultural history of cartography; each contributes to a revisionist history of explorers and exploration; each attempts to read geographical texts as sources for the spatial history of a colony;[29] and each works to maintain something of a double perspective on the history of a colonial space,

---

24. Goetzmann, *Exploration and Empire*; Skelton, *Explorers' Maps*. In the area of South American natural history there is the classic: Von Hagen, *South America Called Them*.

25. For an account of this process, see Bridges, "Historical Role of British Explorers." Historians of cartography have seen their discipline as emerging only lately from a subordinate position with respect to the history of exploration and discovery. See Harley, "Map and the Development of the History of Cartography," 12.

26. Bell, Butlin, and Heffernan, *Geography and Imperialism*; Godlewska and Smith, *Geography and Empire*; Carter, *Road to Botany Bay*; Pratt, *Imperial Eyes*.

27. Nicholas Thomas has criticized the overly general character of much recent theorizing about empire: see Thomas, *Colonialism's Culture*, and Ryan, *Cartographic Eye*, 15. Bayly discusses the importance of attention to the historical specificities of colonial conflict in different regions (as well as synthetic works) in Bayly, *Imperial Meridian*, 14.

28. Ryan, *Cartographic Eye*.

29. Thailand was never colonized by a European power, but Winichakul sees the territorial codification of the region as an indigenous appropriation of European geography in the colonial vein.

remaining attentive to the significance and appropriation of indigenous knowledge in the process of territorial codification. Taken together, this is a fair characterization of the present work as well. Since I have drawn in different ways on the work of Edney, Carter, and Winichakul, I here lay out their arguments and my relationship to them in greater detail.

Matthew Edney's *Mapping an Empire* makes the general point that maps were central to the way the British conceived of British India,[30] but the most interesting part of his book is his demonstration that a particular surveying style—the trigonometric survey—was essential to how the British thought of themselves and their rule over the colony. Edney shows how trigonometric surveying promised to be the "technological fix" that would allow nineteenth-century cartographic practice to approach its (treasured) conceptual ideal. By promising a dense network of mutually correcting control points, closely congruent with the land itself, the trigonometric (or triangulating) survey distanced itself from the imprecision of earlier survey practices and helped rescue the "cartographic ideal."[31] Edney shows that trigonometric rhetoric was more important than trigonometric practice. The institutional framework of the survey of India, the conflicts over cost and approach, the impracticality of nested, multiple-scale triangulated surveys, and the lesser utility of abstract, small-scale triangulations in comparison with local topographical surveys all meant that the ideal of the totalizing trigonometric survey was honored more in word than in practice. But the rhetoric of the Great Trigonometrical Survey afforded the British the comfort of a crucial delusion: "that their science enabled them to know the 'real' India."[32] This "scientistic" ideology not only gave the British a sense of mastery over this distant place, it also embodied the very ideals of British rule. A trigonometric survey involved and implied coordination, uniformity, hierarchical organization, administrative order, and the rule of "transcendent law." It amounted to the geographical assertion that British India was (or could be) a "rational and ordered space, which could be managed and governed in a rational and ordered manner."[33]

Edney's study of the geographical construction of a British colony is thus at the same time a study of a particular surveying style. This book has a similar aim. In addition to examining the geographical construction of British Guiana, I also examine the particularities of the *traverse* survey.

---

30. Edney, *Mapping an Empire,* chap. 1.
31. The cartographic ideal was rooted in the Enlightenment premise that "correct and certain archives of knowledge could be constructed . . . by following rational processes epitomized by mapmaking." Edney, *Mapping an Empire,* 17, 16–30.
32. Ibid., 3.
33. Ibid., 34.

Unlike trigonometric surveys, which were expensive and demanded large and coordinated teams of surveyors, the traverse survey was largely a solitary and nomadic affair in which an explorer, equipped to maintain some positional consciousness, transcribed his route and sight lines onto a cartographic field. The traverse survey, its flaws, and the various "fixes" employed to shore up the reliability of its cartographic products are the focus of chapter 3. I show that the exigencies of the traverse survey led to particular emphasis on landmarks, and in chapters 4 and 5 I detail the process by which such landmarks were constructed and accorded their significance. Having shown that practitioners of the traverse survey continually sought to overstep boundaries, I point in chapter 6 to the contradictions inherent in surveying a boundary by means of a traverse. Edney shows that the trigonometric survey embodied the British ideal of a fixed imperial space. Imperial space, however, is paradoxical. It is, ideally, fixed. But at the same time it is imminently and eminently expandable.[34] The mobile and exploratory traverse survey embodied this other characteristic of the territory of empire.

The solitary nature of traverse surveying makes Paul Carter's *Road to Botany Bay* particularly significant to my work. The great strength of his book was to identify the radical difference between the spatial perspectives of an explorer on the one hand and those for whom exploring was done on the other.[35] He writes: "How could the explorer share his culture's notion of space as passive, uniform and theatrical? How could he pretend that there were no horizons to his knowledge, or that all directions were ultimately the same?"[36] As Carter points out, the terra incognita that lay beyond the expanding frontiers of empire was, above all, a *problem* for imperial administrators. Unknown, inaccessible, heterogeneous, and likely to be volatile, it was the antithesis of the colonial spaces they hoped to see, which were defined, settled, and productive. Carter's insight was to see that travelers looked for, sought out, and in fact "saw" terra incognita. They had dynamic encounters with dynamic spaces; their views were sweat-befogged, personal, and often costly. Their perspective thus shared little with the conventional "stage space" preferred by those seeking to assemble depictions of a stable empire. Before Carter's observations about the distinctive (and, seen from an imperial viewpoint, somewhat *subver-*

---

34. Certeau, *Practice of Everyday Life,* 121. Edney discusses the "extensible" quality of imperial space as well (*Mapping an Empire,* 25). For a discussion of the early modern roots of this tension between "an expanding globe and an enclosing nation," see Cormack, "'Good Fences Make Good Neighbors,'" 640.

35. A weakness, some have felt, is Carter's inscrutable style. See Ryan, *Cartographic Eye,* 16; Seddon, "On the Road to Botany Bay"; Noyes, "Representation of Spatial History"; and Noyes, *Colonial Space,* 11–14.

36. Carter, *Road to Botany Bay,* 62.

sive) perspective of the traveler, it would have been easier to see the whole process of imperial exploration and mapping as a sweeping "strategy" of European imperialism: the "objectification of the gaze" created the "colonial object";[37] the process of imperial geographical exploration (representing the "scopic drive" of European paranoia) constructed a disciplinary cartographic panopticon for its scorned and feared subjects; literary discourse and scientific technology alike were enlisted in a grand *machine à territoire* in which individuals were merely cogs in a coordinated, rational, and centrally controlled system that mapped colonial territories automatically as officials in the Royal Geographical Society and the Colonial Office turned their globes. It is the image that led a previous generation of cartographic historians to depict the history of European expansion as a gradual "unrolling of the map" over (or under?) the non-European world.[38] After Carter such an analysis will not suffice. Within the progressive image of how colonial territory "took shape," he has identified a tension. If before Carter European cartography had been imagined to spread over the globe like a flood of crimson paint radiating from the metropolis, after *The Road to Botany Bay* there is an obligation to recognize very distinct countercurrents in that flood, eddies and backwashes in the tide of empire.

This description of Carter's project makes clear his relevance to a study of imperial surveying and cartography, particularly a case like Schomburgk's work in British Guiana. The traverse surveyor's twin identity—as an explorer of terra incognita and as a surveyor charged with codifying the "stage space" of the colony by his maps—meant that the tensions between the spatial perspective of the explorer and that of the imperial administrator arose at the core of his work. Schomburgk's narratives (and those of other explorers) reveal these tensions, showing traces of the "active engagement with the land" and of the personal risks and anxieties associated with travel in an alien place among alien people. The scholarly approach that insists on reading these traces as mere rhetorical ploys—either as authenticating tropes or as attempts to conceal imperial agency behind a facade of impotence, effeminization, or some sentimental smoke screen—misses something important about these texts.[39] Carter's work offers a different (and better) framework in which to understand these traces. For him, attention to the traces of an active engagement with the land is a point of departure for a history of empire that renounces the imperial point of view. Carter's true travelers, those who engaged an unfamiliar

37. Noyes, *Colonial Space*, 74.
38. Outhwaite, *Unrolling the Map*.
39. An example of this sort of reading would be Pratt's notion of the anti-conquest. Pratt, *Imperial Eyes*, 7 and passim.

space "on the ground," knew the secrets that could not be spoken in the language of empire: that places were unique, particular, and diverse; that they came into view as a result of serendipitous moments; that they took shape in the course of a journey; that they had to be plotted with respect to lines of emotion, experience, and movement, not simply on the homogenizing and bloodless grid of longitude and latitude.[40] Explorers perceived the manifold oversights within the feigned oversight of the imperial gaze.[41] A history that recognizes this, Carter lets on, would be a way forward in the wake of the empire. Such a history would reject the "stage convention" depictions of space that made empire possible and that (he argues) are regularly recapitulated by standard imperial histories.

Paul Carter has written a spatial history of Australia that is thus at the same time an attempt to pay attention to explorers' work in the field. Doing so promises to unsettle the imperial territorial inscriptions their passages made possible. This book has a similar aim. Chapters 4 and 5 examine the notion of the landmark in ways that correspond roughly to Carter's distinction between the perspective of the imperial administrator and that of the explorer. In chapter 4 I demonstrate that landmarks were visible and stable icons of distant territory and that they had a place in an elaborate strategy of imperial representation. In chapter 5 I reexamine the same landmarks and show their distinctive meanings in the context of exploration. The stable, iconic landmark that served, for a metropolitan readership, as a metonymic replacement for the place itself (and thus facilitated the imperial appropriation of distant territory) could constitute, in the context of the exploration, a treacherous obstacle or the symbol of an inaccessible region. In chapter 6 I extend this analysis by turning to the demarcation of the colonial boundary. I show that scrutinizing traces of the explorers' active engagement with the land undermines the stability of the boundary.

Thongchai Winichakul's *Siam Mapped,* a pioneering study of the geographical codification of Thailand in the late nineteenth and early twentieth centuries, has been of particular use in thinking about the durability and political significance of boundaries. Winichakul's study is primarily concerned with understanding the place of a mapped image of Thai territory in the construction of Thai nationhood. Much of his book is devoted to examining the "premodern" or indigenous notions of space in the kingdom of Siam and then tracing the displacements, negotiations, and conflicts that attended the introduction of Western geographical dis-

---

40. Carter's observations can be seen as an extension of Certeau's distinction between "tactics" and "strategies." See Certeau, *Practice of Everyday Life,* xix.

41. Carter's most recent book, *The Lie of the Land,* pushes the pursuit of these key points to considerable lengths.

course, particularly maps. Thailand was never directly colonized by an imperial power, but Winichakul shows the power of the cartographic depiction in installing and communicating European notions of territory. By tracing shifts in the notions of sovereignty and boundary that accompanied the use of Western cartography, Winichakul demonstrates the impact of the Siamese court's adaptation of a new spatial technology. Ultimately Winichakul can show that the projection of anachronistic notions of what "Thailand" was has distorted Thai history and introduced confusion into the construction of Thai nationality. In addition, then, to providing a strong case study of the power of the map in the codification of a possessable territory, Winichakul also brings the power of the map, and the history of cartography, to bear on contemporary geopolitics, on the postcolonial landscape that is a product of colonial spatial practices. He shows that "Thailand" never was what it was imagined to be by those who lament its "dismemberment" during the age of empire.

Winichakul's study of the geographical construction of Thailand is thus at the same time an attempt to understand the legacy of territorial codification by means of surveyed maps and fixed boundaries. This book has in its conclusion a similar, if somewhat speculative, aim. After examining the process of boundary delimitation and demarcation in chapter 6, I turn to a reflection on the later history of the boundaries and suggest that close attention to their making reveals the anachronistic rhetoric that has shaped recent disputes about the limits of Guyana.

A SKETCH OF THE ROUTE

Having positioned myself with respect to several landmarks in the field, I need to lay out the course the present work will follow. This introduction concludes with a brief historical outline of British involvement in Guayana before 1803. Two significant points emerge. First, the colony of Guiana had an extremely ambiguous relationship with Britain. It was a unique place: the only British possession on mainland South America; an "English" region that had retained a great deal of Dutch influence; a continental landmass categorized as part of the West Indies. As if these confusions were not enough, the region had changed hands numerous times in recent memory, suggesting questionable colonial loyalty. Second, the figure of Sir Walter Ralegh and the nebulous figure of El Dorado—the gilt king of a legendary empire—both loomed large in British conceptions of the region.[42] Chapter 2 takes up both of these themes in some

---

42. I follow Armitage (*Sir Walter Ralegh: An Annotated Bibliography*), and Schomburgk too, in spelling Ralegh with no *i*.

detail by examining the place of Ralegh and the myth of El Dorado in the nineteenth-century exploration and territorial codification of British Guiana. The point of departure is Schomburgk's edited edition of Ralegh's 1596 *Discovery of Guiana,* which Schomburgk prepared at the end of nearly a decade of explorations in Guayana.[43] In this sense I begin at the "end" of the most important geographical explorations of British Guiana in the nineteenth century, but I do so in order to return to the beginning of British involvement in the region—Ralegh and the El Dorado quest. I show that both Ralegh's routes and the putative site of El Dorado served as key orienting points for subsequent explorers and even became points of reference in the boundary disputes later in the century. While this material is certainly concerned with British colonial projects in lowland South America, I am after several broader themes at the same time. First, I demonstrate explorers' obsession with each others' routes (a recurring theme of European exploration everywhere) and suggest its significance. Second, I argue that while attention to historical passages helped root British territorial claims, myths proved an unstable (if necessary) grounding for colonial territory.

Myths overwrote terra incognita with significance. Chapter 3 takes up this notion of "unknown land" and examines it in relation to the assignment of the traverse surveyor in British Guiana. This form of geographical labor relied on an explorer's ability to move to and through unfamiliar regions. I show that getting to some appropriate terra incognita was not a straightforward proposition. Not all unknown regions were of equal interest to the multiple parties with stakes in and control over a geographical explorer. The RGS and the colonial governor had different ideas about what constituted the most important regions to map. After opening up the notion of terra incognita to scrutiny, chapter 3 goes on to examine what made a good geographical explorer. I show that the traverse survey constituted a questionable source of geographical knowledge because it was inextricably entangled with a solitary and subjective explorer on a nomadic passage across difficult terrain. Aware of these difficulties, traverse-surveying explorers attempted to "fix" these problems and to enhance the reliability of their surveys. "Fixed points" provided a solution. The quality of a traverse survey depended on the number and quality of the fixed positions an explorer was able to establish, by means of astro-

---

43. *The Discoverie of the Large, Rich and Bewtiful Empyre of Guiana, with a Relation of the Great and Golden Citie of Manoa (Which the Spaniards Call El Dorado) . . . by Sir W. Ralegh, Knight.* For information on editions, see Armitage, *Sir Walter Ralegh: An Annotated Bibliography,* and Nicholl, *Creature in the Map,* 377–78. To avoid confusion and maintain consistency with citations of the edited work, I have adopted throughout Schomburgk's (modernized) spelling of Ralegh's title. For the spelling of "Guayana" here (denoting the larger geographical region) see above, note 8.

nomical observations, dead reckoning of his position, or both. Reliably fixed points allowed an explorer's passage to become a cartographic depiction of the territory. How could a traverse surveyor establish the reliability of his points when they could rarely be tested? I argue that instrumentation was important, but that instrumental rhetoric could be even more so. For instance, invoking Humboldt suggested that the explorer was working within what many thought the most reliable and measurement-centered geographical science of the day. I also show the importance of the tradition of naval navigation in the work of the traverse survey. Not only did the naval tradition provide the instruments and protocols enabling terrestrial positions to be fixed from celestial observations, it also afforded a context in which the reciprocity of map and disciplined passage was well established. Good navigation meant maintaining mapped positions. Correct discipline—of self, of crew—meant that navigators could add to maps. These traditions of discipline were essential for the traverse surveyor.

Here is a truth essential for understanding this book: for all their importance, those fixed points on their own were *useless* for the geographical construction of the colony. As coordinates in the graticule of longitude and latitude, every fixed point had been postulated by the blank map before it was observed on the ground.[44] It did not take an explorer to tell you that the point 2°55′ N by 58°48′ W *existed*. The question was, What stood at that point? (See plate 6 if you can't wait to find out.) To have geographical significance, a fixed point had to be linked to a point on the ground. There had to be something *at* the point. This means that traverse surveyors needed not only to fix points but to create landmarks. After establishing the importance of landmarks to the work of the traverse surveyor, I turn in chapter 4 to a study of the construction of landmarks. The jungles and savannas of Guayana presented difficult and dynamic environments: human settlements were few and transient, cyclical rainy seasons changed the land. I argue that in this recalcitrant environment, landmarking required considerable effort by the explorer, who was obliged to use an array of techniques to represent clear, stable, and visible points and to invest them with meaning in as many frameworks as possible. The best landmark had multiple associations: historical, aesthetic, geological, mythical, and so on. By examining how the publications of the explorer constructed landmarks and situated them in cartographic fields, I show how certain sites became nodes in the geographical construction of colonial

---

44. The use of "graticule" here—meaning both "an imaginary network of meridians and parallels on the surface of the earth or other celestial body" and "a network of lines, on the face of a map, which represents meridians and parallels"—is anachronistic; it took on that meaning in the late nineteenth century. Wallis and Robinson, *Cartographical Innovations,* s.v. "graticule."

territory. Through cycles of reproduction and copying, images of these sites became icons of the colony. Chapter 4 traces these cycles and examines the readership of the texts in question. My aim is to show that landmarks were instrumental in the geographical construction of a colony.

After establishing the function of the landmark in a representational strategy that made the colonial territory appear visible and stable, I turn in chapter 5 to examine the place of the landmark in the experience of the traverse surveyor. I show that landmarks were essential to the explorer's progress through an unfamiliar region: they indicated directions, marked key points in previous passages, and afforded privileged glimpses of the terrain. This attention to the explorer's active engagement with the land exemplifies the tension between the local perspective of the explorer and the overview he was expected to provide on the territory he traveled through. I suggest some ways this tension could be reconciled and show that the explorer's view could be "repatriated" through cycles of representation. I do closer readings of several kinds of landmarks—Amerindian sites, waterfalls—that I suggest had particular significance for explorers. Examining these sites dramatizes the anxieties and negotiations that attended geographical exploration. I ultimately argue that the geographical construction of the colony depended in part on the progressive erasure of these idiosyncrasies of the explorer's experience.

Chapter 6 draws together the ideas introduced in the previous chapters and brings them to bear on the boundary surveys of the colony that Schomburgk conducted between 1841 and 1844. These surveys were traverses, and I show how landmarks and landmarking were employed in the boundary demarcation. More significant, I suggest that Schomburgk, like any successful traverse surveyor, had been trained as a boundary crosser, since this was the particular charge of a geographical explorer seeking terra incognita. I argue that this expansive, nomadic, and boundary-transgressing technique led Schomburgk to demarcate a highly ambiguous boundary. In fact, he demarcated boundaries around boundaries and continually overstepped his previous inscriptions. Understood in this light, the subsequent and enduring disputes over the boundary (it remains a source of international controversy) can be understood not as the effect of an individual motivated by imperial megalomania, as some have suggested, but rather as an unintended consequence of a particular style of survey and all it entailed. As a conclusion, I suggest some implications of this finding for those who care about Guyana, but I also extend the reflection to encompass the postcolonial legacies of colonial boundaries more generally. In the most speculative reach, I ask in closing what the whole of this study of geographical exploration might have to tell us about European obsessions with the possession of nature.

THE DESTINATION: A LITTLE BACKGROUND ON BRITISH GUIANA

British involvement in the regions that would become British Guiana in 1831—and remain under that name until 26 May 1966—began in cycles of misadventure and buccaneering, none more celebrated than the first: Sir Walter Ralegh's ill-fated pair of expeditions to the Spanish Main in 1595 and 1617. Elizabeth's favored courtier, seduced by the golden chimeras circulating about the Spanish possessions in the New World, ended up sacrificing more than simply twenty-three years of his life to his dream of penetrating to El Dorado, the city of fabulous wealth he came to believe lay up one of the rivers draining the region. He sacrificed a son, who fell fighting the Spanish at Santo Thomé. He sacrificed his closest companion and fellow officer, Lawrence Keymis, who, after botching his approach to the Iconuri mine, also botched the first, but not the second, attempt on his own life. Ultimately, of course, Ralegh sacrificed himself. Imprisoned in the Tower and eventually executed, he left as a legacy both an endlessly alluring tale of a star-crossed obsession and the contagious conviction that England could hold a tropical empire.

It is customary to begin the history of Guyana by invoking Ralegh and his pains. The desire to show that his sacrifices were, as one modern historian of Guyana has put it, "not in vain" stands as a subtext (or at least conceit) of several histories of the colony.[45] This Raleghness of Guyana's history is particularly curious given that—although Ralegh wrote up his first voyage in his celebrated *Discovery of Guiana,* a broadsheet of Elizabethan colonial fervor and a work of considerable poetry—none of his exploits touched directly on the regions that would become British Guiana in the nineteenth century. Nevertheless, his *entradas,* by historical convention (and as I suggest in chapter 2, by imperial design), initiate British imperial presence in the region. There is a certain truth to the convention, in that the association of Ralegh with the colony correctly links the British presence in the region with an ambivalent story of squibbed usurpation and questionable legitimacy. Such ambivalence characterized British meddling in the region from 1595 to 1803, when Dutch surrender to British warships yielded a possession that would endure for a century and a half.[46]

The waning of Spanish power in the early seventeenth century gained for their rivals, the Dutch, formal recognition of their settlements at Kykoveral

---

45. For instance Menezes, *British Policy,* 1.
46. The following historical sketch draws on Edwards and Gibbon, "Ethnohistory of Amerindians in Guyana"; Netscher, *History;* Dalton, *History of British Guiana;* Daly, *Making of Guyana;* Farley, "Unification of British Guiana"; Lancaster, "Unconquered Wilderness"; and the work of Menezes.

on the Essequibo and the plantations on the adjacent rivers.[47] From the earliest days of Dutch settlement in the region of the "Three Rivers" the British had made periodic visitations, as plunderers, traders, or some hybrid species. The introduction of sugar (and African slaves) to these colonies between 1658 and 1664 meant they looked lucrative enough to attract the attention of a British naval assault at the second outbreak of Anglo-Dutch hostilities in 1665. They were quickly taken and shortly thereafter retaken in the first of several such exchanges. By the mid-eighteenth century the Dutch West India Company, eager to expand the slave-based plantation cultivation of sugar that had built prosperous little outposts on the alluvial delta soil at the mouths of these rivers, opened free lands on the Essequibo and the Demerara to settlers from any nation; new settlers were absolved from taxation for a decade. The newcomers were primarily English planters fleeing the exhausted soils of Britain's Caribbean sugar islands. Within a few years the English outnumbered the Dutch in several settlements.

The business of these colonies was sugar and, to a lesser degree, cocoa, coffee, and cotton. Although timbering and trade with the Carib and Arawak Amerindians went on, the plantations were the life of the colony and the death of the growing numbers of Africans imported to dike, drain, and clear new plantations, operate the mills, and harvest the cane. The correspondence of the presiding Dutchman of the region in the century, Larens Storm van 's Gravesande, governor of Essequibo and Demerara from 1742 to 1772, indicates that he worked to contest Spanish claims in the interior, commissioning explorers and establishing "posts" upriver in efforts to solidify Dutch territory, but the colony looked to the sea for its livelihood and its future. The slaves, by contrast, had a tendency to look the other way.

In the eighteenth century considerable numbers of slaves escaped to establish communities of "Bush-Negroes" or "Maroons" in the dense jungles and inaccessible savannas of the proximate interior. Expeditions to ferret out these communities and destroy their provision fields were a common facet of life in the colony: all healthy men belonged to the militia, and Amerindians were often recruited as guides and extra hands. Rewards for their services were forthcoming, particularly in periods of direst need: officials bestowed valuable annual "presents" on the Amerindians assembled at the upriver posts, and the colonial government awarded silver-headed batons to the "captains" of participating groups.[48] These

---

47. The name Kykoveral reflected the ultimate imperial desire, to "overlook."

48. The story of the relationship between Amerindians, Maroons, and colonists is a complex one, since the dynamics changed considerably as the colony saw shifts in metropolitan affiliation and slave policy. For an introduction to this material, see Benjamin, "Preliminary Look at Free Amerindians"; Whitehead, "Tribes Make States"; and Menezes, *British Policy*. On the staves themselves, see Menezes, *British Policy,* 48; Rodway, *Handbook of British Guiana,* 28; and Harris and De Villiers, *Storm van 's Gravesande,* 103.

batons appear and reappear in the interior over the next 150 years, and they became syncretistic symbols of leadership within several Amerindian communities. In a looping tale of reciprocating territorial authorities, the Dutch staves, and similar later British models, had to be retrieved from the field (when they could be found) in the late nineteenth century, since they became important evidence in the boundary dispute between Britain and Venezuela. Having deputized the native population to enforce European sovereignty in the bush, Europeans later had to turn back to the Amerindians to sort out the limits of colonial territory.

The Berbice slave rebellion of 1763 and the Dutch reprisals that followed it give some indication of the hardship of the life that drove thousands of runaways into the "bush." Their presence there, more often rumored than real, nevertheless established the character of the interior in the minds of European settlers. The bush came to embody all the anxieties of a slaveholding elite: a free, vengeful, miscegenating population bent on blood, invisible, atavistic. Coupling the presence of dreaded Maroon settlements with rumors of cannibalism by the mobile and well-armed population of exotic Amerindians (whose curare-tipped arrows were a macabre metropolitan fascination) makes it easy to understand why the European population stayed in the open spaces they cleared along the littoral and why, significantly, they adopted as the seal of the colony a ship headed for open water (fig. 1).[49]

Open water brought its own problems. Dutch support for the rebellious American colonies during the fractious years of the American Revolution brought British retribution in the form of four privateers, which captured the three river colonies in 1781 without firing a shot. This British occupation endured only about a year before the French, sending a superior force of eight warships and 2,000 men at arms, made themselves masters of the region. This too was an interlude; the signing of the Treaty of Versailles in 1783 stipulated that Essequibo, Demerara, and Berbice should be returned to the Dutch.

The outbreak of European hostilities in the wake of the French Revolution eventually reverberated through the Caribbean, with high human cost. Between 1793 and 1801 Britain would commit 89,000 men-at-arms to the West Indies, more than half of whom would be buried there.[50] The three river settlements of Guiana changed sides early and often. In 1796 a detachment of 1,300 British troops arrived from Barbados and took the settlements without violence. Surrender terms were gener-

---

49. Figure 1 is reproduced from Rodway, *Handbook of British Guiana*. The seal was accompanied by a suitably mercantile motto: "We give and demand in turn."
50. Marshall, *Cambridge Illustrated History of the British Empire*, 32.

Figure 1. The seal of the colony: trade as motto, means of egress as device. Reproduced from Rodway's 1893 *Handbook of British Guiana*.

ous, and the European population, primarily conservative Dutch royalists and settlers of British extraction, was pleased to divest itself of its connection to the Batavian Republic, widely seen as a French puppet government. So anglicized did the Guiana colonies become through tightened relations with Barbados between 1796 and 1802 that planters greeted the Treaty of Amiens—which returned Essequibo, Berbice, and Demerara to the Dutch—with widespread dissatisfaction. Their disappointment did not last, for Dutch rule was as short-lived as the peace that followed the treaty. An English squadron came about in the mouth of the Demerara the next year, and on 18 September 1803 the Dutch colonies of Demerara and Essequibo signed the Articles of Capitulation and the British flag was raised at New Town (later Stabroek, later still Georgetown); six days later Berbice signed as well. At the London Convention of 1814 Britain paid formal compensation to the Dutch and cemented British control over the three colonies, which in 1831 united to become British Guiana.

At about the same time, an energetic anglicized Prussian living in the West Indies came to the attention of the RGS. Robert Herman Schomburgk's career as a merchant in Richmond, Virginia, had quickly been superseded by his fascination with botany, geography, and the writings of his esteemed compatriot Alexander von Humboldt.[51] After finding his

---

51. Secondary sources available on Schomburgk include Ojer, *Robert H. Schomburgk; Dictonary of National Biography,* s.v. "Schomburgk"; and Rodway, "Schomburgks in Guiana." See also the recent short

way to Tortola, Schomburgk began to ply the trade of surveyor and undertook the mapping of the treacherous outlying island called Anegada. The reefs of Anegada had long taken their toll on West Indian shipping, a toll collected by the semipermanent settlements of pirate-salvagers who made the island home. Schomburgk's surveying exercise won him the praise of the RGS and the attention of the Admiralty, but it won him no friends among the locals, who threatened his life in return for his threatening their livelihood.[52]

In 1835 Schomburgk secured a commission from the Society to conduct a series of explorations in Guayana, and he became the most significant figure in the geographical construction of the colony in the nineteenth century. Although he was not immediately successful in realizing the aims of his employers, Schomburgk quickly oriented himself to his task and conducted a set of expeditions into the interior of the region between 1836 and 1839. He ascended, in order, the Essequibo, Corentyne, and Berbice Rivers and then set out on a circuit of almost 3,000 miles that took him into Venezuela (where he reached the easternmost point attained by Humboldt) and Brazil before returning to Georgetown and, from there, London. He won the RGS Gold Medal in 1840.[53] His publications (in the *Journal* of the RGS,[54] the Naturalist's Library series, and several independently published books) and his correspondence (with Sir Thomas Buxton at the Aboriginal Protection Society, with the Foreign Office, and with colonial governors) drew attention to the colony and specifically to its uncertain territorial limits. With the support of the RGS and the governor of British Guiana, Schomburgk won a Crown commission as "Her Britannic Majesty's Commissioner for Surveying and Marking out the Boundaries of British Guiana." In this capacity he returned to the colony in 1841, accompanied by his brother, Richard, engaged as expedition botanist and funded by the Prussian government. The work of the boundary survey occupied Robert Schomburgk from 1841 to 1844, at which point he returned to London, was knighted, and took up a position in the diplomatic service under the patronage of Lord Stanley.

The importance of Schomburgk's explorations in the geographical construction of the colony is captured best by a late nineteenth-century article that encouraged readers to gauge his merits by perusing the carto-

---

piece by Rivière: "From Science to Imperialism." Other discussions of the explorer's work appear in Daly, *Making of Guyana,* 172–78; Outhwaite, *Unrolling the Map,* 184; Stafford, *Scientist of Empire,* 81–83; and Cameron, *To the Farthest Ends of the Earth,* 18–21. For a detailed account of a contemporary RGS explorer, see Cameron, "Agents and Agencies." Schomburgk's middle name also appears as "Hermann."

52. *Venezuela Arbitration, Schomburgk's Reports,* 48.
53. I follow Menezes (*British Policy*) on the spelling of "Corentyne."
54. Abbreviated *Journal* in the rest of the text and *JRGS* in the notes.

graphic fruit of his labors: "From the various maps accompanying the reports, it may be best seen how the blanks were gradually filled up by one exploration after another, until the final Map left very little to be discovered."[55] In fact it left so little to be discovered that Schomburgk's map constituted the base map over which King Vittorio Emanuele of Italy puzzled out the southwestern boundary of the colony in 1906, leaning over the negotiating table erected in his chambers of state. As the neutral arbitrator in Britain's boundary dispute with Brazil, King Vittorio put his pen down on top of the 9,000 foot prow of Mount Roraima and traced a line to the Takutu and Mahu Rivers, fixing with a calligraphic gesture the geographical boundaries of British Guiana.

That, in a sentence, is the tremendous power of nineteenth-century territorial practices: the explorers, the instruments and techniques they used, the language, images, and maps, the practices of surveillance, control, and legislation, all of these transformed a European terra incognita into a colony of definite form and extent; all of these carved a territory—possessable, remittable—out of lowland floodplains and jungle highlands. With the clarity of hindsight, the "badness" of this imperial enterprise, the destructive homogenization produced by European desire to acquire wealth or virtue or both at the ultimate expense of others, seems well beyond question. The legacy of the enterprise endures. Salient in that inheritance—besides empty mines, scarred minds, and *tristes tropiques*—is a way of thinking about place itself. Former imperial nations have of late, and for better or for worse, given most of these colonial places back to their inhabitants. Of course to say "back" is something of a misnomer, for those geographical constructions—codified, specified, and reified as they were (and for the most part have remained)—never "belonged" to anybody except, of course, those who made them or for whom they were made: imperial nations themselves. This constitutes the most troubling legacy of imperialism. In a world racked by ethnic and territorial strife, where reasoned voices decry the balkanization and tribal warfare of countless postcolonial states, there remains a lingering suspicion that a "postcolonial state" is inexorably a colonial state manqué.

It is in light of this disturbing realization that examining the geographical construction of a colony takes its place as a postcolonial project. The aim, as John Noyes puts it, is to "actualize the traces of resistance to the colonizing project—to develop the hints and associations whereby the colonial text speaks its own impossibility."[56] Such a history, then, will be an intervention of sorts. By exposing the slippery foundations and tippy

---

55. Rodway, "Schomburgks in Guiana," 22.
56. Noyes, *Colonial Space,* 285.

edifice of the imperially defined places, I hope to unsettle imperial territorial installations. Confidence is out of place: "history," Greg Dening wisely points out, "is a poor social reformer," and scholarly subversions offer a thin gruel to those who welcomed me to Guyana with heartier fare.[57] To these and theirs I can offer only dedications.

57. Dening, *Islands and Beaches*, 6.

CHAPTER TWO

Myths and Maps: Making Explorers and Empires

MYTHICAL EMPIRE AND INTERIOR EXPLORATION I: FINDING EL DORADO

Britain's anomalous constitutional relationship with British Guiana under the 1803 Articles of Capitulation (the colony retained its Dutch laws and local political institutions), coupled with Britain's ambivalent history of comings and goings in the region, led to a particular anxiety about what British Guiana was, had been, and should be. This was reflected in, among other things, the confusion in Britain's Parliament over whether Guiana was, or was not, a Caribbean island, a confusion that persisted to the mid-nineteenth century.[1] Guiana was a colony with an identity crisis: an odd, obscure, inscrutable part of mainland South America. History suggested that Guiana had capriciously changed imperial masters without a shot's being fired. British Guiana's history of ambivalence meant a colonial preoccupation with precedents, legitimating passages, key artifacts, and precursors. It was as if "every mere coincidence of historical time and place . . . [could] be interpreted filially, as evidence of legitimate descent."[2]

The filial descent of a British Guiana had to be traced back to the wayward, missing, and ultimately beheaded colonial "father," Ralegh, and his El Dorado *entradas* of 1595 and 1617. No discussion of British exploration and territorial consolidation in the region can begin without addressing Ralegh and the legend he made famous.[3] The Ralegh story did two things. First, it installed a mythical empire at the iconic center of the Guianas.[4] Second, it laid down a set of routes (his own and that of

---

1. References to this misprision include Hancock, *Observations* (1835), 86; BL, Hilhouse MS, 1 and 7; R. H. Schomburgk, *Twelve Views,* "Preface," recto; and CRL, Holmes MS, letter 2.
2. Carter, *Lie of the Land,* 211.
3. A variety of contemporary works examine the El Dorado tale. Among these are Naipaul, *Loss of El Dorado;* Hemming, *Search for El Dorado;* Chapman, *Golden Dream;* Ramos Pérez, *El mito del Dorado;* Quinn, *Raleigh and the British Empire;* Armitage, *Sir Walter Ralegh: An Annotated Bibliography;* Fuller, "Ralegh's Fugitive Gold"; Greenblatt, *Sir Walter Ralegh;* Schama, *Landscape and Memory,* 307–20; Montrose, "Work of Gender"; Coote, *Play of Passion;* and most recently, Nicholl, *Creature in the Map.*
4. The original statement on the significance of myth to exploration is Wright, "Terrae Incognitae." The notion was reexamined in Allen, "Lands of Myth," and more recently, Allen, "Indrawing Sea." Carter has made use of Obeyesekere's work on Cook in arguing that the value of myth lies in its capacity to give

Keymis) that took on particular significance for subsequent explorers. These two things—mythical empire and actual routes—played key roles in the geographical construction of the colony in the nineteenth century.[5] They were important points of reference for explorers in the region, who sought to "find" El Dorado by discovering the source of the myth, and who built their identities around following and surpassing the great Ralegh. Understanding the cycles of priority and posteriority established by explorers—who shored up their own significance by "following" predecessors, even as they traced the routes of those predecessors on new maps—is essential to understanding both the history of exploration and the history of colonial territorial claims.[6]

On his return to London from his Crown-commissioned boundary surveys in British Guiana in the early 1840s, Robert Schomburgk was promptly approached by the Hakluyt Society (an association dedicated to promoting exploration and republishing celebrated travel accounts), which requested that he edit a new edition of Ralegh's *Discovery*. This was to be entirely annotated with the most recent material on the "true course" and place-names of Ralegh's expedition, new observations on natural history, and a reappraisal of the El Dorado myth.[7] Despite being an Englishman by "conversion" rather than by birth, Schomburgk felt himself "strongly biased in favour of Ralegh"; indeed, he even felt a certain kinship with the subject of his biographical and editorial "work of love."[8] As Schomburgk pointed out in the preface to the new edition, Ralegh's text possessed the power to transport him back into the land of his own geographical glory:

> How frequently we are reminded of the pleasures of former days by an accidental word, the perusal of a passage in a book or by the view of a place associated with past enjoyments! . . . Such was my case as I recently perused Sir Walter Ralegh's "Discoverie of Guiana." Every page, nay almost every sentence, awakened past rec-

---

"history an allegorical dimension." Carter, *Lie of the Land,* 210–11. Bayly has identified the importance of "ideas, legends and beliefs" in binding regions in the period 1780–1830. Bayly, *Imperial Meridian,* 14. For other discussions of myth and legend in the process of exploration and territorial codification, see Whitehead, "El Dorado, Cannibalism, and the Amazons"; Lewis, "La Grande Rivière"; and Baker et al., *American Beginnings*.

5. For an account of a similar North American mythical empire and its place in European exploration and settlement, see Baker et al., *American Beginnings*. Particularly relevant are the following essays: D'Abate, "On the Meaning of a Name," and Axtell, "Exploration of Norumbega."

6. For a discussion of the "profound sense of historical awareness [that] guided the European colonial conquest of 'people without history,'" see Prakash, "Writing Post-Orientalist Histories," 353. For a look at the uses of history in Britain's involvement in the Americas, see Sparks, "England and the Columbian Discoveries."

7. Ralegh's *Discovery* has recently been the subject of an extensive anthropological rereading: Whitehead, "Historical Anthropology of Text."

8. Schomburgk, "Introduction," vii.

ollections, and I felt in imagination transported once more into the midst of the stupendous scenery of the Tropics.[9]

This sympathy with Ralegh led Schomburgk to set about an explicit vindication of the veracity of Ralegh's text, coupled with a rehabilitation of Ralegh's character. Schomburgk wrote: "Our object is to view Ralegh chiefly as the father of American colonisation and a promoter of commerce and navigation."[10] This object demanded a considerable apparatus, a running commentary on the original text, which allowed Schomburgk to "prove, from circumstances which fell within my own experience, the general correctness of Ralegh's descriptions."[11] So exuberant was this vindicatory footnoting that the preface apologized for the "overburdening of the text."[12] On certain issues the exculpatory endeavor was easier than on others. Ralegh's much-maligned assertion that he had encountered "a store of oisters upon the branches of the trees . . . very salt and wel tasted" could be raised above scorn by a lengthy footnote detailing the Latin names of the "three species of mollusca chiefly found on mangrove trees," though Schomburgk felt obliged to place himself at gustatory odds with his predecessor, declaring *Ostrea rhizophorea* neither so large nor so delicate of taste as the European variety.[13]

Recasting Ralegh as a scientific and commercial explorer "in embryo" hit a very serious snag in the myth of El Dorado, however. Central to Ralegh's account of Guiana was the seductive story of Juan Martínez de Albújar, which the Elizabethan privateer had wheedled or whipped out of the Spanish governor of Trinidad in 1595. The account—the baroque death confession of a man who claimed to have been the captive of cannibal heathen in the interior of South America, where he had seen an inland ocean in the midst of mountains, a gilded chief, and a golden city (which he almost succeeded in looting)—confirmed Ralegh's highest hopes for the continent that lay to the south.[14] The scattered intimations of Guiana wealth that he and his backers had cobbled together out of seized Spanish correspondence, prying spies, and captured hidalgos took shape around the narrative of an eyewitness. Ralegh painted over the terra incognita beyond the sketchy coast of South America with a beckoning scene: the inland sea of Parima, the golden city of Manoa, and most important, the Indian prince so rich he had himself anointed with powdered gold—*El Dorado*, the golden one.

9. Ibid.
10. Ibid., xxii.
11. Ibid., ix.
12. Ibid., ix.
13. Ralegh, *Discovery*, 3. Schomburgk's comment: ibid., n. 1.
14. See Lorimer, "Location of Ralegh's Guiana Gold Mine."

Schomburgk addressed the matter directly, writing of the captivating magnetism of the gilt cacique and his realm: "Such was the influence of this seducing picture, first sketched by rumour, and then coloured by imagination, that the more victims it drew into its vortex, the more were found to embark in plans for its attainment."[15] Schomburgk went so far as to assert, against all odds and evidence, that Ralegh, "patron of science in general," probably never himself believed in the tale of El Dorado but rather invented and fed it, recognizing that the hardships of settlement and agriculture would be unlikely to attach many diligent men to his expedition.[16] Schomburgk thus construed the myth of El Dorado as the first incentive to travel in the region, a chimerical empire that was also the ultimate destination of the many expeditions that had followed on Ralegh's path. That those expeditions constituted a long litany of failures was well known. The standard mid-nineteenth-century history of the colony described how the auriferous myth exerted a siren call on those approaching Guiana. They "pursued the phantom" to their deaths: "They followed its imaginary track . . . through savage valleys, interminable forests, and perilous swamps, to the south over the dark waters of the Rio Negro and the island-studded Amazon; but the land of promise vanished as they approached, and the further they advanced the more hopeless their pursuit. . . . Their bones whitened the banks of rivers—successive expeditions perished."[17] As for himself, Schomburgk claimed that he had succeeded where all others had failed: he had reached the site of Ralegh's El Dorado.

In this claim he was not, as I will show, alone. In the first half of the nineteenth century El Dorado—and the inland sea of Parima on which it was said to be situated—served as the gravitational center of British Guiana. Each history or account of exploration began with an account of the myth of the interior that had drawn "countless" explorers and treasure seekers into its maw. Explaining the origins of the myth (and locating the sites thought to have given rise to the story) were primary tasks taken on by geographical explorers, whose expeditions could thus drive the sea and the city from the map even as they located the source of the myth. They could complete Ralegh's failed *entradas*.[18] In this sense El Dorado not only proved an orienting point for the process of interior exploration and surveying in British Guiana, it was also a central landmark in crafting the ter-

---

15. Schomburgk, "Introduction," xlviii.
16. Ibid., xxxvi.
17. Dalton, *History of British Guiana*, 124.
18. Leed has suggested the existence of a pervasive "lost-boy complex" among nineteenth-century explorers (*Shores of Discovery*, chap. 8). He traces a set of attempts (Franklin, Stanley, Livingstone, Park) to find lost predecessors.

ritory of the colony: Ralegh was the founder of the British Empire; if the colony encompassed the "actual" site of El Dorado, this fact would anchor an anomalous British possession.

Schomburgk grounded his claim to have reached the true site in the extensive work of his role model and compatriot Alexander von Humboldt, who had found the hydrography of the Guianas and its mythical entanglements worthy material for synthetic investigations of nature in the New World. Writing in volume 5 of his *Personal Narrative,* Humboldt turned his attention to river sources in the Guianas: "The interest we feel in this subject is not that alone which is attached to the origin of all great rivers, but is connected with a crowd of other questions [including] . . . the fable of el Dorado."[19] Having traveled up the eastern branch of the Orinoco as far as the mission town of Esmeralda (where he asked the local inhabitants if they knew anything about a vast inland sea farther east), and having undertaken a "long and laborious study of the Spanish Authors who treat of el Dorado,"[20] Humboldt announced that no lake of such scale could exist in the region. But myth was not arbitrary, and "all fables have some real foundation."[21] Hence an inland sea of some 2,400 square miles[22] and its accompanying legend, which had cost lives on an ill-fated expedition as late as 1775, could not be drained from the map without some account.

Humboldt gathered evidence for the source of the myth during his and Aimé Bonpland's celebrated passage on the Casiquiare canal.[23] The existence of a natural canal connecting the Amazon and the Orinoco, unlikely as it seemed, proved a reality. Humboldt and Bonpland followed the branch of the upper Orinoco to the Rio Negro, a tributary of the Amazon, demonstrating (to general perplexity) that water from one channel could seemingly flow in two directions. While the news that seasonal channels linked "different" rivers was no news to the migrant Amerindians who had used these *itabbos*[24] for generations, marking them with runic petroglyphs, it was a discovery that undermined a fundamental principle of European hydrography: that rivers were confined to their respective river valleys and separated by watersheds. The jungles of South America appeared to defy these simple principles, confronting explorer and hydrographer alike with an anastomosing morass of meandering trib-

19. Humboldt, *Personal Narrative,* 5:312.
20. Ibid., 5:773.
21. Ibid.
22. Brett, *Indian Tribes of Guiana,* 60.
23. Discussions of the controversy over the Casiquiare include Karsen, "Alexander von Humboldt in South America"; Smith, *Explorers of the Amazon,* chap. 7, "Baron von Humboldt"; and Botting, *Humboldt and the Cosmos,* 125.
24. BL, Hilhouse MS, "Geographical Description," 9.

utaries, fluctuating according to seasonal and evolving dynamics. Humboldt wrote: "When immense rivers may be considered as parallel furrows of unequal depth; when those rivers are not enclosed in valleys; and when the interior of a great continent is as flat as the shores of the sea are with us; the ramifications, the bifurcations and the interlacings in the form of net-work, must be infinitely multiplied."[25]

What Humboldt needed to explain was how apparently sourceless rivers could seemingly flow in circles, mingle, and separate. The key, he claimed, lay in understanding how the dynamic force of sedimentation could build up "isthmuses of deposited earth," which separated a stream at its headwaters into two except in "the time of great inundation."[26] An "isthmus" was thus "partly mountainous, partly marshy ground"[27] that lay at the near-contiguous sources of different rivers and would, during the rainy seasons, be entirely submerged. In the traditional mythic geography of the Guianas, Lake Parima lay at the source of both the mighty Orinoco and the Rio Branco (a tributary of the Amazon),[28] and possibly the Essequibo as well; but according to Humboldt it was more likely that what would be found at the sources of those rivers was not a vast inland sea, but rather a vast seasonal floodplain, an "isthmus." Using historical researches in the maps and expedition accounts, Humboldt zeroed in on the source of the myth.[29]

The "classical soil of Dorado de Parima,"[30] Humboldt revealed at the end of volume 5 of his *Personal Narrative,* lay in a boggy isthmus at the quasi confluence and pseudosource of tributaries of all three of the greatest rivers of northern South America: the Orinoco, the Amazon, and the Essequibo. This site on the southern side of the Pakaraima Mountains was "subject to frequent overflowings" and was the location of a shallow lake called Amucu, known through the sketch of a Dutch trader called Nicholas Horstman who had passed the area in the mid-eighteenth century and whose journal Humboldt had been able to consult in a manuscript copy held in Paris.[31] Horstman, originally commissioned by the Dutch to investigate the southern part of the colony, proved to be an enigmatic figure. Instead of returning from his voyage of discovery into

---

25. Humboldt, *Personal Narrative,* 5:315.
26. Ibid., 5:316.
27. Ibid., 5:481.
28. Ibid., 5:375.
29. Humboldt actually separated out two distinct sources for the myth. One had been painted as a plausible common source for the Caura, Caroni, and Paragua Rivers (Paragua "in the Caribee language meaning sea or great lake"; Humboldt, *Personal Narrative* 5:785). The "original" depiction of Lake Parima, however, had placed it farther south and made it longest from east to west.
30. Humboldt, *Personal Narrative,* 5:790.
31. Ibid., 5:593.

the interior—and hence offering the Dutch a cartographic extension of their territory—he kept going, crossing not only the watershed of the river but a political watershed as well.[32] Some rumors suggested he had "gone native" and was leading Amerindian raids in the savannas; others said he had turned traitor and divulged secrets of Dutch emplacements to the Portuguese. What is certain is that after his transformation from commissioned explorer to nomadic "interloper" he had the occasion to meet a Frenchman making his way to Cayenne down the Amazon after some very trying experiences in Peru—Charles-Marie de La Condamine.[33] Horstman sketched him a map of his route (fig. 2), which La Condamine passed on to the French geographer Jean-Baptiste d'Anville, who incorporated the path into his *Carte de l'Amérique Méridionale* of 1748.[34]

Humboldt reconstructed Lake Amucu as the veritable and fabled site of Parima using Horstman's manuscript map and journal. Not only was the spot a marshy isthmus, seasonally flooded, but close attention to Horstman showed Humboldt that the geological substrate of the region seemed to consist of a "micaceous" granite, "full of druses and open veins"[35] of rock crystal. The stage was set: the ground lakelike, the soil sparkling. All that remained was to find the name in the place, to have sufficiently layered it with indications that it was the authentic "source" of the myth of El Dorado. Finding in the writings of a Spanish missionary an account of a small river called the "Parime" that joined the Rio Branco not far from the place where it petered out onto the crucial isthmus, Humboldt considered the case closed. "All names," he asserted, "that figure in the fable of Dorado are found in the tributary streams of the Rio Branco." For good measure he also pointed out that "Parima seems to denote vaguely great water" in one of the "Caribee dialects."[36] Multiple lines of inquiry—cartography, etymology, ethnography, and hydrography—intersected on a swampy corner of the savannas south of the Pakaraima Mountains, marking the point as the confluence of myth and the epicenter of European desire in lowland South America. The Amerindian village at that point was called Pirara, after the local creek. Humboldt wrote: "Here, in a river called Parima, and in a small lake connected with it, called Amucu, we have basis enough on which to found the belief in the great lake bearing the name of the former; and in the islets and rocks of

---

32. De Villiers, "Foundation and Development of British Guiana," 13.
33. The dramatic story of the La Condamine expedition has been well told in a number of places. See, for instance, Pratt, *Imperial Eyes*.
34. Figure 2 has been reproduced from *Brazil Arbitration, British Case Atlas*, 8.
35. Humboldt, *Personal Narrative*, 5:792.
36. Ibid., 5:795.

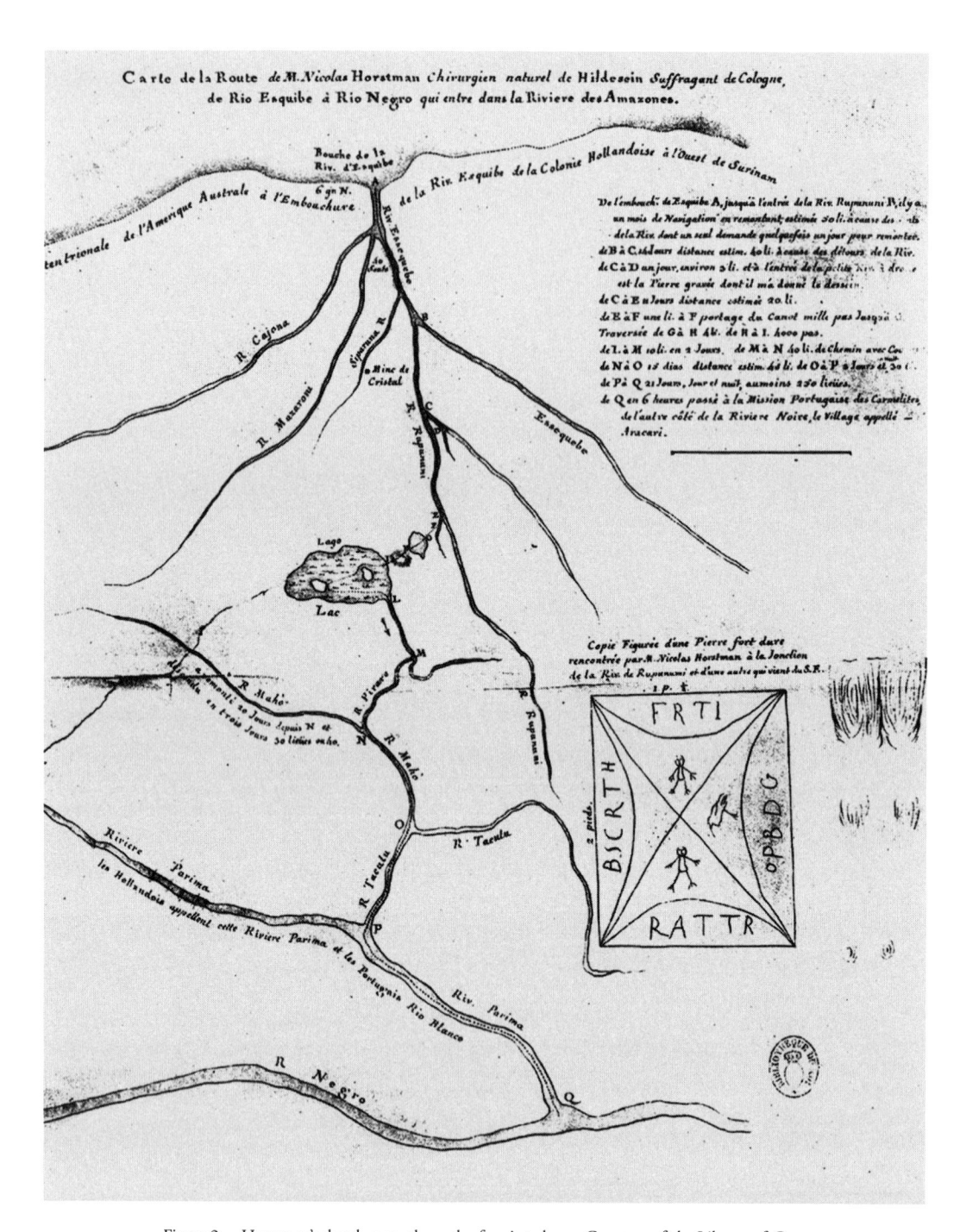

Figure 2. Horstman's sketch map: the path of an interloper. Courtesy of the Library of Congress.

mica-slate and talc which rise up around the latter, we have materials out of which to form that gorgeous capital, whose temples and houses were overlaid with plates of beaten gold."[37]

Though "the myth of El Dorado . . . wandered from West to East," Humboldt wrote, he had "followed it" as it had been "tacked on to different localities," until it came to rest on the small swampy Lake Amucu.[38] Myth had become place: "Dorado, like Atlas and the Hesperides, has gradually stepped out of the realm of poetry to take up a position in that of systematic geography."[39] Through Humboldt's work Horstman became, in a sense, the first European to "find" El Dorado, but it was not clear that he ever knew of his good fortune. His voyage—into the interior and never back out—made him an equivocal founding figure, not least for the British at the end of the nineteenth century, whose attempts to defend territorial claims to the southern savannas depended on establishing the extent of earlier Dutch possessions. Commissioned by the Dutch, Horstman's path traced out Dutch claims. But only if he stayed Dutch. If he had joined forces with the Portuguese, then his route (or at least the part of it traced subsequent to his switch in allegiance) could be construed as a Brazilian claim. Both sides marshaled evidence of his affiliations in the hopes of substantiating their territorial claims on his ambiguous passage.[40]

Humboldt himself never stood on the ground that his researches had made into a sacred site in the geography and history of South America. He found El Dorado in manuscripts in Paris. Others, however, were in a position to reach the point on the ground.

MYTHICAL EMPIRE AND INTERIOR EXPLORATION 2: REACHING EL DORADO

Getting to El Dorado had been a theme of English travelers in the Guianas long before they read Humboldt's findings. Charles Waterton, an eccentric naturalist whose *Wanderings in South America* recorded his experience botanizing and collecting specimens in the British colony (where he managed his family plantation from 1806 to 1816),[41] enticed his readers into the pursuit of the legend: "According to the new map of South America, Lake Parima or the White Sea ought to be within three or four days' walk from this place. On asking the Indians whether there was

37. Cited in Brett, *Indian Tribes of Guiana*, 60.
38. Humboldt, "Preface," viii.
39. Humboldt, "Concerning Certain Important Features of the Geography of Guiana," 6.
40. *Brazil Arbitration, British Case*, 27–28 and passim.
41. Blackburn, *Charles Waterton*; Waterton, *Wanderings*, biographical sketch.

such a place or not, and describing that the water was fresh and good to drink, an old Indian who appeared about sixty said there was such a place and he had been there."[42]

One of Waterton's proposed journeys had the search for the lake as its entire purpose.[43] John Hancock, another early nineteenth-century explorer who made his way into the interior of British Guiana in 1810 (and to whom I will return), made reference to the mythic site to underline the breadth of his peregrinations in the interior when he wrote, "The attention of Europeans was long excited by the fables of El Dorado and the Lake Parime, where the writer has travelled."[44] Back in London, members of the RGS heard a number of reports from travelers who claimed to have reached the sacred site in the early 1830s. Captain James Edward Alexander, reporting on his West Indian voyages in the Society's second *Journal,* declared: "My purpose was now to proceed up the noble Essequibo river towards the El Dorado of Sir Walter Raleigh."[45]

The following year, in 1833, the Society heard a paper titled "Journal of a Voyage up the Massaroony in 1831," whose author had made the most studied attempt to reach the site of Lake Parima and to account for its not being a lake: "On consulting the maps and observations of Humboldt, I was immediately convinced that the Massaroony must be the natural drain of the intermediate space between the Cuyuny and the Essequibo, and by giving it a south-west direction it would intersect that undiscovered region, the Eldorado, or Great Golden Lake of geographical fable."[46]

The writer, William Hilhouse, had been a resident of the colony of British Guiana since 1815,[47] rising to prominence by organizing the Amerindians of the interior into an armed force that helped suppress the John Smith slave rebellion in 1823.[48] By the early 1830s he had became the local colonial expert on matters related to the bush. Whereas a decade earlier Waterton had wandered about looking for Lake Parima without the benefit of having read Humboldt's *Personal Narrative,* Hilhouse knew the book well.[49] Familiar with the *itabbos,* or flooding rivulets that laced the savannas—as well as with the hypothesis that mica and quartz had been the golden deception of Ralegh, his predecessors, and his follow-

42. Waterton, *Wanderings,* 114.
43. Ibid., 174.
44. Hancock, *Observations* (1840), 10.
45. *JRGS* 2 (1832): 65. See also Alexander, *Transatlantic Sketches.*
46. *JRGS* 4 (1834): 26.
47. CRL, Long Collection, scrapbook H.
48. The subject of Costa, *Crowns of Glory.* See also Menezes, *British Policy,* 211.
49. For Hilhouse's praise of Humboldt, see BL, Hilhouse MS, "Journal," 3 recto.

ers—Hilhouse used these elements to substantiate his own claim to have reached El Dorado via the Mazaruni River.

Like Humboldt, Hilhouse was persuaded that no Lake Parima remained in existence; he based this claim on the testimonies of Amerindians he had come to know in his official colonial capacity of "Quartermaster General to the Indians." Hilhouse initially followed Humboldt in ascribing the rise of the fable to the seasonal flooding of the *itabbos,* but in his two-month excursion up the unexplored Mazaruni, a large tributary of the Essequibo flowing from the southwest, Hilhouse found what he declared to be the "key of Eldorado."[50] In the company of twenty-three Akawoi Amerindians whom he and his companion had hired as guides and paddlers,[51] Hilhouse followed the river through a maze of low islands to a large waterfall, Teboco. There the cataract dropped through a narrow granite pass. Climbing to the top of Teboco Falls, the party saw a stagnant river, "with little more current than some lakes."[52] The surrounding cliffs were precipitous and stony, showing clearly their geological strata. He described it this way: "At the point of Teboco, the granite assumes a regular formation for the first time, and is ever afterwards found in strata, at an angle of about five degrees above the horizon; its apex being nearly northward. It forms the base of all the cliffs to a height from six hundred to about one thousand feet, when a perpendicular and cubical formation of quartz is the general superstructure to about fifteen hundred feet higher."[53]

The barren rock formations gave Hilhouse the impression of the dry bed of an ocean, and remarking that the Teboco Falls broke through a fissure in the granite fifty feet high on each side, Hilhouse was convinced that he had discovered the deeper meaning of El Dorado: "If, at any former period the horizontal stratum of granite at the pass of Teboco was unbroken, the still water must have been raised at least fifty feet above its present level, and a vast lake, ten or twelve miles wide by one hundred and fifty or two hundred miles long, would be the natural consequence." If it were so, he continued, "El Dorado need be no fable."[54] Moreover, a quartz crown like a shining beacon would have stood over the sea. Hilhouse christened the summit "Raleigh's Peak" (fig. 25, lower right corner). He

50. *JRGS* 4 (1834): 40.
51. Ibid., 27.
52. Ibid., 32.
53. Ibid., 33.
54. Ibid., 32. It is possible that Hilhouse got this idea from the suggestive view of Lake Guatavita in Humboldt and Bonpland, *Vues des Cordillères.* The lake (itself associated with the El Dorado legend) had a distinctive geological fissure that served as a spillway and limited the depth of the water. See Botting, *Humboldt and the Cosmos,* 148.

was the first explorer personally to drive the cartographic fantasy from the map: he had found El Dorado and abolished it entirely.[55]

To move El Dorado over to Teboco Falls was to suggest that Humboldt had perhaps not found the true site after all. It is little surprise, then, that Humboldt did not think particularly highly of Hilhouse's catastrophist account for the origins of El Dorado. The Prussian sniffed at the latter's "dearth of astronomical observations" and somewhat sketchy details on heights and distances. Humboldt had considerably warmer things to say about the work of his protégé Robert Schomburgk, who in 1836 became the first European to stand on the shores of Lake Amucu *with the knowledge of what that meant.* On the first page of the preface to his edition of Ralegh's *Discovery,* Schomburgk wrote that he had "traversed . . . the site of the gorgeous capital of El Dorado with its sea-like lake."[56] He wrote about the experience in more detail in an earlier publication, the large-format illustrated book he published in 1841, *Twelve Views in the Interior of Guiana.* The third plate in that book depicts the approach to the Amerindian village of Pirara (3°38′ N, 59°16′ W), and nearby Lake Amucu (plate 1).[57] The text orients the reader to the mythic significance of the place. El Dorado was "another Scylla and Charibdis," "a waste of human life unparalleled in the history of imaginary schemes," and nothing less than "a device invented by Satan to lure mankind to destruction."[58] Only after these solemnities did Schomburgk reveal that he was bringing readers to the very threshold of this magnetic point. Referring to the researches of "the most eminent traveler of our age," who caused the erasure from maps "of those last vestiges of that delusive bubble, El Dorado," Schomburgk explained that Pirara had been demonstrated by Humboldt's "deep reasoning" to be the "classical soil" of the myth. Placing himself before the scene, Schomburgk invoked the sublimity of the landscape: it possessed power which "that traveler alone can conceive who treads a *terra incognita* connected with such associations."[59]

---

55. "The result of this expedition proves that there is no alpine or other lake in this region which is the site of El Dorado or Parime" (BL, Hilhouse MS, "Journal," 9 recto).

56. Schomburgk, "Introduction," vii.

57. Plate 1 and subsequent illustrations from the *Twelve Views* have been reproduced from the copy held by the Yale Center for British Art, cataloged as 720 fol. A.

58. Schomburgk, *Twelve Views,* 7.

59. Schomburgk did, however, invoke Hilhouse's grander cataclysmic theory, adopting it to Humboldt's locus classicus at Pirara. He wrote: "The geological structure of the region leaves but little doubt that it was once the bed of an inland lake, which by one of those catastrophes of which even later times give us examples, broke its barriers and forced a path for its waters to the Atlantic. May we not connect with the former existence of this inland sea the fable of El Dorado and lake Parima? Ages may have elapsed, generations may have been buried and returned to dust; the nations who once wandered on its outer banks may be extinct and exist no longer even in name; still the tradition of the lake Parima has sur-

REMAPPING AND THE MAKING OF EXPLORERS

Here we need to pause and consider this most interesting idea: a terra incognita that, though by definition unknown, is nevertheless recognized to be a historical source. The seeming paradox captures many of the functions of mythical places in the history of exploration and geographical construction of colonial territory. Through the mediating account of natural history, the eye of the explorer not only could survey what was before him but could suffuse the place with deeper meaning that gave significance and historicity to his own presence. By constructing the place as the epicenter of all previous expeditions, Schomburgk's act of seeing Pirara completes the journeys of Keymis, Ralegh, and the thousands of nameless precursors who died in pursuit of the point where Schomburgk stands. He has walked over the banks they whitened with their blanching bones; he has survived where they failed. The "fact" that the Parima lay firmly at Schomburgk's feet confirmed his identity as an explorer and allowed him to substantiate the prophecy of the great Humboldt.[60] The explorer and the narrative of the exploration thus became critical bridges, linking discontinuous empires (Amazonian, Britannic), completing the place itself; the narrator stands at the head of a long lineage of those who have sought to possess the site. The rhetorical power of this maneuver can hardly be overstated. While Schomburgk can, with effusive grace, declare himself to be a geographical and colonial pilgrim—merely following the steps of a giant, and the very progenitor of British imperial power at that—he at the same time elevates British Guiana from an entirely marginal status in that empire to its very historical center. This moves his own place from that of an explorer at the margin to that of an explorer who has traveled to the symbolic and historical center of terra incognita.

The map was a key instrument in this negotiation of precedent. Schomburgk requested permission from the council of the Hakluyt Society to "illustrate Sir Walter Ralegh's Journey up the Orinocco by a Map."[61] When the Society ratified the inclusion of the map (fig. 3),

---

vived these changes." Schomburgk, *Twelve Views*, 8. The reference to a recent cataclysm may be to Mea Kaumen. See Oldroyd, "Historicism and the Rise of Historical Geology." Another possible reference would be Lyell's discussion of the bursting of glacial lakes in the Alps: Lyell, *Elements of Geology* 1:192ff.

60. It is an exemplary instance of the traveling rhetoric identified by Leed, in which the explorer becomes responsible for "the conjunction of a dreamed unconscious landscape with an observed reality in the present time." Leed, *Mind of the Traveller*, 139. Schomburgk's contemporary, Alexander William Kinglake, offered a wry commentary on the ubiquity of the trope: "If one might judge of men's real thoughts by their writings, it would seem that there are people who can visit an interesting locality and follow up continuously the exact train of thought that ought to be suggested by the historical associations of that place." Kinglake, *Eothen*, 94.

61. Schomburgk, "Introduction," xi.

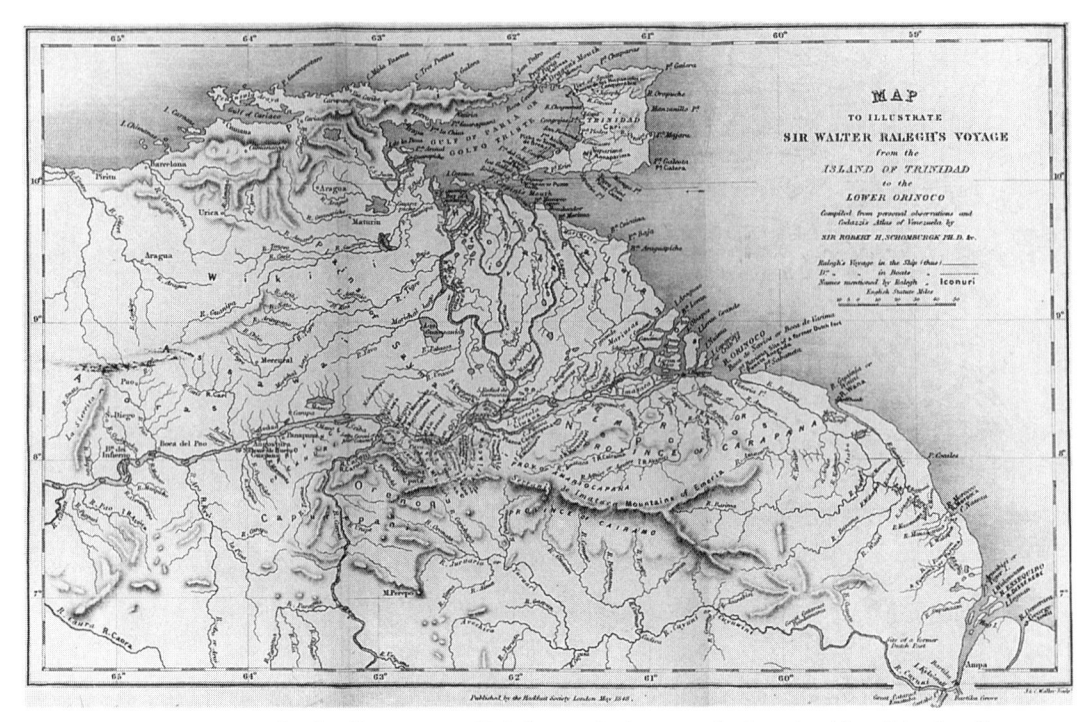

Figure 3. Schomburgk's remapping of Ralegh: a metaleptic cartography. Reproduced from Schomburgk's edition of the *Discovery;* courtesy of the New York Public Library.

Schomburgk had the privilege of tracing Ralegh's route across his own map, a map he drew "in great measure from personal observations made during eight years' rambles through Guiana." Schomburgk made Ralegh walk over Schomburgk's Guiana; a small dotted line marks Ralegh's "course" on Schomburgk's map. This graphic maneuver merits closer attention. It creates a shifting relationship between Schomburgk and his precursor. Initially it seemed that Schomburgk was merely invoking an illustrious predecessor in order to increase his own stature and authority. But the remapping of Ralegh's route demonstrates an additional element. Schomburgk's conceit—his portrayal of himself in Ralegh's wake—is actually just one half of an elaborate rhetorical posture that simultaneously cedes precedence to Ralegh (in order to invoke his authority) while overwriting his text with hundreds of notes and redrawing his path on a map whose legend boldly proclaims that it is Schomburgk's. This latter process actually *strips* Ralegh of his geographical authority. Schomburgk has construed himself as a follower of Ralegh, but Ralegh has been reconfigured and surpassed by being placed on Schomburgk's map. Schomburgk has invoked Ralegh's authority only to subsume that authority entirely.

The relationship is not one of succession (in which authority falls to the successor), nor is it one of fidelity to precedent (in which authority remains with the predecessor). Instead, the relationship is circular. What is generated is a subtly tautological ring of legitimization. Ralegh legitimizes Schomburgk; Schomburgk exculpates and restores Ralegh. Schomburgk walks Ralegh's path, but Ralegh walks on Schomburgk's map. Schomburgk has drawn geographical authority from an antecedent, even as he has, to a great degree, stripped that antecedent of geographical content. I have borrowed a term from classical rhetoric to describe these cycles of priority and posteriority on which the geographical authority of explorers was built: *metalepsis,* "founded on the relation between the antecedent and the consequent, between what goes before, and immediately follows."[62] To give to the term the explicitly cyclical significance I am after, I must stretch its traditional meaning, where it denotes exclusively retrospective tropes.[63] As I use the term it refers to invocations of history that are deployed in order to authorize even as they are stripped of their authority and content.[64]

Schomburgk's relationship to Ralegh affords a choice example of this trope, but this is no isolated occurrence of a curious but insignificant relationship between two explorers and their texts. It is rather an exemplar of a scenario that is replayed repeatedly in the annals of imperial exploration: later travelers cycle back over their predecessors by editing their texts and redrawing their routes on fresh maps, all the while borrowing orienting points, geographical authority, and ennobling historical context from them. One thinks immediately of Cook's presiding legacy for subsequent explorers in the Pacific and Australia, of the role of remappings in the geographical construction of colonial Africa. For Guiana, Schomburgk himself would be followed on, corrected, and annotated by Barrington Brown in the 1860s and 1870s, and later in the 1930s by Walter Roth, who would translate and remap the Schomburgks. As I will show in the next chapter, Robert Schomburgk both employed and usurped the authority of his predecessor Hilhouse. Through cycles of remapping Schomburgk negotiated a relationship with Humboldt as well. Powerful as this trope proves, it is nearly imperceptible, smuggled in in footnotes and dotted lines.

The myth of El Dorado served a metaleptic role in the writings of ex-

62. *Oxford English Dictionary,* s.v. "metalepsis"; Blair, *Lectures on Rhetoric,* xiv.
63. A variety of meanings for the term have been suggested. See Lanham, *Handlist of Rhetorical Terms,* 65. The *OED* states that "in many English examples the use appears vague." For a similar stretch of the meaning of metalepsis, see Miller, *Topographies,* 21.
64. For a discussion of the formation and malleability of canonical texts (and the potential value of such an approach to Science Studies), see Schaffer, "Contextualizing the Canon."

plorers who visited Guiana. Invoking El Dorado conjured up a wealth of weighty historical associations, but as I have shown, *what El Dorado was* was left open to a myriad of interpretations and reconstructions. Writers and explorers emphasized the power of the myth, but they did so as a prelude to providing an account of how they found its source. The issues of authority that were at stake are evinced by the testy spat that arose between Schomburgk and the author of a book titled *El Dorado*, published in 1844 in New York. The author, Jacob Adrian Van Heuvel, claimed to have been awakened to the importance of the fabled city, and Ralegh's voyage, "by a visit I made, some years hence, to a part of British Guiana."[65] Van Heuvel, a Yale graduate and law student, was retracing his roots. His father had been born in Demerara, and the younger Van Heuvel claimed the elder had been governor.[66] In *El Dorado* Van Heuvel focused on the roots of the legend rather than on his family tree. He reported having made his way some distance up the Mazaruni on his voyage in 1819 or 1820 and having parleyed, through the intercession of an unnamed "Agent," with the "Charibee chief, Mahanerwa," who happened to be biding his time in a nearby camp, waiting for an epidemic in Georgetown to subside before approaching the town.[67]

"Mahanerwa" (now written Mahanarva)[68] was a celebrated figure in the early history of the British colony. Reputed to be the cacique of a large and strong Carib tribe, he arrived in Georgetown in 1810 seeking "presents" from the British, giving rise to a spate of colonial tales about the interior (to which I return at the end of this chapter).[69] Van Heuvel's assertion that he had an audience with the chief was a very strong claim to historical and geographical knowledge of the terrain.[70] In the words of another colonial author, "He was the most intelligent and correct of all the Indians I ever met with."[71] The encounter with Mahanarva thus served as a crucial authenticating episode in Van Heuvel's text. He not only had privileged access to the geographical origin of the El Dorado story, but he also portrayed himself as traveling back in time as he trav-

65. Van Heuvel, *El Dorado*, introduction.
66. Dexter, *Biographical Sketches of the Graduates of Yale College*, s.v. "Vandenheuvel." No "Van Heuvel" (or "Vandenheuvel," the original family name) appears on the "List of Commanders and Governors of Essequibo, Demerary & Berbice up to 1814." Netscher, *History*, 151–53.
67. Van Heuvel, *El Dorado*, 82.
68. Menezes, *British Policy*, 23 and 182.
69. For the most extensive work on these relationships, see Whitehead, "Carib Ethnic Soldiering," and Whitehead, "Ethnic Transformation."
70. The Schomburgks never met him, but they wrote about visiting the ruins of one of his camps. Schomburgk, Brown, and Im Thurn all claimed to have traveled with descendants of Mahanarva.
71. Hancock, *Observations* (1835), 22n.

eled up river, which allowed him access to the temporal as well as spatial fountainhead.[72]

Approached by the men who were to lead him to the Carib chief, Van Heuvel wrote: "They were nearly naked; and not only were their bodies painted, but their heads were profusely covered over with a paint of scarlet brilliancy; their faces marked with black streaks across their cheeks, their eyebrows painted, strings of shells around their necks, and war clubs hanging at their wrists."[73] This apparition (and the flashes of gold ornaments)[74] did not fail to trigger the author's awareness of his position on the shore of an ancient order. He continued, "I looked at them with historical recollections connected with their nation, with a powerful sensation; and the scene . . . has ever remained fresh in my mind."[75] Having construed his encounter as *the* Carib encounter,[76] Van Heuvel reenacted the El Dorado quest, commencing an "inquiry . . . concerning the existence of lake Parima, its character, etc." The answer came quickly. "Leaning out of his hammock, with a stick in his hand," Van Heuvel wrote, Mahanarva "marked on the sand of the floor the issuing of the river Parima from the lake, the situation of the Branco," and other details of "the geography of the region concerning which I made my inquiries."[77]

Based on this exchange Van Heuvel concluded that Humboldt had been misguided in his assertion that all references to El Dorado and Lake Parima could be traced to the floodings of little Lake Amucu, lying in the marshy plain between the Rupununi and the Takutu. Instead, Van Heuvel conjured again a vast sea to the west, perhaps 250 miles long, situated in the terra incognita to the southeast of the Orinoco. He acknowledged that much of this "body of water" was likely a "temporary inundation," referring to the hydrography of Humboldt, but he questioned just how temporary it might be. After all, he reasoned, given that Guiana had two rainy seasons and two dry, "each of three months," and that when it rained it did

72. Leed (*Mind of the Traveller,* 207) examines the work of Fabian (*Time and the Other*) in the context of the history of travel. Leed claims that European eighteenth- and nineteenth-century "philosophical travel" is distinguished by the assertion that "differences of space can be read as differences of time." The trope as applied to rivers—upstream, back in time—is explored in Schama, *Landscape and Memory,* "Water."
73. Van Heuvel, *El Dorado,* 35.
74. Gold crescent-shaped ornaments were linked to the El Dorado legend from the very earliest accounts. For an account of gold in the region and a reappraisal of the El Dorado legend, see Whitehead, "Mazaruni Pectoral."
75. Van Heuvel, *El Dorado,* 35.
76. For the historical context of this encounter, see Hulme and Whitehead, *Wild Majesty.*
77. Van Heuvel, *El Dorado,* 36–37. For accounts of the exchange of the "geographical gift," see Bravo, "Ethnological Encounters"; Bravo, "Accuracy of Ethnoscience"; Belyea, "Amerindian Maps"; and Sparke, "Between Demythologizing and Deconstructing the Map."

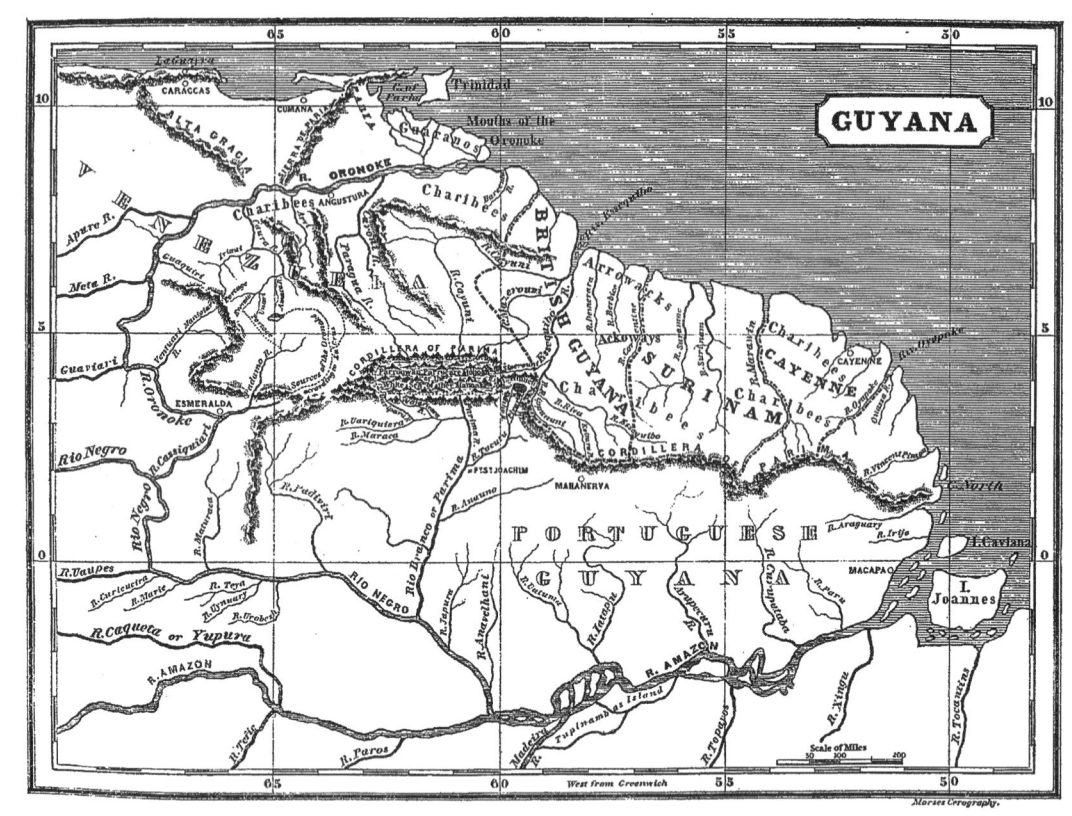

Figure 4. Van Heuvel's map of Parima: a rival claim to El Dorado. Reproduced from Van Heuvel's 1844 *El Dorado*.

so "continuously," it seemed a safe proposition that "water must fill the savannah" for half the year at least and quite probably more. If the sea was there more often than not, by what cartographic convention did Humboldt and subsequent geographers omit it from the map?

This reasoning underlay his cartographic remedy. Using the "large map of Arrowsmith"—the very same one that Schomburgk had set out to improve on for the RGS—Van Heuvel traced over it the scratchings he claimed to have seen in the sand on the banks of the Cuyuni and presented a "sketch map" as the frontispiece to his book (fig. 4). Here the "White Sea of the Manoas" is apparently restored, in orientation and size not differing greatly from the early representations of Jodocus Hondius and Jan Jansson (figs. 5 and 6).[78] On closer inspection the great instauration of Parima by Van Heuvel is less decisive; his depiction does not share the confident iconography of those earlier maps. It is as if Van Heuvel

78. Figure 5 has been reproduced from *Venezuela Arbitration, British Case Atlas,* 1. Figure 6 is from the author's collection.

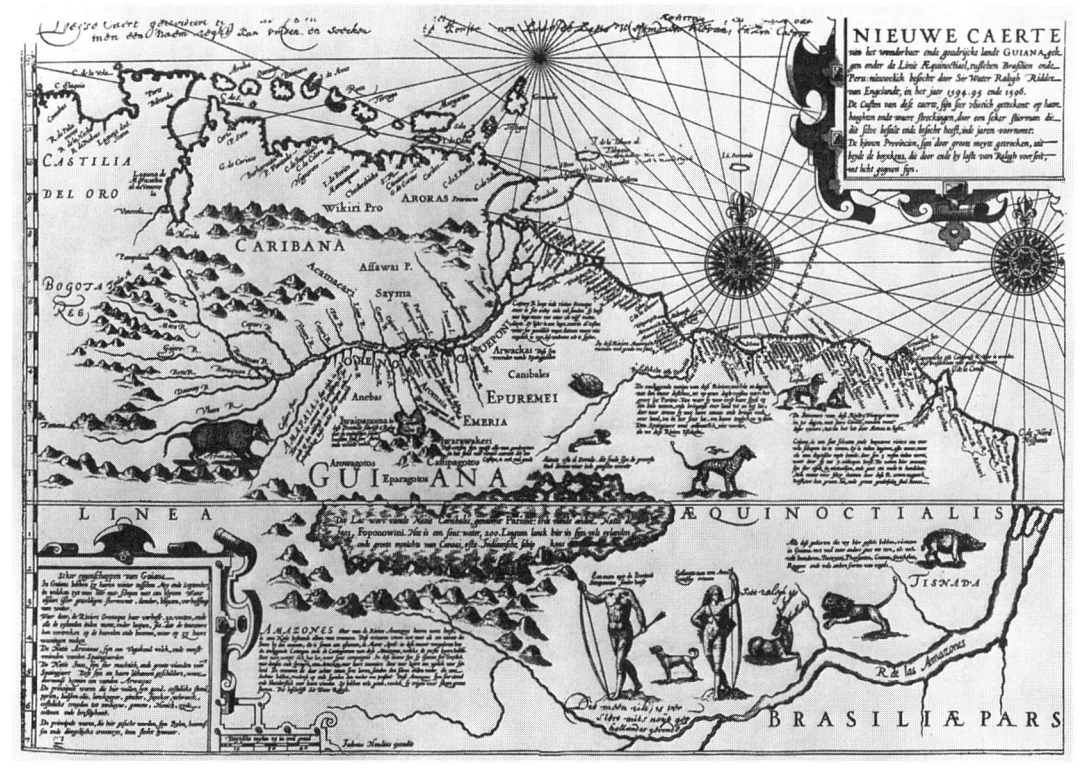

Figure 5. The Hondius map of Guiana, 1599. Courtesy of the Library of Congress.

could not entirely decide how to go about representing a seasonal sea (fig. 7). Within the bounding ring of the "Cordillera of Parima" lies an area of nervous visual vibration. The barred sign for "water" competes with tufted bursts presumably designating marsh; throughout the "body" of the White Sea run jagged ridges. This superimposition of cartographic signifiers for water, vegetation, and rock is minutely overwritten by a concatenation of place-names—"Parroowan, Parrocare, Monoan or the White Sea of the Manoas"—as if by sheer onomastic recitation the place might somehow force itself onto the page, the place, and the mind.

A contemporary guidebook for map drawing recommended using the guidelines of a sort of cartographic mimesis, particularly in depicting bodies of water: "The waviness or curvature of lines is meant to represent the water ... under that condition that smooth water presents when slightly disturbed, as by throwing a stone into it." Along a rocky coast, the author continued, these lines should be "proportionately bold and irregular," whereas along a level coast they would be "proportionately placid."[79] The visual density of Van Heuvel's White Sea, nebulously circumscribed by a

79. Wilme, *Handbook for Mapping*, section 6.

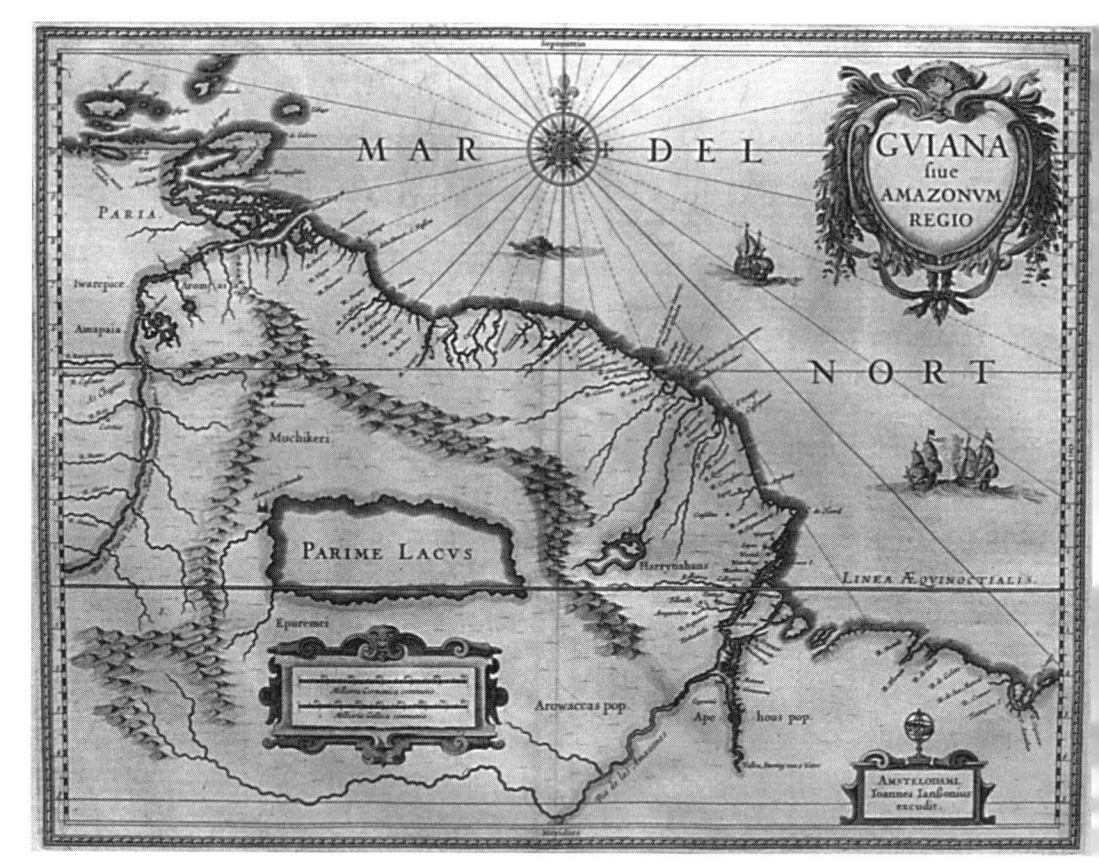

Figure 6. The Jansson map of Guiana, 1647. Collection of the author.

dotted line, suggests nothing so much as the condition of smooth water disturbed by centuries of speculation, rumor, and suggestion. His map offered a mimetic representation of Parima that stretched cartographic convention in its fidelity: the location was confusing, mixed. With no temporal axis, the map was trapped in semiotic indecision.

Schomburgk had no patience for interpretive niceties of cartographic ambiguity when he came upon a copy of Van Heuvel's *El Dorado* while he was living in Richmond after his return from his boundary expeditions. He was busily preparing the Hakluyt edition of the *Discovery,* and even if Van Heuvel was not literally restoring the inland sea but rather sketching a slightly more suggestive symbol for a Humboldtian isthmus, the little-known American was still intent on moving the source of El Dorado. A reassertion that the authentic Lake Parima might lie to the west, at the headwaters of the Parima, was an assertion that Lake Amucu— whose marshy shores Schomburgk had by this time celebrated in word, hallowed in image, and defended (as will be seen) in a military expedi-

Figure 7. A detail from Van Heuvel's map of Parima: the ambivalent inland sea. Reproduced from Van Heuvel's 1844 *El Dorado*.

tion—was nothing more than another savanna mudhole. It was to suggest that an obscure New Yorker of dubious scientific and scholarly qualifications was making a bid to inherit Ralegh's mantle, to be the "discoverer" of El Dorado once and for all.[80]

Castigating the "pompous title" of the work, Schomburgk set out in a long note to his edition of Ralegh's *Discovery* to refute Van Heuvel's book, which Schomburgk claimed was "illustrated by a map on which the lake Parima figures in its whole extent." Without acknowledging the ambiguity of that figuration, he insisted that the author "has fully restored [the inland sea]."[81] Humboldt himself had praised Schomburgk for having "brought light to bear" on these regions that "prior to his expedition lay, so to say, in the dark," and with this authority behind him, Schomburgk made short work of Van Heuvel's credentials to reinvent an inland sea.[82] "Mr. Van Heuvel," he wrote witheringly, "visited the *coast regions* of Guiana,

---

80. In fact, Van Heuvel had achieved some fame for founding his own utopian empire. He squandered a family fortune buying up land near Ogdensburgh, New York, where he subsidized his tenant farmers and became an eccentric local hero. The town was renamed Heuvelton in the early 1830s. Hough, *History of St. Lawrence*, 204.
81. This quotation and the previous one are from Schomburgk, "Introduction," liv.
82. Humboldt, "Preface," viii.

without *penetrating into the interior.*"[83] Not only was the author thus not a true explorer at all, but he was wanting in the linguistic and geographical expertise necessary to invent and erase lakes: "His conclusions respecting this lake rest only upon what he learned from some Indians, whose language he did not understand, and upon the maps . . . of the last century."[84] Robert's brother, Richard, was equally dismissive.[85]

Robert Schomburgk did not stop at characterizing Van Heuvel's work as a "step backwards in geographical knowledge." He went so far as to call for a veritable censure of El Dorado fantasy and hydrographic speculation: "All who take interest in that science ought to aid in preventing the dissemination of such absurdities."[86] The suggestion was that Van Heuvel's "seed," sown abroad, might beget a new generation of pursuit and desire, might establish (or reestablish) a new golden empire, *Van Heuvel's El Dorado*. Schomburgk took advantage of the opportunity to address the British Association for the Advancement of Science at Cambridge in 1845 to give a talk "On the Lake Parima," and he mentioned the recent retrogression of geographical knowledge represented by Van Heuvel.[87] Schomburgk would not allow his new place as the keeper of the El Dorado tradition to be usurped by anyone else. His journeys were not to be traced on Van Heuvel's map, but Van Heuvel's travels were to be retraced on his, where they would be found to be short and shallow.

REMAPPING AND THE MAKING OF TERRITORY

In the previous section I showed that explorers paid attention to cycles of priority and that they remapped each others' geographical labors in order to shore up their own identities. In this section I extend this observation and show that these same cycles of historical awareness could be used to ground the territorial claims of a colonial power. The act of claiming territory is charged with a paradox to which I return in the final chapter: namely, that the act of taking possession always implies a previous *lack* of possession. This is not some poststructuralist paradox but an issue raised by nineteenth-century territorial disputants: in the boundary controversy over British Guiana's southwestern limits in the 1890s, Brazil used just this argument to call into question British claims to the upper Essequibo. The Brazilian case posed the question: If Schomburgk raised a flag at the

83. Schomburgk, "Introduction," liv. Emphasis mine.
84. Ibid.
85. M. R. Schomburgk, *Travels,* 2:262.
86. Schomburgk, "Introduction," liv.
87. Schomburgk, "On the Lake Parima," 51.

sources of the Essequibo in 1837, didn't that imply that they were *not* British territory before he did so?[88]

The strategies of exploration and mapping that have concerned us in this chapter—those that relied on (and extended) metaleptic cycles—afforded a means to avoid this foundational problem. They generated a shifting relationship between priority and posteriority, eliding the foundational moment: a subsequent explorer was "following" a predecessor, but having stripped that predecessor of geographical content—by remapping his routes and landmarks—the subsequent explorer had control over those seemingly prior routes and points. Rather than inaugurating imperial legitimacy in a single, tenuous (temporal or spatial) point, the circular structure of metaleptic cycles closed the authority of past and present into mutually reinforcing hoops to bind territorial claims. Below I examine two episodes in the consolidation of the territory of British Guiana and suggest that explorers' attention to Ralegh and El Dorado played a role in shaping and deepening territorial claims. The regions discussed—Pirara village and Barima Point—became points of contention in the arbitrations later in the century. As chapter 6 shows, the settling of British Guiana's territory was a complicated affair, and it would be impossible to argue that considerations of myth and history drove the process entirely. At the same time, it would be equally wrong to overlook their significance altogether. El Dorado and Ralegh's routes mattered to those who set about defining what British Guiana was in the nineteenth century.

Links between Pirara and El Dorado were part of the issue at stake in 1841 when British troops were to be sent to occupy the village and protect it from the encroachment of Brazilian forces. This fiasco of a military expedition (detailed in chapter 6) was precipitated by the actions of a renegade missionary who had established a mission in the area in 1839. But the path to military action was laid by Schomburgk's zeal for defining the boundaries of the colony in such a way as to include Pirara and hence (among other things) to protect the mission from Brazilian slavers' *descimientos*.[89] A recent microhistory of the "incident at Pirara" by Peter Rivière offers a detailed reconstruction of the events that led up to the colonial government's (and the Colonial Office's) decision to dispatch, and then hastily to recall, a costly expeditionary force to occupy Pirara

---

88. The implications were subtle: Brazil claimed that Schomburgk, as a merely scientific explorer, had no authority to claim territory. Therefore his taking possession was invalid but was not insignificant, in that it meant that the region did not at the time fall within the British territorial domain. For a further discussion of the question, see *Brazil Arbitration, British Counter-case,* chap. 6, 120. I return to this in the conclusion.

89. "Descents," the term (of reproach) used for the raids by slavers and military press-gangs.

during Schomburgk's boundary survey of the region.[90] Although this action enabled the missionary Thomas Youd to reestablish his mission after it had been sacked by Brazilian slaving parties, Rivière is ultimately at something of a loss to explain exactly why any right-thinking imperial administrator would bother with such a project.[91] Recognizing the indecisiveness, backtracking, and confusion that surrounded the expedition, and the haste with which it was recalled, Rivière sees the entire incident as an instance of imperial "absent-mindedness." For Rivière the incident demonstrates that "at least parts of an empire may be the result not of any grand design, but rather of the unintended outcome of a number of individuals going about their lives with Britain absent-mindedly looking on."[92]

This may be so, but is the incident at Pirara the best example? A certain view of Pirara makes it seem like an impossibly lonely outpost in a second-tier colony, a poor place to make a border stand, even if, lying at the intersection of two river systems, it could conceivably be a point of some future commercial or strategic significance should trade between the Essequibo and the Amazon develop (Humboldt's "isthmuses" were, by definition, "portages" as well).[93] But this chapter has shown that Pirara was assuredly not just any "part of an empire." As a result of the work of Humboldt and Schomburgk, Pirara possessed mythic stature in British imperial geography and imagination. These explorers' work attached to the thatched village a symbolism that placed it at the heart of British Guiana at least, and possibly of British imperialism as well: Pirara was El Dorado; El Dorado was Ralegh's destination; Ralegh was the founder of the British Empire. On reaching the site for the first time, on 24 May 1838, Queen Victoria's birthday, Schomburgk did not fail to raise the "British Union" for what he was certain was the first time in this "corner of her dominions."[94] But in his ruminations on Ralegh, Schomburgk represented Pirara not as a corner of empire, but as its cornerstone. While the site itself was so barren and deserted that the commander of the military expedition was reported to have exclaimed shortly after landing, "It must be the last place God Almighty made!"[95] the illustrator on the 1841 expedition looked across the barrenness and seemed to see, in the smoldering

90. The subject of Rivière, *Absent-Minded Imperialism*.
91. Ibid., 170, 173–74.
92. Ibid., 177.
93. Schomburgk emphasizes this fact in the conclusion of his *Description of British Guiana*, 150–51, writing that (with the construction of a short canal on the upper Essequibo) "inland navigation might be extended to Santa Fe de Bogata, and even to the Pacific on the West, and to Buenos Ayres on the South."
94. *JRGS* 10 (1841): 176.
95. Goodall, "Diary December to June," 51.

savanna fires that illuminated the night sky, a flickering mirage of the imperial metropolis. He wrote, "The part in the extreme distance put me in mind of a view of London from Hempstead Heath."[96] His watercolor of the departure of Brazilian stragglers depicted the triumphant rising sun of British imperial rule.

Schomburgk's view of Pirara was the reverse of the dreaded "calenture" that had terrified Ralegh and his contemporaries. That disease of the long sea journey was supposed to cause men to mistake the sea for a green field; they ran off the ship to their deaths. The view of Pirara was a view over green fields that had to be seen at the end of a long journey as a glistening sea.[97] That Schomburgk could look out across the savanna and "see," using his knowledge of geography and hydrography, an ancient inland sea confirmed that Pirara was the destination of Sir Walter Ralegh, the seat of the "Empyre of Guiana" that he had claimed in the name of Queen Elizabeth. At the same time that this view of Pirara established Schomburgk's identity as an explorer of the first rank, it also confirmed the place of the territory in the British Empire. If Pirara was (once) a sea, then a small Amerindian settlement became nothing less than the *fons imperium*, the ultimate destination of the very founder of the British Empire. Pirara became the source of British imperial desire itself.

Moreover, if the low savannas of the interior of Guiana were once the bed of an inland sea, then the flat-topped plateaus became, in a geological past, the "islands" they so strongly suggested.[98] I have already alluded to the island ambiguities in the British conception of British Guiana;[99] if those highlands were once islands, then Guiana itself was once part of the archipelago of the Lesser Antilles, part (in a geological sense) of the British West Indies, not an inexplicable toehold on an Iberian continent.[100] Given that Sir John Barrow—the president of the RGS, Hilhouse's correspondent, and the man who gave Schomburgk his commission—was described by a confidant as "that great covetor of islands,"

96. Ibid.
97. *OED*, s.v. "calenture." See also John Donne, "The Calme," in Gardner, *Metaphysical Poets*, 55 n. 6.
98. *JRGS* 6 (1836): 255. Webber cited Wallace on the theory of Guiana as a former archipelago: "The greater part of the valleys of the Amazon and the Orinoco as well as of La Plata must have formed part of the ocean, separating the elevated lands into groups of islands." Webber, *British Guiana*, 61. For Schomburgk on "island-like" forms in the savannas, see RGS, Map MS, "Plan of Annay." Dettelbach mentions that the British Empire had the archetypal form of "many distant islands connected by threads of naval power to one central island." Dettelbach, "Global Physics and Aesthetic Empire," 258. The special territorial integrity of islands is treated in Holdich, *Political Frontiers and Boundary Making*, 22.
99. See note 1.
100. For a discussion of the relation between geology and territorial claims in the colonial setting, see Stafford, "Annexing the Landscapes of the Past."

seeing Guiana as islands placed it closer to the concerns of the most powerful director of British exploration.[101]

No one on the Pirara expedition, nor any of those who approved it, could have been unaware that they were treading on and claiming, in the words of Richard Schomburgk, "soil rich in fable and myth." He went on: "At my very feet [lay] the '*mar de aquas blancas,*' the 'mar del dorado,' the 'city of Manoa, glittering with gold' whither the boldest adventurers of Spain, Portugal and England had wandered ever since the 16th century, including the four expeditions undertaken between 1595 and 1617 by the great yet unfortunate Walter Raleigh."[102]

These associations, written up in Robert Schomburgk's *Twelve Views* and accompanied by the picturesque view of Pirara itself, were fresh in the mind of the colonial governor, Henry Light, who organized the thirty-five-man military expeditions shortly after receiving his three colored editions of the work.[103] Officials in the Foreign and Colonial Offices too could not have been oblivious to the mythical resonances of Pirara, a place they could know only from Schomburgk's view, whether colored or no. In this sense the myth of El Dorado cannot be overlooked in examining the strategically inexplicable British military expedition to "hold" the scruffy huts of Pirara against the threat of Brazilian encroachment in 1841. Pirara made British Guiana El Dorado, and El Dorado made British Guiana British.

Linking Barima Point to Ralegh's passage (and thereby solidifying British claims in that region) required more subtle invocations. In the introduction to his edition of Ralegh's *Discovery,* Schomburgk described how he had, more or less, followed in Ralegh's footsteps: "As Her Majesty's Commissioner to survey the boundaries of British Guiana, I explored in 1841 that wondrous delta of the Orinocco: on that occasion I encamped at the Punta Barima." In the course of this sentence Schomburgk takes the reader to Ralegh's starting point, at the Orinoco delta, and then, with a quick colon, brings specific attention to a point within that delta, on its easternmost branch, where Schomburgk describes making camp. Schomburgk has followed Ralegh into the mouth of the Orinoco and brought readers' focus to Barima Point. This is somewhat curious, since Ralegh

---

101. Cameron, "Agents and Agencies," 19. As Richard Drayton has pointed out, Barrow's concern with islands must be understood as part of a broader imperial strategy to seize and hold geographically "strong" points, which controlled access to larger regions, particularly bodies of water. Interest in Guiana in the period can thus be fitted into a pattern of British maneuvers that led to the annexation of Singapore, Fernando Pó, the Falklands, Aden, and other areas.

102. M. R. Schomburgk, *Travels,* 1:307.

103. Schomburgk, *Twelve Views,* subscription list. Light made arrangements in October 1841. See Rivière, *Absent-Minded Imperialism,* 88. The role of the *Twelve Views* in constructing an image of British Guiana is the subject of chapter 4 below.

himself was never, as Schomburgk's own remapping makes clear, within 150 miles of Barima Point. Why did Schomburgk elide that geographical point with Ralegh's presence?

The answer lies in understanding what Schomburgk was doing camping on a sandbar of the Barima River on 11 May 1841. In fact, he was setting up boundary marks. After doing so on Barima Point, he and his party crossed over to the neighboring spit, which they christened "Victoria Point" before cutting a crowned "VR" into a tree and hoisting the Union Jack. Patriotic enthusiasm on Barima Point was both a strategic move of some importance and, for long thereafter, a hotly contested issue in the boundary dispute between Britain and Venezuela. In the plan Schomburgk had submitted to the British government in 1839 seeking the formation of a commission to lay out the boundaries of the colony, he had tried to pique the attention of the governor and the Foreign and Colonial Offices by pointing out the strategic significance of the mouth of the Orinoco, the main thoroughfare for trade in northern South America. The area around Barima Point, he had noted, "merits the greatest attention on account of the political importance of the Orinocco."[104] English writing on the subject styled the sandy troolie swamp[105] of Barima Point as nothing less than "The Dardanelles of the Orinocco."[106] It was perhaps this heated sense of the importance of the claim, substantiated in Schomburgk's memo by reference to some fragmentary evidence of a former Dutch post,[107] that led Lord Palmerston to sanction Schomburgk's boundary expedition immediately rather than making its work contingent on the results of negotiations with Venezuela. Better, he thought, should conflict arise, that the Venezuelans be forced to object to a surveyed claim and thus be the plaintiffs bearing the burden of proof.[108]

It took less than six months for Venezuela to lodge a formal complaint, carried to Georgetown by diplomatic agents, against what they perceived to be an infringement on Venezuelan territory; it took more than fifty years to reach formal accord on possession of the point, and the precise location of the boundary has remained a subject of dispute to the present day.[109] Schomburgk's marks were disavowed by Britain within a year of his making them, after an embarrassed Foreign Office conceded that back in 1836 the British chargé d'affaires at Caracas had lodged a formal

---

104. *Venezuela Arbitration, Venezuelan Case,* "General Summary," 1:22.
105. So called because of the preponderance of troolie palm, *Manicaria saccifera*.
106. See, for instance, *Venezuela Arbitration, Venezuelan Case,* "General Summary," 1:23.
107. British territorial claims ostensibly conformed to the limits of the Dutch possession in 1803.
108. *Venezuela Arbitration, Venezuelan Case,* "General Summary," 1:23. See also Rivière, *Absent-Minded Imperialism,* 67–68.
109. For a recent Venezuelan statement, see Tamayo et al., *Guayana Esequiba*.

request that the Venezuelan government place "a conspicuous beacon on Cape Barima" to protect British shipping in the Boca de Navíos. The request amounted to an admission of Venezuelan sovereignty. The governor of the colony informed the Venezuelan emissaries that Schomburgk's marks were "merely" geographical, for the purpose of "identification," and that "the erection of the flag was just an ebulliation of spirits on the part of one of the Indians" in the party.[110] Somewhat vague assurances were offered, and confirmed by Schomburgk, that the flag had been "pulled down or blown down before the party left the river."[111] In 1844 Lord Aberdeen, then in charge of the Foreign Office, offered Venezuela the whole of the Barima River, making a boundary offer that would have given up much of the territory that Schomburgk had surveyed and delineated as British.[112] This offer (later notorious, since Venezuela would do less well under the subsequent arbitration) still stood when Schomburgk sat in Surrey penning the preface to his edition of Ralegh's *Discovery*.

Seen in this light, Schomburgk's preface, weaving together the threads of Ralegh's *entrada* and his own encampment on the Barima Point, must be understood as a subtle invocation of the historical depth of British territorial claims to the mouth of the Orinoco. In the first paragraph of the book Schomburgk drew Ralegh's presence over Barima Point and the Amacura River, and in doing so Schomburgk wrote a defense of the flag that he (doubtless) set over the point he had just christened "Victoria." Schomburgk's preface asserted the depth of British claims in the Orinoco delta, and it did so at a crucial moment.

Schomburgk's meticulous editing and remapping of Ralegh's *Discovery* was a key source of information about Guiana and British relations with Guiana. It came to light in the midst of fraught diplomacy concerning British claims on the territory. The text insisted on the legacy of British presence in a disputed region, and it was commissioned and read by individuals with influence on colonial policy.[113] The president of the Hakluyt Society, Schomburgk's employer, was none other than Sir Roderick Impey Murchison, the "Goldfinder" himself, whose geological speculations had led to a predictive theory for the scientific pursuit of future El Dora-

---

110. Webber, *Centenary History*, 206.
111. Ibid.
112. Ibid., 213.
113. Stafford has pointed out that the Hakluyt Society "emphasized the causal links between exploration and world power while it served geographical scholarship." Stafford, "Geological Surveys," 14. Murchison explicitly saw this "historical intelligence bureau" as a way to retrieve lost information of potential importance to British expansion.

dos[114] and whose trumpeting of geographical exploration in the service of imperial ardor has earned him the title "architect of imperial science."[115] Murchison had also been among the earliest members of the Raleigh Club, that private association of traveling men out of which the RGS took shape.[116] Although Stafford shows that his interest spanned public spectacle and backroom policy (with varying degrees of success), his links with the Colonial Office were strong,[117] and in the 1850s Murchison "helped the official mind establish its negotiating stance over boundary delimitations for . . . British Guiana."[118] Less than two years after the publication of Schomburgk's edition of Ralegh's *Discovery,* Great Britain withdrew Lord Aberdeen's offer of 1844 and reasserted its right to Barima Point and Victoria Point.

There is no guarantee that Schomburgk's edition of Ralegh's *Discovery* directly affected colonial policy on British claims in the Orinoco delta.[119] But the book took up its place on the shelves of the imperially minded subscribers to the Hakluyt Society's series on British exploration and precipitated a wave of Ralegh interest among influential military men like Vice Admiral Sir Charles Malcolm and Major General John Briggs, who sat on the Society's council.[120] The text became the most up to date and reliable source on the disputed region, and by shaping the attitudes of a set of British colonial enthusiasts, it deepened British claims to the territory in the Orinoco delta in the late 1840s. Remapping old routes anchored the empire.

114. Stafford, "Geological Surveys," 12.
115. Stafford, *Scientist of Empire.*
116. On the Raleigh Club and the origins of the RGS, see Livingstone, *Geographical Tradition,* 158–60.
117. Ibid., 81. The RGS secured permission for Schomburgk to do scientific work on his boundary surveys.
118. Stafford, "Geological Surveys," 18.
119. As Curtin has put it: "Officials in the Colonial Office did not, in fact, write down all that they believed to be true about the world in which they lived or the regions they governed. Beyond the world of dispatches, there was also a world of unstated assumptions." Curtin, *Image of Africa,* vii. At the same time, one cannot overlook other (important) factors that bore on British policy concerning Guiana. Richard Drayton has drawn attention to the influence of trade considerations in the period: Britain enjoyed a lucrative cotton export to Brazil, despite maintaining protective tariffs on Brazilian sugar, an arrangement up for renegotiation in 1842. Given this situation, avenues for contraband trade may have seemed both desirable in themselves and useful as sticks to carry to the bargaining table. Palmerston's direct effect on the decisions concerning Pirara (and Guiana generally) should not be underestimated. See relevant correspondence, Ousely to Palmerston, 16 June 1840, PRO, FO 13/172.
120. Schomburgk addressed the Geological Society on the subject of Guiana gold on 4 December 1844. The decision to send James Sawkins and Charles Brown from the Jamaica Geological Survey to Guiana was an extension of this interest. Murchison was instrumental here. Stafford, *Scientist of Empire,* 84–86.

THE OTHER MYTHICAL EMPIRE: MAHANARVA

Schomburgk's explorations and writings fixed the site of El Dorado firmly to the village of Pirara. But El Dorado had long been more than a place: it was a promise as well. From the beginning the tale had made Guiana a proleptic empire, an Elysium that lay *ahead* both spatially and temporally.[121] It never quite happened. Those European colonists who made British Guiana their home knew only too well that they were living in a region unceremoniously dubbed the "land of mud" by their West Indian neighbors. In the early nineteenth century the story of a different mythical empire arose in the colony, a story that served as an ironic commentary on Guiana's unfulfilled promise: the proleptic empire was converted into a prolapse empire.

This countermyth grew out of the colonial tales of Mahanarva's first arrival at the coast in 1810. Mary Noel Menezes, a reliable modern source for ethnohistory of British Guiana in the nineteenth century, writes that "the British had their first brush with the Caribs when the Carib chief Mahanarva strode into the capital city and threatened to unleash his Indians unless they were appeased by the usual presents they had been accustomed to receive from the Dutch."[122] This version of the Mahanarva story is itself inflected by Mahanarva mythology, but it came into circulation early. Waterton told the story this way:

> The misinformed and timid court of policy in Demerara was made the dupe of a savage, who came down the Essequibo, and gave himself out as the king of a mighty tribe. This naked wild man of the woods seemed to hold the said court in tolerable contempt, and demanded immense supplies, all of which he got; and moreover, some time after, an invitation to come down the ensuing year for more, which he took care not to forget. This noisy chieftain boasted so much of his dynasty and domain that the government was induced to send up an expedition to his territories to see if he had spoken the truth and nothing but the truth. It appeared however that his palace was nothing but a hut, the monarch a needy savage, the heir-apparent nothing to inherit but his father's club and bow and arrows, his officers of state as wild and uncultivated as the forests though which they strayed.[123]

This tale, featuring colonists duped by the wiles of a savage, was picked up and retold exactly by Richard Schomburgk, and missionaries later embroidered it further for moral effect, describing how Mahanarva had slain a Macushi slave on the floor of the Court of Policy to make clear his disappointment on learning that the slave trade had ceased.[124]

---

121. For an interesting look at utopias and frontier spaces, see Porter and Lukermann, "Geography of Utopia."
122. Menezes, *British Policy,* 23.
123. Waterton, *Wanderings,* 115.
124. Brett, *Indian Tribes of Guiana,* 134. The incident is retold in Menezes, *British Policy,* 182. I follow Rivière (*Absent-Minded Imperialism*) in the spelling of Macushi.

The reality of the affair, if it is to be gleaned from the manuscript minutes of the Court of Policy in the National Archives of Guyana, appears to have been considerably more subdued. According to these sources Mahanarva himself had come to the coast—seeking his annual presents in the Dutch tradition—only after the "envoy" he had sent six months earlier had been roundly ignored in the colony, having, in the euphemistic condescension of colonial rhetoric, "so little the appearance of what he pretended to be."[125] If he was indeed the envoy of a "great chief," he was encouraged to bring the great chief himself back when he came. So Mahanarva came to Georgetown, hardly with the flourish suggested in the accounts above, but rather in the wake of his spurned ambassador. In fact, if the Court of Policy records can be trusted (and there is no reason to think they would not address his arrival as a security threat if it was seen as such), Mahanarva seemed more interested in selling a few Macushi slaves he had brought with him than in terrorizing the colony. The risk of his making attacks on the "indians settled in the backlands" seemed real, as did the possibility that if he could find no one to buy his human wares, he might feel compelled to kill them rather than trouble with dragging them back up the river and around to Brazilian outposts to sell.

So far was any sense of immediate threat from the minds of the court, though, that in their discussions of the matter the conversation turned to whether Mahanarva and his people should be encouraged to settle nearby, so that their "alliance for the purpose of internal defense [might] be sought."[126] This was ultimately deemed unwise, since the members of the court felt that bringing a new tribe into the regions of the backlands could only disturb the relations between the various groups the colonists already depended on for support in controlling runaway slaves. The court agreed to offer Mahanarva the presents he had come seeking, in return for a pledge that he would make no war on the Amerindians in the backlands or those in the deeper south, would abandon all slaving, and would agree to be accompanied back to his region by an expedition commissioned by the governor to take a census of his fellow tribesmen. The head of that expedition was John Hancock, and he and his party would be the first Europeans since Nicolas Horstman to make a passage to the Rupununi savannas that would be preserved for historians in a map and a narrative.

Hancock's report did not suggest that Mahanarva was an indigent pretender—as Waterton, Schomburgk, and others claimed—or that he was an Aztec prince, as he is supposed to have asserted. It seems rather that what Hancock discovered was consistent with what the court had ex-

125. NAG, MCP, 29 October 1810.
126. Ibid.

pected and with Mahanarva's own protestations. Hancock wrote: "On entering the Macoosy country we summoned a meeting of all of the Chiefs of that part in the name of His Majesty, as authorized by the Governor of Demerara. . . . They unanimously agreed that Mahanarva was not only Caqui [sic] of the caribees, but was acknowledged by them and all the other tribes."[127]

In light of this report and the minutes of the Court of Policy, corrections are in order: Mahanarva did not terrify the colony, did not take it by storm or hold it in awe. Nor does it seem likely that he ever presented himself as a truculent imperial prince. An aside on the matter from Hilhouse, the most knowledgeable official in the colony concerning the affairs of the Amerindians, called the whole event, including the Hancock expedition, merely an "affair connected with the suppression of the indian slave trade," which does not suggest that the colony was gripped with fear.[128]

If Mahanarva did not present himself as the ruler of a threatening empire in the interior, as colonists suggested long after, the elaborate reading of the Mahanarva episode by Wilson Harris, Guyana's modern literary lion, seems unsalvageable. For Harris the Mahanarva encounter, gleaned from "little more than footnotes in the history books," presents a charged microcosm of colonial encounter. He places the tale at center of his Mittelholzer lecture, "History, Fable, and Myth in the Caribbean and Guianas," where Mahanarva becomes the *beau-ideal* of the shamanic trickster: "Mahanarva's 'lie' gives us an insight (if we begin to free ourselves from dogmatic morality) into the trickster womb of the shaman. When Mahanarva claimed that his fighting forces were intact we now know from insights gleaned into our own psyche and into the so-called savage mind that he was compensating in himself losses his people had endured over centuries."[129] Harris's sinuous prose winds to a consideration of Mahanarva as a half-feminine, half-warrior archetype comparable to Pallas Athena and perhaps even Saint Peter—both paradoxical progenitors of autochthonous peoples.

The only problem with this expansive (and imaginative) reading of

127. *Brazil Arbitration, British Counter-case,* 104. For a discussion of the role of such expeditions in constituting (rather than merely identifying) tribal authority, see Whitehead, "Carib Ethnic Soldiering," and Whitehead, "Ethnic Transformation." Commissioners like Hancock saw what authority Mahanarva possessed in European terms, and in doing so and writing their perception of his political power, they *created* a Mahanarva *chief* in the interior after the models of tribal sovereignty they imagined.

128. BL, Hilhouse MS, 9.

129. Harris, "History, Fable and Myth," 23. The passage ends, "[Mahanarva] became the womb of the tribe in certain respects that are analogous to traces of mythology—ancient Greek, Persian, Mithraic as well as Christian—in which stones and rocks become charged with architectural latencies, inner rooms, etc., and therefore give birth to numinous tenant [sic]."

Mahanarva is that it is based on bad history—based, in the ultimate irony, on a colonial fiction about the events themselves. Harris's errors can be summed up as follows. First, it is not correct that in 1810 "little need existed any longer to guard the escape routes of African slaves."[130] If anything, in the wake of the abolition of the slave trade (in 1807) the need was as intense as ever. Second, it was not the case that "slavery had been or was on the point of being abolished" in 1810. A quarter century of slavery was still ahead. Third, as I have shown, there is no archival evidence that Mahanarva "claimed a considerable fighting force lay at his command in the Bush, which would constitute a threat to Stabroek." Fourth, Harris's depiction of European treachery (he claims that a "scout was dispatched—unknown to Mahanarva—to reconnoiter the position") is false. The minutes of the Court of Policy record that the interpreter was asked to explain to Mahanarva the need for an expedition to census his settlement.[131] There are strong reasons to suspect that Mahanarva himself, or at the very least some of his companions, guided Hancock and the government expedition up the Essequibo.[132] Finally, as I have shown, there is no evidence that Hancock found "Mahanarva's ancient command had shrunken to rags." The best evidence against this story lies in the minutes of the Court of Policy two years later, on the occasion of Mahanarva's return to the coast. In response to his renewed request for presents, "His Excellency assured the chief" that "any article that could be had at this moment in the colony" would be procured for him and his people. "The chief having been satisfied therewith," he departed.[133] It is true he was not given all the items he wanted, but he was not scorned out of the court based on Hancock's reconnoiter of the previous year.

These facts offer strong evidence that Mahanarva concocted no elaborate ruse in an attempt to defraud or terrorize the colony. The Mahanarva myth (not the one he himself is supposed to have propagated, but rather the one propagated by Harris and others), revealed as a *colonial* myth, bears with difficulty the weight of the analysis Harris would like to offer. Not only does the suggestive notion of Mahanarva as parthenogenic

---

130. This quotation and those that follow are from Harris, "History, Fable and Myth," 22.
131. NAG, MCP, 29 October 1810.
132. Dates provide the best evidence. Hancock and his commission left Georgetown in November 1810; Mahanarva addressed the court of policy on 29 October of the same year. Given the approach of the rains, it would have been necessary for the expedition to leave quickly and make good time to reach Mahanarva's settlement before the rivers rose. Without an Amerindian guide who knew the route, the group would not have been able to make the trip as quickly as it did. In addition, there was little (if any) precedent for the expedition, and without the arrangements for a guide and promises of food and hospitality (both of which Hancock states the commission received), no one would have been willing to embark on such a novel trip on such short notice.
133. NAG, MCP (petitions), 1812.

propagator of a phantom race fall by the way, but so too, it seems, must fall Harris's still more interesting reading of the Mahanarva encounter as a symbol of the curious entanglements of colonial territory:

> Mahanarva's "lie" to the governor brings into play a fateful—however subconscious—erosion of the character of conquest. The shroud which was parted [by Hancock's expedition] gave the governor a view of his hypothetical kingdom. There were no Carib fighting forces lying in the bush, either to threaten Stabroek or alternatively to secure the interior for the English Crown, but instead a chasm of losses—the primeval fall-out of a broken people.[134]

This "discovery," Harris points out, had a subversive effect. Imperial surveillance had undone itself, and the governor, by "finding out" the ruse of the shaman, was at that moment converted into a territory- and authority-conjuring shaman himself: "The statecraft of the European nineteenth-century representative of the Crown came into intimate rapport with the trickster-shaman of aboriginal allies. It was a marriage of alien and yet conspiring functions (trickster to statesman)." If the Caribs were a ruined people, without extent or power, if they could not hold their territory and enforce its limits, then over whom did the British preside? Who were their subjects? To whom were they paying their tributary "presents"? What was British territory, and who was enforcing its limits—both the limits of their coastal plantations and the limits of their territorial aspirations in the interior—against the Maroons, Portuguese, and Spanish?[135] Understood in this way, the Mahanarva myth suddenly becomes a remarkably clear instance of the kind of "inappropriate" realignments that emerge in European efforts to articulate their cultural and scientific institutions in the colonial context.[136] The "institution" in question here is nothing less than territory itself, and territorial authority. The "realignment" results from the disconcerting proximity of alien concepts—of sovereignty, boundary, authority—that unsettle and displace Western ideas when they must be played out in non-Western contexts.

What can be made of this (wonderful) reading if the Mahanarva story, as told by Harris (and others), is not true? An answer depends on recognizing the deeper truths expressed by the colonial myth. Although I have shown that the story does not conform to the events, as best they can be reconstructed from the documentary sources, the myth of Mahanarva-trickster remains as true as all myths, in that it constitutes "the answer to an unspoken question about a matter of great import."[137] The question it

---

134. This quotation and those that follow are from Harris, "History, Fable and Myth," 23.
135. Whitehead, "Tribes Make States."
136. Prakash, "Science 'Gone Native' in Colonial India," 172.
137. Scholes, *Structuralism in Literature,* cited in Mentore, "Relevance of Myth," 11.

addresses, as Harris rightly observes, is indeed one about the nature of colonial territory and territorial authority. The longevity and currency of the tale have a significance beyond the issue of its accuracy. By inventing, telling, and retelling this story about the interior of Guiana, colonists, travelers, and imperial agents alike were telling a story about colonial territorial anxieties. The tale of tragicomic disappointment revealed their suspicion that the proleptic empire of the interior was in fact a fallen empire. The promise of the interior had withered: the Mahanarva myth was the anti-El Dorado.

Not only did the tale subvert the burdensome mythology of Guiana's boundless promise of future wealth, it also reflected the colonists' anxieties concerning their tenuous command over their actual possessions. Colonists' control over their plantations and slaves was very much contingent on relations with and between Amerindians, and in this sense Mahanarva and his bedraggled Macushi captives posed a very real threat to colonial property. Reinventing the threat as that of a terrifying warrior tribe allowed the colonists to displace the true anxiety of territorial control, drowning a chuckle at themselves in a hearty guffaw at the cheek of a sneaky (but ultimately impotent) old Indian.[138] This story depicted an empty interior and recast the space and its alien inhabitants as an extension of the plantation backlands, a sorry scene of ragpicking Amerindians willing to split wallaba shingles all day for a dram of rum. The myth of Mahanarva pulled the interior up to the familiar "backdam" of plantation life, narrated the interior as a dependency, and let the colonists laugh at themselves for having ever imagined that it could have been a place of promise *or* a threat.[139] The emperor had no clothes; the Amerindian empire had long since fallen. There was nothing to fear in the interior, and nothing to hope for.

The truths that this fiction concealed—the collapse of the boundless promise of the interior, colonists' tenuous grasp over their slaves, colonists' authentic anxieties about the interior and what it held—are inscribed in the myth itself.[140] Wilson Harris, then, proves right, if for the wrong reasons. The story of Mahanarva is indeed a story about the need

---

138. Of similar processes Greg Dening writes: "It was a translation into entertainment of ethnographic moments in which the European strangers confronted the otherness of the Pacific island natives, tried to describe that otherness and in that description possess them." Dening, *Mr. Bligh's Bad Language*, 271.

139. "Backdam" (literally, the back dam) denoted the portion of a plantation that was nearest the "bush," or uncultivated areas. By extension, it could be used to mean the bush itself. For a discussion of the term in modern usage, see Roopnaraine, "Behind God's Back," 58ff.

140. Thomas does a similar reading of a South African colonial fiction, *Prester John*. He shows that the focus on tensions between whites and blacks concealed deeper concerns about the divisions between Dutch and British colonists. Thomas, *Colonialism's Culture*, 143–53.

for "spatial re-assembly" and the need to "salvage the muse of authority," though it is not merely a story *about* those needs. The story itself was an attempt to *address* those needs. As a myth it was first and foremost an intervention. In telling the story of the interior as a collapsed empire, colonists reassembled the space and salvaged their own authority.

EPILOGUE: THE AMBIVALENCE OF MYTH AND METALEPSIS

This chapter has argued not only that metaleptic cycles defined how geographical explorers thought about their routes, destinations, and identities, but that those same cycles played a crucial role in grounding British territorial claims. El Dorado was a place explorers sought, but it was also what British Guiana, in some sense, was. Ralegh's routes were paths that Schomburgk could follow, and could redraw on his own map, but in doing so he deepened the roots of British claims in the eastern Orinoco. I have also suggested that mythic empires play an equivocal role in the codification of colonial territory. The multiple Mahanarva myths laid out in the previous section offer an example. Through the story of Mahanarva's fallen "empire" the threatening interior was transformed into the hinterlands of the colony. But in the process the hinterlands became a farcical space. The same story illustrated the still greater paradox that resulted from rooting the myth of El Dorado at the center of the colony's geography and identity.

The result of the cycles of myth, remapping, and metalepsis detailed in this chapter was that by the mid-nineteenth century British Guiana had been incorporated into imperial writings under the rich and promising sobriquet "El Dorado." El Dorado proper had been erased, displaced, and replaced. It had been attached to swamps, micaceous hills, and granite passes. But it remained fixed as a figure of speech. The missionary Rev. W. T. Veness published his vision of the colony's future in London in 1867 as *El Dorado, or British Guiana as a Field for Colonization*.[141] A late-century colonial booster echoed the writings of many when he exhorted his countrymen to prove Ralegh right by *building* the empire he had sought, given that it could not be found; only then could outsiders "regard us once more as the El Dorado that Sir Walter Raleigh once called our land."[142] The cost of the limitless colonial prospects this figure could still evoke was the chimera it had always been and could not but remain. In the nineteenth-century reappropriation of the story of El Dorado, the romance of the fruitless quest was fixed to Guiana more firmly than any

---

141. Veness, *El Dorado*.
142. Tinne, "Opening up the Country," 34.

prospect of colonial prosperity. The heart of Guiana, and the telos of the colonial venture, might be called a promising El Dorado, but that implied as much that it could not be reached (and might not exist at all) as that it was a rich prize.

Nothing illustrated this better than the toponymic insubordination of members of the beleaguered West India regiment who stumbled out of their boats at Pirara landing in February 1842 only to discover the forbidding emptiness of the savanna they had been sent to "secure." The frères Schomburgk might step out of the boats onto the sacred soil of the "*mar de aquas blancas*" and ruminate on this marsh as the epicenter of South American myth and the lodestone of interior exploration, but the men of the West India regiment, in a gesture so elusive as to have been passed over entirely by commentators, named their mud-walled fortification on the outskirts of the precious Pirara *Fort New Guinea*. Had no one told them this was Guiana? What was the joke? The only clue lies in a forgotten meaning of "guinea": from its earliest coinage in English, "guinea" was not infrequently used as an adjective, "loosely . . . as a designation for an unknown and distant country."[143] It meant, in a way, "no place." What one wished to find in Guiana was, iconically, not there: Guiana itself embodied nowhere.

Such was the ambivalence of myth in the formation of colonial territory. Metalepsis, too, proved a slippery technique for codifying colonial geographies. Stripping predecessors of their content while claiming their authority was a game difficult to control. Remapping predecessors could undermine as well as root the possessive force of their passages. The story of Sir Walter Ralegh's own map of northern South America demonstrates this well. When Schomburgk sat down in 1848 to trace Ralegh's route on the Schomburgk map that served as the frontispiece to the Hakluyt Society edition of the *Discovery*, he had no map of Ralegh's to work from. In the *Discovery* Ralegh had written to his dedicatee that such a map was forthcoming; but, he continued, "I shall most humbly pray your Lordship to secret [it], and not to suffer it to passe your own hands."[144] Secreted it was, so successfully that Schomburgk offered a footnote to the edition he edited, reading, "It appears he never executed this map, or if he did so, it has been lost."[145]

The latter supposition appears to have been correct. Unknown to Schomburgk, in 1849, the very year after he wrote those words, the British Museum acquired a vellum roll that was cataloged as "a map of Guiana

---

143. As in guinea pig, guinea fowl, etc. See *OED*, s.v. "guinea."
144. Ralegh, *Discovery*, 26.
145. Ibid., 26n.

circa 1660."[146] Not until sometime after 1861 did a handwritten slip of paper find its way into the catalog,[147] reading "by Sir Walter Raleigh post-1596. Holograph."[148] If the slip is correct, British Library Additional MSS 17940a is the earliest known cartographic representation of Guiana by an English hand (fig. 8).[149] Given that it should thus have been a celebrated document, its anonymity posed a problem. A late nineteenth-century keeper of the British Library's map room (and researcher on British Guiana's boundary arbitrations) puzzled over this when he asked why "this celebrated map of Guiana . . . was allowed to remain in its present manuscript form." Ultimately he concluded that this could be counted not as "evidence of the lack of enterprise of English cartographers" of Ralegh's day, but rather as proof of their good sense, since "the complete lack of truth in the chief feature of the map, the great lake in the center" may have "militated against its publication."[150]

This is unlikely, for, as I have shown, the presence of the lake did not militate against the publication of dozens of subsequent cartographic depictions of South America running right up to the early nineteenth century. Nor can it be claimed that it was not the practice of English explorers of the period to see their charts through to publication. The earliest printed maps of Guiana, published in 1599, were acknowledged in the legend to have been derived from the chart of a seaman who claimed to have been on the voyage and who had doubtless sold his sketch to the engraver.[151] Maps offered a powerful graphic tool for substantiating exotic (and suspect) geographical discoveries. Yet the author of the *Discovery*, whose need to substantiate the truth of his account surpassed that of any of his contemporaries (given his numerous detractors), insisted that his map was to be kept "inward" rather than turned "outward." If the map could substantiate the *Discovery*, then why was it to be *secreted*? Why not trumpeted in broad publication like the narrative of the *Discovery* itself?

---

146. Nicholl states 1650 (*Creature in the Map*, 346). This is at odds with Skelton, "Raleigh as a Geographer," De Villiers, "Famous Maps," and a variety of other sources.

147. Nicholl, *Creature in the Map*, appendix 2.

148. See Friederichsen, "Sir Walter Raleigh's Karte von Guayana"; De Villiers, "Famous Maps"; and Skelton, "Raleigh as a Geographer."

149. I have written an essay on the history of this identification, but space does not permit its inclusion. In brief, the identification was made by Johann G. Kohl in the early 1850s. Not until Kohl's work on the cartography of the Americas received broader attention (in connection with the Columbus centennial events as well as the boundary disputes themselves) did it come to be widely accepted.

150. De Villiers, "Famous Maps," 181.

151. The earliest printed map of Guiana was that of Jodocus Hondius (1599). De Villiers claims that the British Museum copy is a unique exemplar. The Hondius map was quickly copied by Theodor de Bry, who published his version in Germany later in the same year. It is Skelton ("Raleigh as a Geographer," 141) who affirms that these printed maps were likely made from a draft sold to the engraver by a sailor.

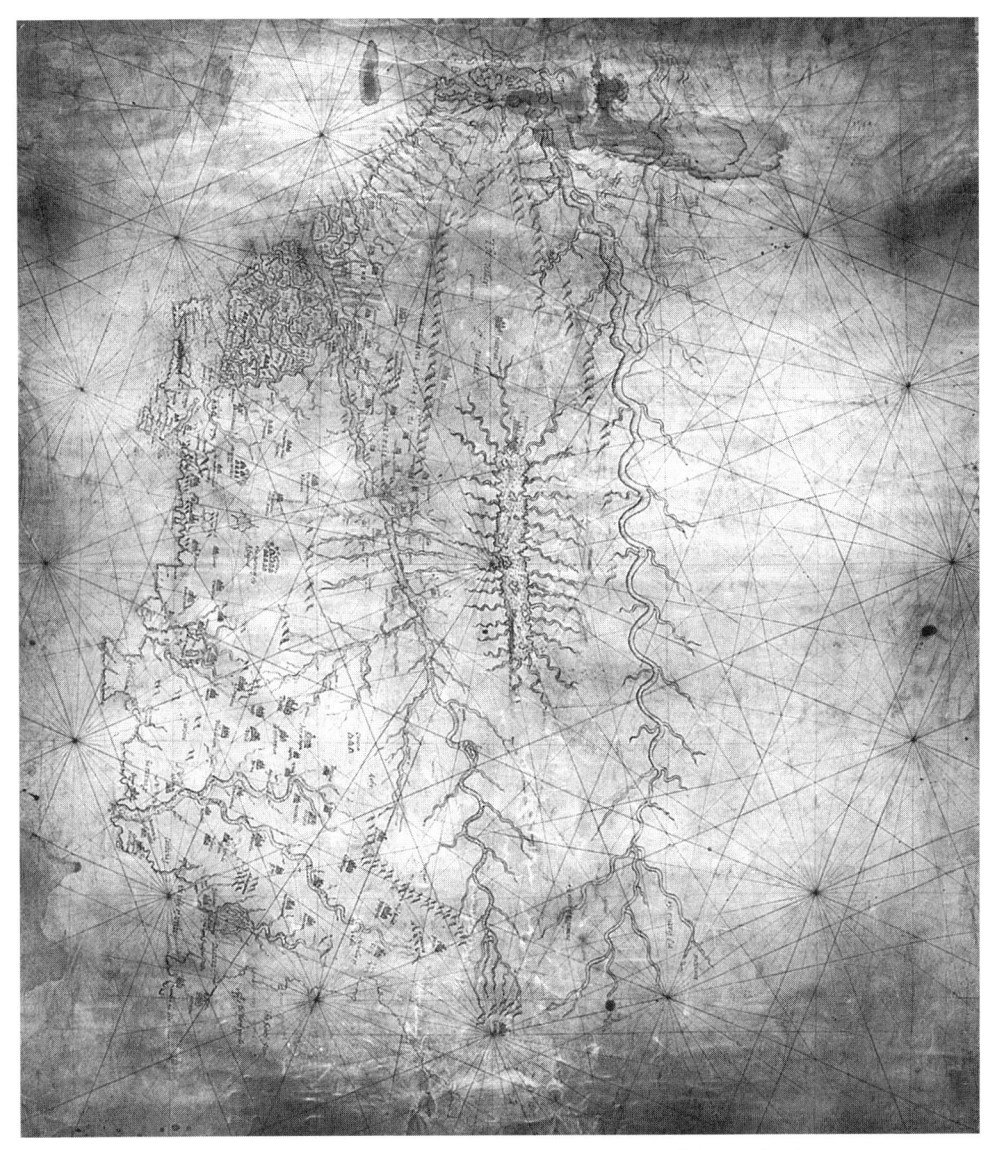

Figure 8. El Dorado on paper: British Library Add. MSS 17940a. Courtesy of the British Library.

The answer lies in the power not of a map, but of the absence of a map. The rhetorical device by which an author "emphasizes something by pointedly seeming to pass over it" is called *occupatio*.[152] It is a clever device. The map that does not appear in the *Discovery* fails to appear to the

---

152. Lanham lists as synonyms *praeteritio, paralepsis, occultatio,* and *parasiopesis.* Lanham, *Handlist of Rhetorical Terms,* 68. Recent work on the rhetorical tropes of imperial representation includes Lewis, "Rhetoric of the Western Interior," and Spurr, *Rhetoric of Empire.*

accompaniment of considerable fanfare. The warnings of secrecy, the explicit withholding of the map from public view, underline the fact that the map is not present. Lord Howard is warned that if he should let slip his cartographic charge, it is *the place itself that will be lost*. Right in the text of the *Discovery* Ralegh wrote of the map that "by a draught thereof all may bee prevented by other nations":[153] the copying of the chart threatened the swallowing up of the very place itself. The inescapable implication is that those who had access to Ralegh's map had special access to Guiana, that the map possessed Guiana in a way that the narrative did not, that *the substance of the place of Guiana was held in the map*. When it comes to the distribution and handling of the chart, there is no delight in "discoverie" but rather an anxiety about secrecy. The map is not included with the *Discovery* because the map would give the place away. The map could give away the place because the map possessed the place. By claiming that a map of his route would entirely reveal the place, Ralegh turned his journey into a journey of possession, a journey that "got" Guiana and brought it back so completely that he could give it away *as a territory*.

All this introduces a considerable irony: whereas Schomburgk's edition of Ralegh probably bolstered British claims to the Barima Point region, Ralegh's *own map* was used against those same claims in the boundary arbitration at the end of the century. In 1886 President Cleveland's British Guiana-Venezuela Boundary Commission assigned to the Spanish international jurist Severo Mallet-Prevost the task of evaluating the "cartographical testimony of geographers" concerning territorial rights in the disputed region. His report, included in the *Venezuelan Counter-case* in 1889 and amply illustrated in the accompanying folio atlas of facsimiles, began with a brief explanation of the importance of cartographic evidence: "Maps perform the function of pictorially expressing the views of the particular geographers or map makers who have been instrumental in bringing about their publication. They furnish us, therefore, with the opinions of a particular class of experts."[154]

Salient among these experts was Ralegh himself. While acknowledging that the earliest maps were of only limited weight in examining a territorial conflict rooted in issues of "effective occupation" and the extent of Dutch claims in the eighteenth century, Mallet-Prevost nevertheless insisted that to understand the meaning and development of later maps it was "essential first to examine Raleigh's map of 1595." The reason for this, he argued, was that "that map, for the first time, made familiar the word 'Guiana.'" As a secret document, of course, the map itself would

---

153. Ralegh, *Discovery*, 26.
154. *Venezuela Arbitration, Venezuelan Counter-case*, 1:267.

have made the word familiar to few people indeed. But Ralegh's writing did make the word Guiana familiar in Britain, and this map, Mallet-Prevost argued, showed what Ralegh thought Guiana was. "Guiana" in Ralegh's chart, he explained, was the watershed of that creeping lake positioned at the center of the vellum sheet, Manoa, the Lago del Dorado. Mallet-Prevost considered this depiction of Guiana—those letters inscribed across the South American interior for the first time[155]—charged with significance. Read by him they indicate that "this 'Guiana' was a region in which the world at that time firmly believed, but which neither Ralegh nor any other European had ever seen: it was a mythical land, unexplored, unknown, *free from the political control of any European nation.*"[156]

In other words, Ralegh's map indicated that "Guiana" was not a territory at all, not seen, controlled, or located. This was underscored by the failure of Ralegh's expeditions. Ralegh's *entrada* was recast by the Venezuelan delegation as a nonexploration. His very inability to "reach" Guiana amounted to proof that the region was inviolate and that the routes to it were controlled by Spaniards. This amounted to proof of uninterrupted (and precedent) Spanish control over the region. In the *Venezuelan Case,* submitted to the arbitrators, Mallet-Prevost wrote: "Of all foreign adventurers Raleigh was the most famous. His expeditions were the best planned and the best manned. His ambition, ability and resources were greater than those of any foreigner who ever attempted to penetrate into the interior of Guiana, *and his failure to accomplish this was the most signal of which any record remains.*"[157]

For Venezuelan officials the nonpenetration by "the most brilliant commander of England" was inscribed in his chart, which did not stake a territorial claim but rather inscribed a cartographic statement of belief. Unreached, unseen, "Guiana" could not have been possessed; Ralegh's map depicted not a place, but a nonplace, not a territory but a nonterritory. Ralegh's map as a nonterritory bolstered Venezuelan claims, hence the placement of a facsimile of Additional MSS 17940a in the prominent anchor seat at the beginning of the *Venezuelan Case Atlas.*

In the end Ralegh was right in the anxiety he expressed to Lord Howard about the power his chart would have in conveying the substance of Guiana. But it held that power in a way that neither of them could have expected. Ralegh's chart was ultimately deemed to contain the nonplace of Ralegh's Guiana. The map, it was argued, did not hold the place as

---

155. He notes the possibly precedent "Cartas de Indias" from the Archive of Seville but offers no further commentary on it.
156. *Venezuela Arbitration, Venezuelan Counter-case,* 1:272. Emphasis mine.
157. *Venezuela Arbitration, Venezuelan Case,* 1:44. Emphasis mine.

much as it held that the place was not held. It could not so much give the place away as give away that in 1595 Guiana was not Ralegh's, was not England's. So faithfully did the text of the chart transcribe Ralegh's encounter with South America that it held within it the story of the *entrada* Ralegh had failed to make. It held this knowledge in its geographical inaccuracies—which would be replotted on other maps to show where Ralegh had not been—as much as in its symbolic configuration: rivers go into the interior, and rivers come out of Lake Parima, but they do not meet; you cannot get *there* from *here*.

The power of a map—to install places and cause them to vanish, to make and unmake explorers and empires—is nowhere better exemplified than in the curious history of Additional MSS 17940a. It conjured a nascent English Guiana out of a handful of letters written across a sheepskin, and it lay at the starting point of an argument to wrest portions of that Guiana out of the British Empire three hundred years later. No one seemed to recognize this power better than Ralegh himself, who "secreted" a "plott of Guiana" on his person right to the end. That "plott"—not impossibly 17940a itself—is found listed in the "inventory of such things found on the body of Sir Walter Rawleigh, Knight, the 15th day of August, 1618," along with a "Guiana idol of gold," a sample of "Guiana ore," and a few personal effects.[158] It was these few items Ralegh had elected to take with him in his hasty and ill-fated attempt to flee England and the death sentence that awaited him.[159] Had he chosen to secrete the map because he thought it might save him? That he could use it to "give" Guiana to the French in return for asylum and patronage? That it would help substantiate the Guiana gold he carried as well? Is it not almost as likely, though, that he secreted the map out of fear it would expose his fatal failure? That it would show where he had not been, show how the body of Guiana had slipped through his fingers?

158. Schomburgk, appendix to *Discovery*, 228.
159. For a full account of the circuitous path to Ralegh's conviction, and its relation to his second voyage to Guiana, see Harlow, *Ralegh's Last Voyage,* and Coote, *Play of Passion.*

CHAPTER THREE

Traversing Terra Incognita:

Getting There and Making Maps

ELUSIVE INCOGNITAE

Schomburgk's sympathetic attachment to Ralegh stemmed as much from common inadequacies as from common geography. The bond between the *Discovery*'s author and its nineteenth-century editor was strengthened by Schomburgk's multiple early failures to reach the interior of Guiana. Schomburgk and his kit arrived in the colony from Tortola in early August 1835. His first task was to introduce himself to locals who had expertise on the interior, with an eye toward planning his first expeditions. At the top of his list of contacts was William Hilhouse, whose correspondence with Sir John Barrow and the RGS in the early 1830s drew attention to British Guiana and its potential as a strategic point in British access to South America. These writings were instrumental in the Society's decision to commission a Guiana expedition.[1] Hilhouse was one of only seventeen corresponding members of the RGS throughout the world in 1836, and he provided unofficial assessments of Schomburgk's fitness, qualifications, and progress during his early work for the Society.[2]

---

1. Hilhouse's *Indian Notices* circulated at the RGS in 1833. See RGS, Correspondence, Hilhouse to RGS, 1 July 1831. In it Hilhouse had written of the region's "scientific and political importance" and claimed "Demerara justly calculated to become the key introduction to the new world, with all her vast unexplored field of information, wealth and power." Hilhouse, *Indian Notices*, 130. These lines would have been read in the context of statements made by George Canning, the British foreign secretary, just a few months before: "Spanish America is free and if we do not mismanage our affairs sadly, she is English." Miller, *Britain and Latin America*, 1. The RGS's early sponsorship of a major (and costly) set of expeditions up the rivers of British Guiana affords an interesting parallel with the Society's early work in Australia. Barrow had links to the West Indies through his Admiralty mentor Lord Macartney, former governor of Grenada, who like Barrow had strongly opposed the return of the eastern Dutch possessions (like Surinam) captured during the war. It has recently been argued (Cameron, "Agents and Agencies," 19–24) that Barrow's enthusiastic support for the 1836 RGS-sponsored expedition to Australia can be traced directly to Barrow's desire to solidify British interests in the East Indies, and particularly to his wish to protect the northern shore of Australia from foreign (French or Dutch) occupation. If a major river emptying onto the northern shore of Australia could be found, it might provide the means for a convenient and lucrative "opening up" of that region's unexplored interior. Barrow likely perceived the potential for an "imperial beachhead" in Guiana, as he did in Australia, particularly if one of the large rivers in the colony was found to be, or could be made, navigable.

2. RGS, Correspondence, Hilhouse to RGS, 17 August 1835.

Hilhouse articulated Schomburgk's geographical assignment in one of those letters. He was to "prosecute his discoveries . . . 'till the whole of the Terra Incognita changes its names on our maps."[3] Hilhouse considered himself qualified to evaluate this process of exploration because he had engaged in it himself. When he and Schomburgk sat down in early April 1835 to discuss the interior—and how Schomburgk might make his way to the terra incognita of the deep south—Hilhouse was the reigning European authority on matters of exploration in British Guiana. Not only had he recently published several memoirs of his Guiana expeditions in the *Journal* of the RGS, he had also produced a short book on the Amerindian tribes, a memoir on the ichthyology of the colony, and also, significantly, several maps.[4] Although he had managed to alienate the governor and other colonial officials by his persistent and outspoken objections to Amerindian policy in the colony (losing so much clout that he was eventually barred even from visiting Amerindian settlements), he continued to have considerable sway at the RGS as a local informant.[5] His initially favorable, if reserved, judgment of Schomburgk's fitness was an important vote of confidence in 1835; Hilhouse's later unfavorable assessments of Schomburgk's early failures nearly cost him his commission.

The substance of that April meeting can be summed up easily: where to go and how to get there. Getting to terra incognita proved more complicated than the recent arrival from the Virgin Islands had suspected. It was not just that the "team of horses" he had alerted the Society that he would require would be entirely useless for negotiating the jungle tracks and whitewater rivers that stood between the coast and the "unknowns" of the interior.[6] Hilhouse could help him arrange for a boat and a crew. But once he had those, where did he point them? Which rivers actually led to some terra incognita? Which ones petered out into marshes before they could reach the Sierra Accarai, the putative watershed with the Amazon basin and, depending on who was consulted, the southern boundary of the colony as well? Hilhouse had already mounted the Mazaruni, and the Cuyuni went nearly due west, when what was wanted was to go south. The Essequibo, the Demerara, the Berbice, and the Corentyne all headed that way, but little could be confirmed concerning the upper reaches of any

3. RGS, Correspondence, Hilhouse to RGS, 30 October 1839 (listed in Menezes as July).

4. Note that ichthyology was not irrelevant to interior exploration: see my discussion of the "pacu" fish and its use as an indicator of particular geographies (article forthcoming in *Ethnohistory*). For a bibliography of Hilhouse, see Menezes' supplementary notes to Hilhouse, *Indian Notices*, 138–48.

5. His struggles with the governor as well as his financial difficulties may explain why he did not receive the RGS commission for the Guiana expeditions. See Menezes' introduction to Hilhouse, *Indian Notices,* and Menezes, *British Policy,* 101–2, 106–7.

6. RGS, Correspondence, Schomburgk to RGS, 8 August 1835.

of them. There was terra incognita at the headwaters of each, but not all terrae incognitae were created equal, as Schomburgk would soon discover.

A roster of previous passages was essential to selecting an appropriate route. The Demerara had been ascended by Charles Waterton, who had crossed over to the Essequibo above the post and then made his way by the Rupununi to Fort São Joaquim; Horstman and Hancock had taken similar paths. This meant the Essequibo below the Rupununi, the Demerara considerably above the post, and the Rupununi as far as the Brazilian fort, were unknown only in a limited sense. It was true that Horstman carried nothing more than a compass,[7] and though Hancock had made use of a sextant,[8] his sketch map (to which I will return) was little known and considered of questionable reliability.[9] As for Waterton, he was a self-proclaimed "wanderer," not a geographical explorer. Cartographic data collected on these rivers could thus be said to be poor, though none of them could be truly called terra incognita. What was the right river for auspiciously beginning a career as an interior geographical explorer?

In helping Schomburgk work out the answer to this question, Hilhouse had a powerful graphic tool: his *General Chart of British Guiana* (fig. 9).[10] A copy of this chart would have been the field on which Hilhouse and Schomburgk plotted his approach to the interior.[11] A closer look shows that the chart is a rich source for the "perspective" on the interior at the beginning of the nineteenth century. Though evidently laid out on a relatively simple geometric projection—in which longitude and latitude have been projected as perpendicular straight lines[12]—Hilhouse's

---

7. *Brazil Arbitration, British Case*, 25.

8. See manuscript note on figure 16.

9. Hancock's work was little known outside the colony, as Hilhouse noted: "The public have never benefited one iota from their expedition, nor has anything been added by it to science or political information." BL, Hilhouse MS, 9. Hancock is an interesting character. Persuaded of the wisdom of Amerindian botanical remedies, he studied with *piai* (shamans) and advertised a book, *A Treatise on Inflamations and Fever, Its Doctrines, Nature and Treatment and on False Hypothesis in Medicine (Founded on the More Successful Methods Pursued by Certain Aboriginal Natives of North and South America in the Cure of Disease)*, which appears never to have been published. It was advertised on the endsheets of Hancock, *Observations* (1835). A flavor of his ethno-medico-botanical interests can be garnered from his publication "Observations on Certain Resinous and Balsamic Substances Found in Guiana," *Edinburgh Journal of Science*, n.s., 1, no. 2 (April–October 1829): 223ff. His "scientific" stature clearly came under some scrutiny for his membership in the "Eclectic and Verulam Philosophical Society of London," an organization he defended as "misunderstood." Hancock, *Observations* (1835), "Preface." Hancock fits well into Grove's discussion of late eighteenth-century colonial doctors. Grove, *Green Imperialism*, 380ff.

10. Published by James Wyld, "geographer to the king," in 1828 and again in 1836. Figure 9 has been reproduced from BL, Maps 84010.1.

11. As indicated by the legend of Schomburgk's 1835 proposal map, discussed below.

12. It is the same projection he used in preparing the map to accompany his "Memoir on the Warrow Land of British Guiana." *JRGS* 4 (1834): 321. For a contemporary account of the construction of this projection see the 1814 textbook by Jamieson, *Treatise on the Construction of Maps*, section 9, para. 409, "Perspective Representations of the Globe," exercise 1.

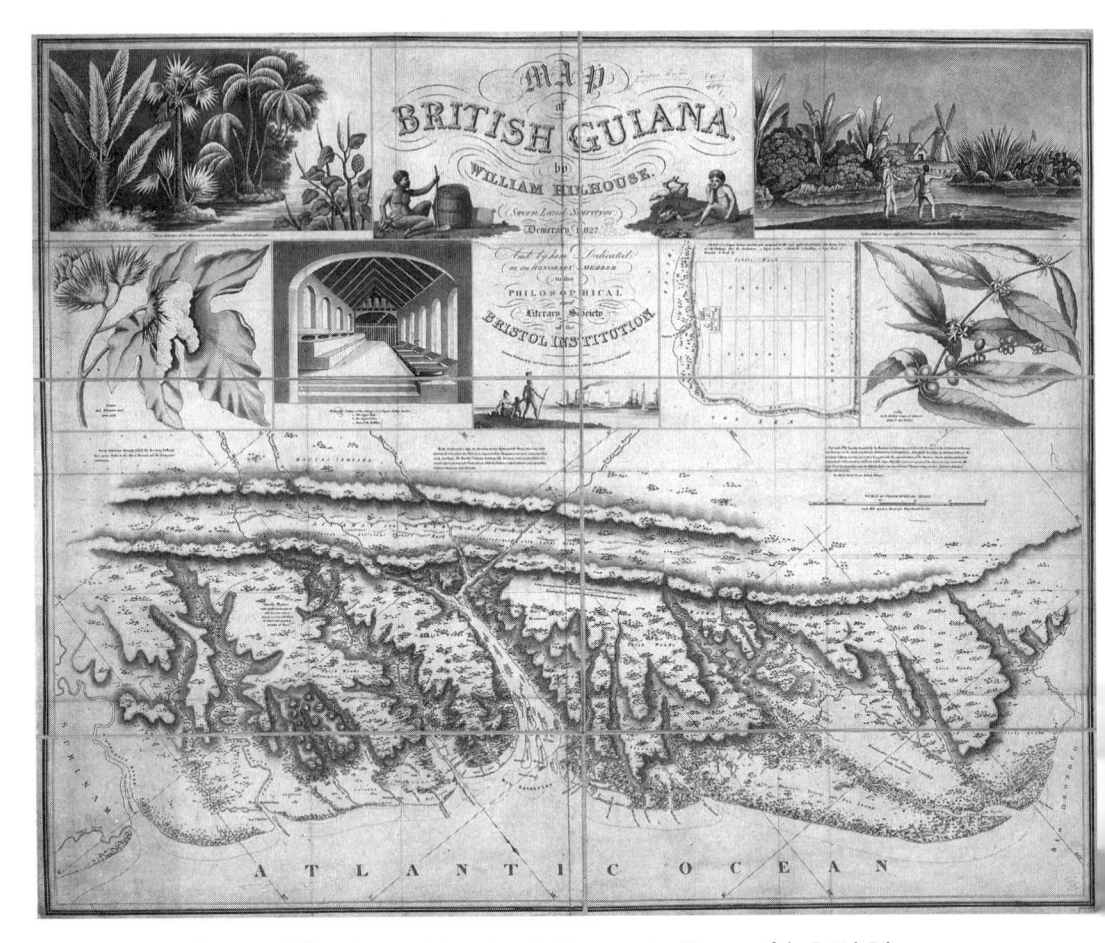

Figure 9. Hilhouse's general chart: the colonial perspective. Courtesy of the British Library.

chart conveys the impression of a view from a point high above the coast down into the interior. Close examination of the image explains this effect. First, the conventional orientation of north and south has been here inverted, meaning that south is to the top of the sheet; the flip has the effect of putting the deep south on the horizon line.[13] This inversion, though, is not alone responsible for the chart's drawing the viewer's eyes to this southern horizon and, hence, *into* the interior of the colony from the coast. The forty-five degree rotation of the graticule of the chart with respect to the frame of the image means that the lines of latitude and longitude have been redrawn in such a way as to give the image a deceptively perspectival character. The rivers—dominant images because of the dark

13. This was not unique. See, for instance, Thomas Walker's 1799 "Chart of the Coast of Guyana" (BL, Maps K 124.40).

hachures that define the drainage areas—have been drawn disproportionately wide at the mouths, rapidly tapering to thin lines. This has the effect of exaggerating the sense of perspective, by apparently foreshortening the dominant image in the field while emphasizing the visual *direction* by means of what appear at first glance to be dark arrows pointing south. The placement of the symbolic trees (the "bush") in elevation, complete with shadows, further emphasizes the perspectival effect. The chart draws the eye into the interior while presenting an overview of the whole colony.

Hilhouse's *General Chart* made one thing very clear. The only way into the interior was via the Essequibo River and its two major tributaries, the Cuyuni and the Mazaruni. The Corentyne, Berbice, and Demerara peter out on reaching the first of the two parallel ranges that mark the "backdam" of the colonial littoral. Hilhouse advised Schomburgk that his first expedition should be up the Essequibo, and that he should attempt to trace it to its sources. Hilhouse's chart, with its beckoning perspective, made the approach look deceptively easy. All one needed to do was follow the river, which threaded its way to the south. In his letter to the RGS, Hilhouse called Schomburgk's Essequibo trip, condescendingly, "his training expedition."[14]

LOCAL REORIENTATION AND DISORIENTATION

The effect of Schomburgk's meeting with Hilhouse, and, more generally, the effect of Schomburgk's local orientation when he arrived in the colony in 1835, can be dramatized by comparing his two earliest proposal maps for his Guiana expeditions, one done before and the other after April 1835. Juxtaposition of these manuscript maps, one submitted to the RGS from Tortola in late 1834 or early 1835 (fig. 10) and the other submitted to the governor of the colony in the summer of 1835 (fig. 11), reveals a significant shift in Schomburgk's conception of the object of his expedition.[15] This shift of scale—from the grand circuit depicted in the RGS proposal map to the local expedition within the framed boundaries of British Guiana—betrays a tension in Schomburgk's work and a conflict between the aims of his multiple masters, who had very different terrae incognitae in mind for him to map.

The RGS proposal map shows Schomburgk's conception of his expedition as a ranging, pan-national "Scientifick Journey to South America," as its legend states. The note on the map that it was "compiled from

---

14. RGS, Correspondence, Hilhouse to RGS, 17 August 1835.
15. Figure 10 has been reproduced from RGS, Map MS, "Ven S/S2." For information on figure 11, see note 16 below.

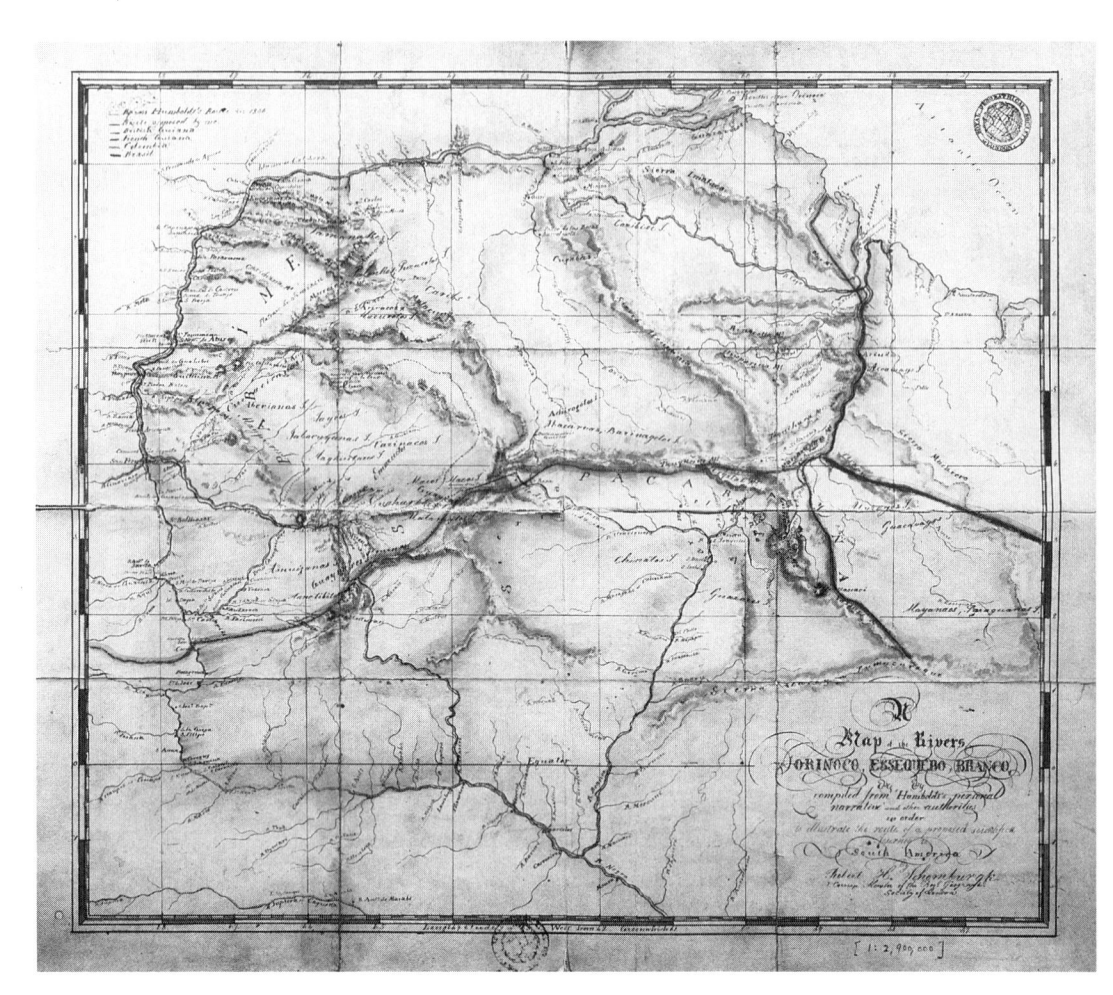

Figure 10. Schomburgk's first proposal map: grand plans. Courtesy of the Royal Geographical Society.

'Humboldt's personal narrative'" illustrates that Schomburgk saw his work in the context of that of his celebrated countryman. This proposal map also dramatizes that, originally, Schomburgk's primary interests did not lie within the territorial limits of the colony of British Guiana. The proposed journey passed along the edge of British Guiana before heading off for the Venezuelan highlands and the Rio Branco. The primary aim of the passage depicted was to link the "observation" points in Guiana to those made by Humboldt in "Colombia" while "fixing" the sources of the Orinoco, which Humboldt had been unable to reach. In short, British Guiana itself was not the object of Schomburgk's major expedition as he proposed it to the RGS, except incidentally as he passed by on his way to the terra incognita of lowland South America.

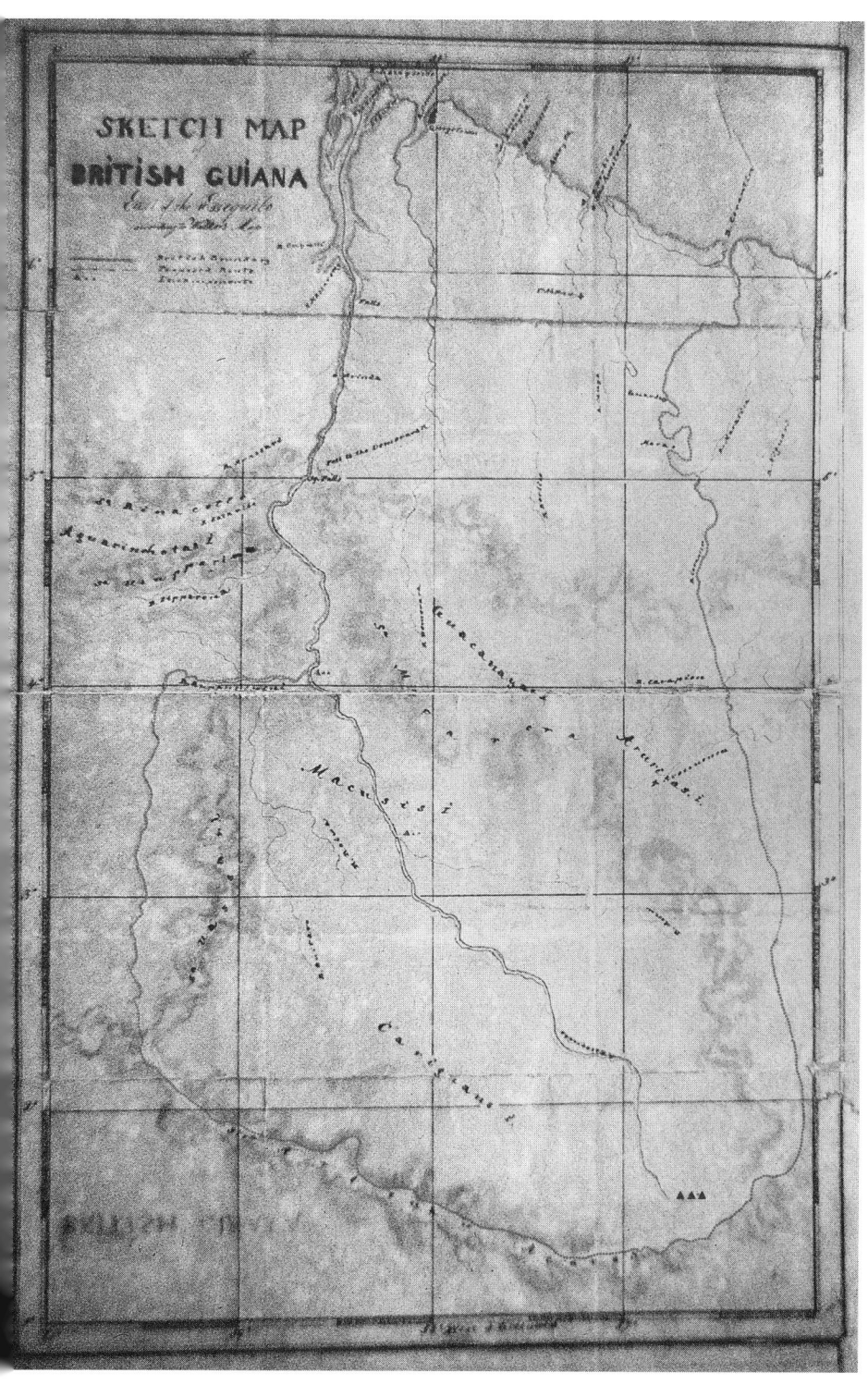

Figure 11. Schomburgk's second proposal map: a circumscribed view. Courtesy of the Caribbean Research Library, University of Guyana.

But the other proposal map (recently discovered in the Archives of the University of Guyana) attests to a spatial realignment of the explorer on his arrival in the colony. This manuscript "Sketch Map of British Guiana East of the Essequibo" represents the colony as a cartographic unit, occupying the frame of the map, bounded by speculative dotted lines.[16] The legend indicates that the map has been prepared from the local work of Hilhouse and J. E. Alexander, among others.[17] The eastern branch of the Essequibo bears a striking resemblance to that laid down by Alexander in his submission to the RGS *Journal* in 1832, where the river culminates in a forked tongue just below the second line of latitude (fig. 12). At that hypothetical source the optimistic Schomburgk expectantly placed a small blue triangle on his proposal map, which the accompanying key identifies as an "encampment." Equally optimistic, a clear, straight, dotted blue line traces the route up the river he had settled on with Hilhouse.

In comparing these two proposal maps we see Schomburgk's geographical and cartographic attentions refocusing on the "plot" of the colony and its local terra incognita exactly at the time of his arrival in the colony itself. The explanation for this shift lies in the politics and economics of his expedition. Schomburgk arrived empty-handed and in need of donations from the colonial chest. He quickly discovered that, with abolition under way, a financial crisis loomed over the plantocracy and the government; there was very little local interest in the exact geographical coordinates of the sources of the Orinoco. If the plantocrats could be interested in the interior at all—and from the outset that looked doubtful, given their preoccupation with an impending labor crisis, competition from East Indian and Brazilian sugar, and a host of other practical issues[18]—what would interest them was the extent and potential resources of the colony they inhabited. Moreover, Schomburgk suddenly found himself among locals, some of whom knew the interior quite well. Some had made "bush expeditions" to hunt runaway slaves in the days before abolition; others had worked the whole of their lives as land surveyors defining plots of land in the bush, sometimes at considerable distances from Georgetown. Surrounded by this company, Schomburgk's rhetoric of a "journey of discovery" into the terra incognita of South America, which had seemed appropriately virile while writing letters to the RGS from his elegant island chambers in Tortola, was quickly toned down: he

---

16. CRL, RBR Maps, accession no. 109883. Archivists now report this map lost.
17. Ibid., legend.
18. For an analysis of the plantocracy's attitude toward the interior in this period, see Lancaster, "Unconquered Wilderness," 13–26.

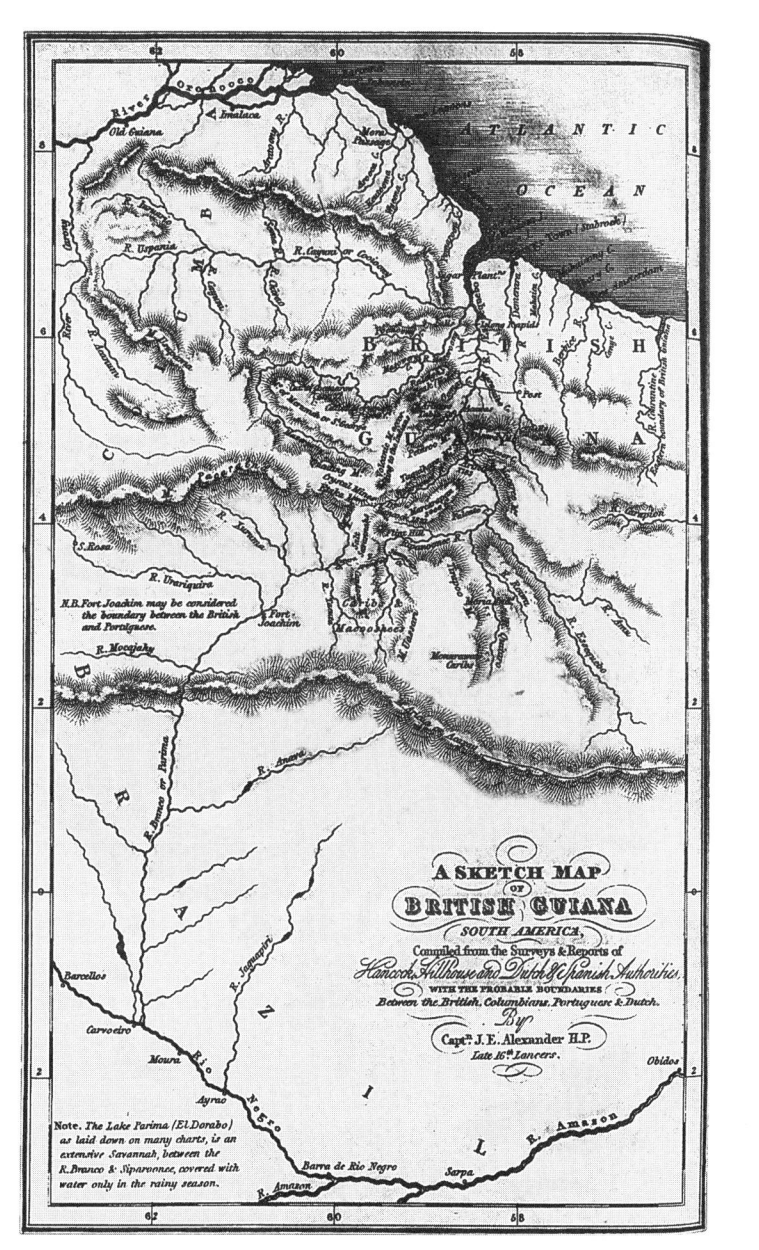

Figure 12.   Alexander's map for the Royal Geographical Society. Reproduced from the Society's *Journal*, 1832.

Figure 13. Detail from Hilhouse's general chart: plotting colonial control. Courtesy of the Library of Congress.

now explained that he would be making a "tour" of the interior of the colony.[19]

Hilhouse's chart, while it beckoned into the interior, also provided a graphic representation of the character of colonial concerns of the sort that Schomburgk would have encountered on his arrival. Hilhouse was a plantation owner as well as a sworn land surveyor, and his perspective on the interior, and the colonial possession of the interior, was inseparable from the importance he attached to control by the propertied order on the coast. Hilhouse not only decorated the elaborate heading of his printed chart with images depicting plantation order, he even inscribed his role in achieving that order directly on the surface of the chart itself. Closer attention to the writing over the Berbice districts reveals a small caption reading "Extirpated Camps of Bush Negroes" (fig. 13). As military leader of mercenary Amerindian bounty hunters, Hilhouse very likely conducted the expeditions that effected the extermination.[20]

The clearest statement of Hilhouse's preoccupation with the planta-

19. RGS, Correspondence, Schomburgk file, proposal to Carmichael-Smyth, 1835.
20. For details on Hilhouse's militia work, see Roth, "Hilhouse's Book of Reconnaissances."

Figure 14. Detail from Hilhouse's general chart: jungle and swamp made legible. Courtesy of the Library of Congress.

tion life of the colony is embedded in the images that make up the top half of the chart—the "legend," which is a storyboard for the evolution of the colony itself.[21] The images in the upper left and upper right corners amount to a "before and after" image of colonial exploration. On the left is a depiction of the perspective of the riverine explorer: plants encroach, the way is obscure, the viewer is left with the sense of pushing between the palm fronds for a glimpse of what lies ahead (fig. 14). On closer examination, however, this image proves to be considerably more than a glimpse of the bush. In fact, it is a *key* to guide the eyes of the interior explorer concerned for the welfare of the colony. Rather than a "picture" of the bush, the image is a schematic diagram. It presents a field, but not a field of rushes; rather, it is a field of *signs:* each tree and bush is labeled with its name, and the image itself is called "plants indicative of the alluvion, or best description of lands for all culti-vation." The image indicates how to identify the "wilderness" bestsuited to being cleared for growing sugar and the other staples of plantation cultivation.

As if to reinforce the immanence of this exploitable fertility, Hilhouse included a second and parallel image on the other side of the sheet. Out of the troolie swamp could come, under the proper conditions, this

---

21. For an exemplary account of the "reading" of map legends as critical elements in the "map as a text of possession," see Clarke, "Taking Possession: Cartouche as Cultural Text."

scene of colonial productivity (see the upper right corner of fig. 9). The mill turns in the background, wringing the sweet juice from the cane; a plume of smoke shows that the juice is being refined into Demerara crystals. In the foreground the well-toffed master, his overseer, and dog observe the enthusiastic cutting of a troop of barefoot slaves. The meandering flow of the creek has been transformed into the orderly and productive flow of the plantation ditch, permitting the easy passage of cane to mill. Hilhouse declared that the ubiquity of such canals "reduces the labour of Negros one fourth," affording the plantation owner in Guiana greater productivity.[22] The images below this tableau further underline the role of order in delineating colonial space: the architectural regularity of a sugar boiling house stands next to a rectilinear plan of a plantation.[23]

Planning his routes into the interior on Hilhouse's chart, Schomburgk would have seen the interior framed by these images of plantation life. The owners of those plantations framed his preparations in a concrete sense. In Georgetown they constituted more than his links to Georgetown society (the local greats who invited the new arrival to meals and shooting excursions); some of them sat on the Court of Policy and thus held the purse strings of the expedition. In this environment, confronted by the immediate need to secure funding from individuals with a very limited "scientifick" interest in the interior, it is not difficult to understand how Schomburgk rapidly refocused his interests on the geographical body of the colony itself. Although his statements of purpose to the board of the RGS had been flavored by the ideas of sweeping Humboldtian geography, those RGS sponsors had so far offered him a slim £50 advance. The governor of the colony, on the other hand, Sir James Carmichael-Smyth—from whom Schomburgk was, according to his RGS directives, to "receive further instructions"[24]—had offered him, over the grumblings of the Court of Policy, £150.[25]

The issue of to whom, exactly, Schomburgk owed his primary loyalty quickly became a sticky issue of explorer etiquette, one that he navigated with even less success than he did the rapids he encountered on his first push up the Essequibo. He was traveling under a double commission, si-

---

22. Hilhouse, "Remarks and Observations," 2.
23. It is interesting to consider the way the spatial layout of a plantation paralleled the geographical conception of the colony itself: a proper sugar plantation had its own complex hydrography, configured for the circulation not only of water but of boats as well; to be a good possession it had to be circumambulated and have its boundaries recorded on a map. It is possible to construe the work of geographical explorers like Schomburgk as extending the cadastral preoccupations of local plantation surveys to the scale of the colony as a whole.
24. Rodway, "Schomburgks in Guiana," 2.
25. RGS, Correspondence, Schomburgk file, proposal to Carmichael-Smyth, 1835.

multaneously in the pay of the RGS and of the colonial governor. One might presume that this cooperation implied a unanimity of purpose, but it was not at all so. The geographical objectives of his two sponsors proved to be strikingly different, if not explicitly opposed. Much has been made of the "practical" orientation of the early RGS, but the "utilitarian appeal" of the Society's projects and the agenda of promoting the "social value of geography to the service of empire"[26] did not always translate into geographical goals shared with local colonial officials, as the story of Schomburgk's early work reveals.

It was considerably easier to draw a line to the sources of the Essequibo on a proposal map than it was to get to the point on the ground. Schomburgk aimed to be cost effective. He resolved to use a crew of coastal boatmen only as far as the first Amerindian village, where, he understood, "money is of no value." This would allow him to secure a native crew by trading "items of nominal value."[27] In addition, the lush vegetation he presumed he would find at the headwaters would be something of a botanical El Dorado. A large number of new and unusual botanical specimens, which could be sold to his ten subscribers at £2 a hundred,[28] promised to "add materially to the fund of an intended larger expedition."[29] Schomburgk set out on 21 September 1835 on the first leg of the expedition that was supposed to take him to the sources of the Essequibo before the rainy season in late March. By 23 October the party had reached the confluence of the Rupununi, but they had yet to come across a populous village of Amerindians willing to offer their services as paddlers to replace Schomburgk's expensive crew. These boatmen had escorted him from the coast at the cost of nearly £3 a day or, in the economy of the expedition, no fewer than 150 botanical specimens, dried, pressed (and transported to London) per diem. To cover the whole sixty-odd days during which he retained their services, Schomburgk would have needed to collect a weighty 9,000 specimens. He sent back exactly 170.[30]

Nor did he get to the sources of the Essequibo. Late in the season, unable to secure paddlers, Schomburgk was stopped by a fourteen-foot waterfall on the upper Essequibo. Leveled by fever, he had to be carried part of the way back from his first expedition in his hammock. His first report

---

26. Livingstone, "History of Science and the History of Geography," 291.
27. RGS, Correspondence, Schomburgk to RGS, 8 August 1835.
28. RGS, Correspondence, Schomburgk to RGS, 15 September 1833. Rivière ("From Science to Imperialism," 2) suggests that, in addition to these ten (German) subscribers, Schomburgk also had twelve English ones, paying a slightly higher rate.
29. RGS, Correspondence, Schomburgk to RGS, 8 August 1835.
30. *JRGS* 6 (1836): 284. A passage in RGS, Correspondence (Schomburgk to RGS, October 1835) suggests that he had found more but had not been successful in shipping them back to Europe.

to the RGS made much of the christening of the "Great Cataract" (which he named after King William IV)[31] and included several maps (including a sketch of the fortifications of the Brazilian Fort São Joaquim on the Takutu), but the directorship of the RGS was not pleased. The committee reviewing his work read a chilly letter from Hilhouse that concluded: "He was stopped by a fall of 20 feet! 2000 feet would not have stopped me!"[32] Taking the lead from their local adviser, the committee ruled that "no accession to our geography of the least importance in that part of South America has been derived from Mr. Schomburgk's journeys" and advocated a suspension of his funding.[33]

This unfavorable report set in motion considerable debate over Schomburgk's next river, a debate that demonstrates most clearly how different terrae incognitae embodied different interests. Schomburgk wrote the Society that he intended to explore the Corentyne, the eastern boundary of the colony, and by the time its negative reply reached him it was too late to change his plans. The Corentyne had been the governor's idea, and Hilhouse thought it was a bad one. He wrote the RGS: "By pointing out the Corantyn river the Governor has, I think, more eye to his own merit in dictating the survey of the boundary river than in the general interests of scientific discovery."[34] In the wake of the failure of this second expedition, Schomburgk admitted in a confidential letter to the Society that, though he had tried to sound enthusiastic about the Corentyne, this was because he knew the governor had permission to open his mail to the Society.[35] Pleading to the RGS secretary his willingness, fidelity, and sense of isolation, Schomburgk would later confess, "I tell you now, that the Lieutenant Governor's wish amounted to almost an order to explore the Courentyne."[36]

Barrow, for one, had expected as much. When he saw Schomburgk's Corentyne proposal he penned in an internal memorandum that "the Society takes no interest in the Boundary of the colony."[37] Governor Carmichael-Smyth did, however. The upheavals of emancipation help explain why. In 1836 the colony was halfway through the turbulent years of "apprenticeship," the four-year transition from slavery to full emanci-

---

31. In doing so he patched up an earlier toponymic faux pas: in 1832 he had tried to dedicate his chart of Anegada to the king and was informed by the secretary of the RGS that charts were not considered appropriate to dedicate to the sovereign. RGS, Correspondence, Schomburgk to RGS, February 1833.
32. RGS, Correspondence, Hilhouse to RGS, 12 April 1836.
33. RGS, Correspondence, Schomburgk file, "Report of Committee of the Guyana Expeditions."
34. RGS, Correspondence, Hilhouse to RGS, 10 October 1836.
35. RGS, Correspondence, Schomburgk to RGS, 22 April 1837.
36. RGS, Correspondence, Schomburgk to RGS, 12 May 1837.
37. RGS, Correspondence, Schomburgk file, Barrow's memorandum of 13 October 1836. Was he concerned that the firm delineation of boundaries would merely constrain any future expansion of British territorial interests in the region?

pation.[38] When rumors of emancipation began to circulate in the colony in 1833, the first year of Carmichael-Smyth's governorship, planters anxious for their "property" began to talk of smuggling their slaves across the ten-mile-wide mouth of the Corentyne at night and slipping them ashore by launch into the Dutch settlement of Nickerie, where there was no meddlesome talk of liberation. Carmichael-Smyth lost no time in declaring that he would personally see to the hanging of the first planter who tried it.[39] Such high-handedness, coupled with the confusion and anxiety over emancipation, led to bad blood between the plantocracy and the governor, who shortly thereafter brought libel charges against several colonists for scathing editorials in the local press.[40]

If emancipation threatened to precipitate a night migration over the Corentyne from west to east, the potential for an equal and opposite reaction was also recognized. Surinam had a long history of slave runaways, and if freedom lay across the Corentyne there was reason to think that many might try crossing to the protection of the British flag, affording a much-needed boost in the midst of the labor crisis.[41] Just what areas fell under that protection, then, amounted to a very timely question during the chaos of apprenticeship.[42] A mapped survey would offer an occasion to set the Union Jack along the course of the river. A formal pageant of British colonial authority would provide just the show needed to dissuade discontented colonists from meddling in the expatriation of slaves, while drawing the line of the "safe zone" that might encourage Dutch slave defections. With this agenda in mind, Carmichael-Smyth at first warmly, and later more heatedly, encouraged Schomburgk to make the Corentyne his second expedition.

It was a fiasco. Although Schomburgk was able to claim, in defiance of Hilhouse's prediction (and his *General Chart*), that the river was not a mere alluvial rivulet, it had not proved a highway to the interior either.[43]

---

38. Note that apprenticeship was originally planned to last six years.
39. Webber, *Centenary History*, 166.
40. Ibid.
41. Such flights by Dutch slaves occurred in this period. In the NAG several reports from James Shanks, the superintendent of rivers and creeks for Berbice, record his having instructed the post holder on the Corentyne to look out for a party of "Surinam Negroes" and encourage them to go down to the coastal plantations, where they would find work. It appears that rumors were afoot that the post holder had made arrangements with the Dutch to sell them back to their pursuers. NAG, Loose reports, Shanks to Light, [1839].
42. In the interests of concision, I have omitted a lengthy discussion of the ways abolition affected land surveying and the legislation around territorial claims in the colony. This period saw a number of significant changes in land tenure and in the related enterprises of surveying and surveillance (see my forthcoming article "Abolition and the Cadastral Ideal").
43. Hilhouse's prediction that the Corentyne would not reach the Sierra Accarai hinged on an argument about the species of fish living in the river, knowledge he claimed to have learned from Amerindian informants. See above, note 4.

A pair of cataracts—christened, in a gesture of toponymic reconciliation, the Smyth and Barrow Falls—blocked Schomburgk's route. A near mutiny among the Amerindians he called his crew did not advance the cause. They refused to show him the portage.[44] Aware that his exploring career hung in the balance, Schomburgk attempted an impromptu expedition up the neighboring Berbice. He called this terra incognita, but it was difficult to conceal that the marshy, meandering stream obviously did not lead to the deep south. Nevertheless, Schomburgk's narrative of the Berbice expedition salvaged his commission. In it he described a dramatic encounter with a party of Carib slavers whom he claimed to have overawed and persuaded to desist from their trade in innocents.[45] This sort of abolitionist legwork won metropolitan allies, but the real salvation of the Berbice expedition (and ultimately Schomburgk's career itself) lay elsewhere: narrating the events of the first day of the new year, 1 January 1837, Schomburgk described the moment that would later become something like the apotheosis of his exploring career: "Some object on the southern point of the basin attracted my attention; I could not form any idea of what it might be, and I hurried the crew to increase the rate of their paddling; in short time we were opposite the object of our curiosity—a vegetable wonder!"[46] The water flower before them, a gargantuan lily, became the ad hoc telos and turning point of the expedition narrative; here lay a meet reward for all their toils. Wrote Schomburgk, "I . . . felt myself rewarded."

And that was only the start. This "most beautiful specimen of the

44. A report from Superintendent Shanks in the NAG suggests that Schomburgk likely decided not to go above the falls because of the hazard posed by the uncooperative Amerindians, rather than because he could not find the portage (as he claimed). Shanks wrote: "It may not be inopportune, to state here that the Indians made no difficulty in showing us the path around the falls at which point Schomburgk failed, and which was certainly most surprising considering that the path is within fifty yards of one of the falls and in an open place." Shanks also denied that Schomburgk had surveyed above the falls as he claimed. That survey was the basis for his dismissal of Hilhouse's claims concerning the alluvial character of the stream. Shanks wrote: "But how he could have arrived at that fact [that the river was nine hundred yards wide above the fall], I cannot understand—the Indians, several of them the same that were with him, declaring that the gentleman did not go beyond the falls." NAG, Loose reports, Shanks to Light, n.d. [1840].

45. The story of the Carib-Dutch slave party is plausible. Documents in the NAG confirm that the Corentyne post holder, N. De Wolff, was considered a rogue, whose Dutch ancestry and sympathies brought him under suspicion for involvement in illicit trade of slaves, Amerindians, timber, and rum. In addition to making a pointed allegation of smuggling in 1840, his superior, the local superintendent James Shanks, not only accused De Wolff of periodically abandoning his post and enticing local Amerindians to cut timber for him in return for illegal liquor, but also intimated that he had been involved in the sale of a young Amerindian boy. In calling on the governor to replace him, Shanks stopped just short of calling the post holder seditious, writing, "[He shows] in a most marked manner his contempt of your reporter's authority, in fact, he seems to think himself beyond the reach of all interference." NAG, Loose reports, Shanks to Light, 25 February 1840.

46. *JRGS* 7 (1837): 320.

Botanical World"[47] would eventually be named after Queen Victoria, become a centerpiece of the Great Exhibition, precipitate the building of elaborate new tropical aquatic greenhouses for its propagation in the British Isles, and amount to nothing less than a metropolitan sensation (see chapter 4). If anything saved Schomburgk's career in January 1837, it was that flower. Though in fact previously "discovered" several times by other South American explorers, the *Victoria regia* was discovered better by Schomburgk, who quickly spied in its plump and aromatic blossoms the promise of royal patronage and the rescue of what was well on its way to being his third expeditionary debacle. The timing was good. In May 1837 Schomburgk sent off an elaborate description and sketch of the flower to the RGS. In June King William IV died and the young Victoria acceded to the throne. The flower, or rather a large oil painting of it, was a choice offering to have on hand for the coronation of the Society's new patron (plate 2).[48]

Schomburgk also made clear in his correspondence with the Society that "a misunderstanding," rather than cataracts, had stopped him from executing the geographical expedition deep into the interior that he had originally proposed. He pointed out that the governor had to grant passes to enable anyone to pass the "posts" established at the upper edge of settlement on each river.[49] This made the governor the ultimate keeper of terra incognita. Beholden to him for money and obliged to pass all letters through his mail, Schomburgk explained that he thought he was supposed to go where the governor suggested. As for the explorer's own interests, he wrote: "When I first planned the expedition it NEVER entered my views to dedicate to British Guyana more than a superficial investigation."[50] His aim had been to cross over into Brazil and beyond, but he claimed he had been told it was not allowed.

Such was the tangle of politics and the discordant visions of what was the terra incognita that mattered and how one got there. An orientation to the local scale of the colony—the product of local colonial concerns and authority—nearly cost Schomburgk his commission with the RGS, which wanted those in its employ to send back properly geographical observations and maps rather than preoccupy themselves with local colonial reconnaissance. Schomburgk was given one more chance to demonstrate that he understood what the RGS was after. On 12 September 1837 he

---

47. Copy that appeared on the broadsheet printed to advertise the Guiana Exhibition of 1840. Exemplar held by the Lilly Library, Indiana University.
48. Plate 2 is an 1848 lithograph of the painting of the flower on display at the Royal Gardens, Kew. See *Floricultural Cabinet and Florists' Magazine* 16, no. 15 (March 1848): E.
49. RGS, Correspondence, Schomburgk to RGS, 22 April 1837.
50. Ibid. Emphasis in the original.

embarked on what would be a circuit of almost two years and 3,000 miles, an expedition that would win him the Society's Gold Medal in 1840.

MAPPING TERRA INCOGNITA: THE TRAVERSE AND ITS WEAKNESSES

Reaching the right sort of unknown regions demanded that an explorer negotiate committees, history, and budgets as well as rivers. But getting to someplace suitably incognita did not complete the task. The obligation of a geographical explorer was to return from a cartographic blank with a map. Schomburgk understood this clearly after his early reprimands from the RGS. His copious notes on Amerindian behaviors had been edited out of his reports before they saw publication in the Society's *Journal,* and he had been exhorted to concentrate on "geography" rather than on descriptive accounts of his travel.[51]

Schomburgk had sent back maps, including three multiple-sheet charts of the rivers he had ascended and a sounding chart of the mouth of the Corentyne. And while no one objected to the quality of the surveying represented in these manuscripts, such large-scale river charts—where a mere seventy geographical miles became eleven feet of chart—were not exactly what the *geographical* society had in mind. The more extensive such observations became, the more difficult it became to distinguish hydrography from geography. Vice Admiral Frederick William Beechey, later a rear admiral and the first naval professional officer at the Board of Trade and president of the RGS, acknowledged that river surveys were a murky business. Of river surveying he wrote in the *Admiralty Manual of Scientific Enquiry:* "It becomes difficult to avoid infringing on what properly belongs to geography. The two sciences are indeed here so nearly allied that it is scarcely possible to avoid encroaching on the sister branch."[52] The RGS clearly hoped Schomburgk would soon submit a map of some land, preferably one with much larger coverage. The chart of the Corentyne mouth was forwarded to the Admiralty, and the river charts (much reduced) were incorporated onto a small plate of Guiana prepared by John Arrowsmith, which was used to print the maps accompanying Schomburgk's printed reports in the *Journal* (fig. 15).[53]

Close examination of Schomburgk's manuscript charts reveals the kind of surveying he did. Each of his river charts is a route survey (a traverse)

---

51. RGS, Correspondence, Schomburgk file, "Report of Committee of the Guyana Expeditions"; RGS, Correspondence, Schomburgk to RGS, 13 August 1837.
52. Herschel, *Manual of Scientific Enquiry,* 66.
53. The large charts (not shown) would include RGS, Map MS, "Guyana S 28."

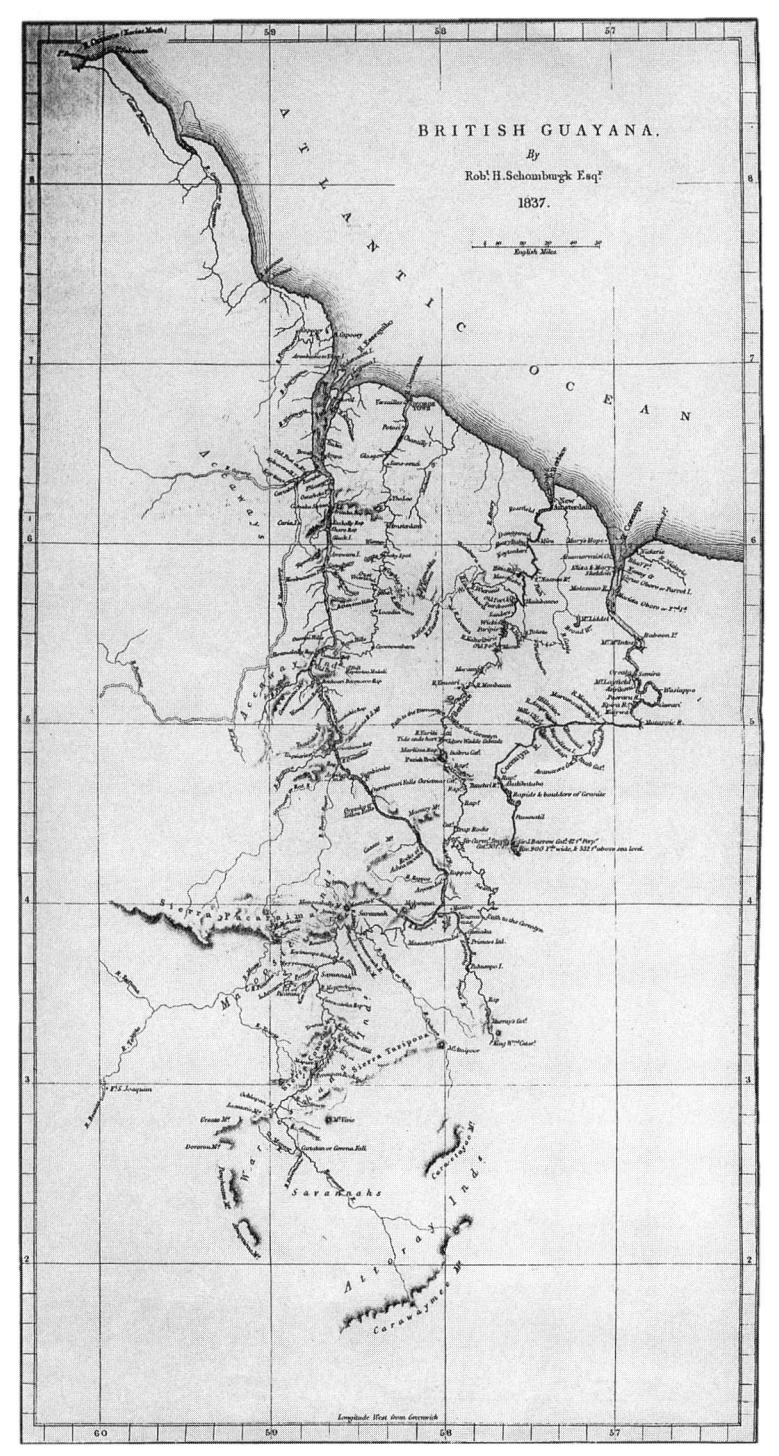

Figure 15. Arrowsmith's map for the Royal Geographical Society. Reproduced from the Society's *Journal*, 1837.

in which the map incorporates a mixture of points fixed by astronomical observations and others fixed by means of bearings (compass and sextant) to distant objects taken en route. On the Corentyne chart, for instance, points established astronomically are labeled, and latitude and (where relevant) longitude are given. In his reports Schomburgk included lists of the astronomical observations from which these points were derived. Dead (or "deduced") reckoning was used to fill in the course of the river between these points, and evidence for Schomburgk's calculations (from elapsed time to distance along compass-determined vectors) can be seen in the erasures on the map: on portions of the river rectilinear grids oriented with the river's direction have been penciled in and used to establish bearings with respect to the river itself. They are nearly invisible, but in numerous places pinpricks through the surface show that Schomburgk has transferred his working sketches (now lost) onto the finished manuscript by means of these graticules.

These palimpsestic elements of Schomburgk's early cartographic work only confirm what might be surmised about his surveying practices. Matthew Edney, writing on surveying and mapmaking in British India, describes the traverse survey as the "ubiquitous form of survey in all of Britain's possessions in the eighteenth century."[54] While improved portable instruments—rugged chronometers, artificial horizons,[55] and more precisely graduated sextants[56]—made it possible for surveyors working in this tradition to supplement their direction-and-distance observations by more (and better) observations of latitude and longitude, enhanced traverses remained, into the twentieth century, the only possible surveying style for explorers expected to cover large areas with a minimum of institutional support.[57]

The technique had its roots in army and naval traditions. On the military side, the colonial traverse constituted a version of the *reconnoiter,* a robust task involving a brisk incursion into alien territory to grasp the essential features of unfamiliar terrain. According to one instructor in the art, the reconnoiter called for "boldness, perseverance, and a Schmalcalder compass."[58] The aim was to establish the "lay of the land" in a single passage so that officers could think about the battle *field* itself, not

---

54. Edney, *Mapping an Empire,* 92.
55. Day, *Admiralty Hydrographic Service,* 56, and Brewington, *Navigating Instruments,* 54.
56. Cannon, *Science in Culture,* 97. Bosazza and Martin write: "By 1821 [sextants] had been developed to being vernier instruments with telescopes, and users knew how to resilver mirrors in the field. In most cases they read to 20″." Bosazza and Martin, "Geographical Methods of Exploration Surveys in the 19C," 32.
57. For an detailed account of these "exploration surveys" in Africa, see Martin, *Maps and Surveys of Malawi.*
58. Jackson, *Course of Military Surveying,* 94.

merely the way there and back.[59] Anxious to distinguish the vigorous and comparatively elite practices of military surveying from its more insipid cousin, the civil survey, one author went to some length to point out that the difference between the two lay in instrumentation: "The former [civil surveyors] have almost uniformly adopted a mode of surveying without any instruments save the land-chain, while we, on the contrary, throw aside the chain and adhere to our instruments."[60] Hobbled by their chains, civil surveyors were portrayed as topographical pedants: the quaint tradition of the chain-dragging surveyor was suitable only for the open, pastoral, settled, and essentially "fixed" territory of the English countryside. "How could it be practiced in the jungles of India?" asked a military surveyor with an eye for colonial interiors.[61] After all, "in order to run the necessary lines the surveyor is obliged to cut his way through copse, wood, plantation, etc." Practitioners of the traverse preferred to associate themselves with the traditions of nautical navigation, whose instruments (sextants, almanacs, chronometers) they appropriated and whose predicament (voyaging in a trackless expanse) they portrayed themselves as sharing.[62]

This instrumental advantage gave traverse surveying scope and reach and made it the obvious technique for the geographic explorer working in the terra incognita of the colonies. That an enhanced version of the eighteenth-century traverse remained current through the nineteenth century is evinced by E. C. Frome's handbook for colonial surveying, published in 1850, which dedicated a chapter to the conduct of a good traverse.[63] This longevity does not mean that explorers and geographers were oblivious to its flaws. They were not. The traverse, even supplemented by increasingly accurate celestial observations, remained a distinctly second-best (if not actually suspect) source of geographical data. A synthetic trigonometrical survey—done by a team, composed out of a net of precise triangles measured out over the region by means of a cumbersome theodolite—was always to be preferred where resources allowed. The erasures on Schomburgk's river charts show one reason why. Traverse surveys demanded juxtaposing multiple frames of reference to produce a cartographic arti-

---

59. Edney argues that during the later Enlightenment, "the essence of military science was geometry, geography, and their point of contact: mapping." Edney, "British Military Education," 18.

60. Jackson, *Course of Military Surveying,* 156.

61. Ibid., 157.

62. In describing the task of the colonial surveyor, E. C. Frome wrote of the passage into the interior: "The traveller then finds himself in the same predicament as at sea, having little beyond his dead reckoning to trust to for the delineation on paper of his day's work. In this position he must look to the heavens for his guide; and hence the necessity of his becoming himself, or having with him, a good and rapid observer." Frome, *Outline of the Method,* "Colonial Surveying."

63. Ibid.

fact. This meant the map had to incorporate different kinds of observations: traverse data's lack of uniformity often led to inconsistency, and in this way the traverse survey fell far short of the "cartographic ideal."[64] It afforded no reliable system for correcting error and could never be systematic. Most damning of all, the traverse was inextricably entangled with the idiosyncrasies of the traveler and the contours of the irregular earth. Some places were easy to see, others were tougher. Some regions were cloudy, others were clear. Some places were flat, others were mountainous. Observers too had their characters, and given that they made brisk passages, they were rarely able to produce enough observations of a given place that statistical methods could be employed to weed out errors.

The shortcomings of the traverse survey "point to the core of the epistemological dilemma of modern mapping: no matter how accurately and precisely the world's structure is measured, that structure is created through the surveyor's and the geographer's experiential perception."[65] Dependence on itinerant eyes and the peculiarities of route and weather meant the traverse survey sat awkwardly in the cartographic pantheon erected by Enlightenment philosophes.[66] European geographers employed a number of techniques to shore up the lapses of praxis that threatened the privileged place of the map as the paradigm of an archive of knowledge. At the end of the eighteenth century, increasingly elaborate systems for correcting, compiling, and collating geographical data emerged. But these patchwork solutions for averaging and combining diverse material could not solve the core dilemma of mapping.[67] Only an entirely different surveying technique—triangulation—promised such a solution.

Exigencies of cost and personnel meant that trigonometric surveys were far beyond the reach of individual explorers like Schomburgk and colonial backwaters like Guiana. Nor was Schomburgk part of a centralized survey administration that would see to the compilation of redundant field observations. But while Schomburgk had no choice but to conduct traverse surveys, some traverse surveys were better than others. Hilhouse made certain Schomburgk understood as much when they met in April 1835. Hilhouse made no effort to hide his scorn for the traverse survey of John Hancock, who had made a map of the interior of Guiana on his trip up the Essequibo to reconnoiter Mahanarva's dominions. Hilhouse may have possessed a tracing of this original manuscript map (fig. 16).[68] But per-

64. Edney, *Mapping an Empire*, 95.
65. Ibid.
66. Ibid., 17, 49–53.
67. Ibid., 96–104.
68. Figure 16 has been reproduced from *Brazil Arbitration, British Case Atlas*, 19.

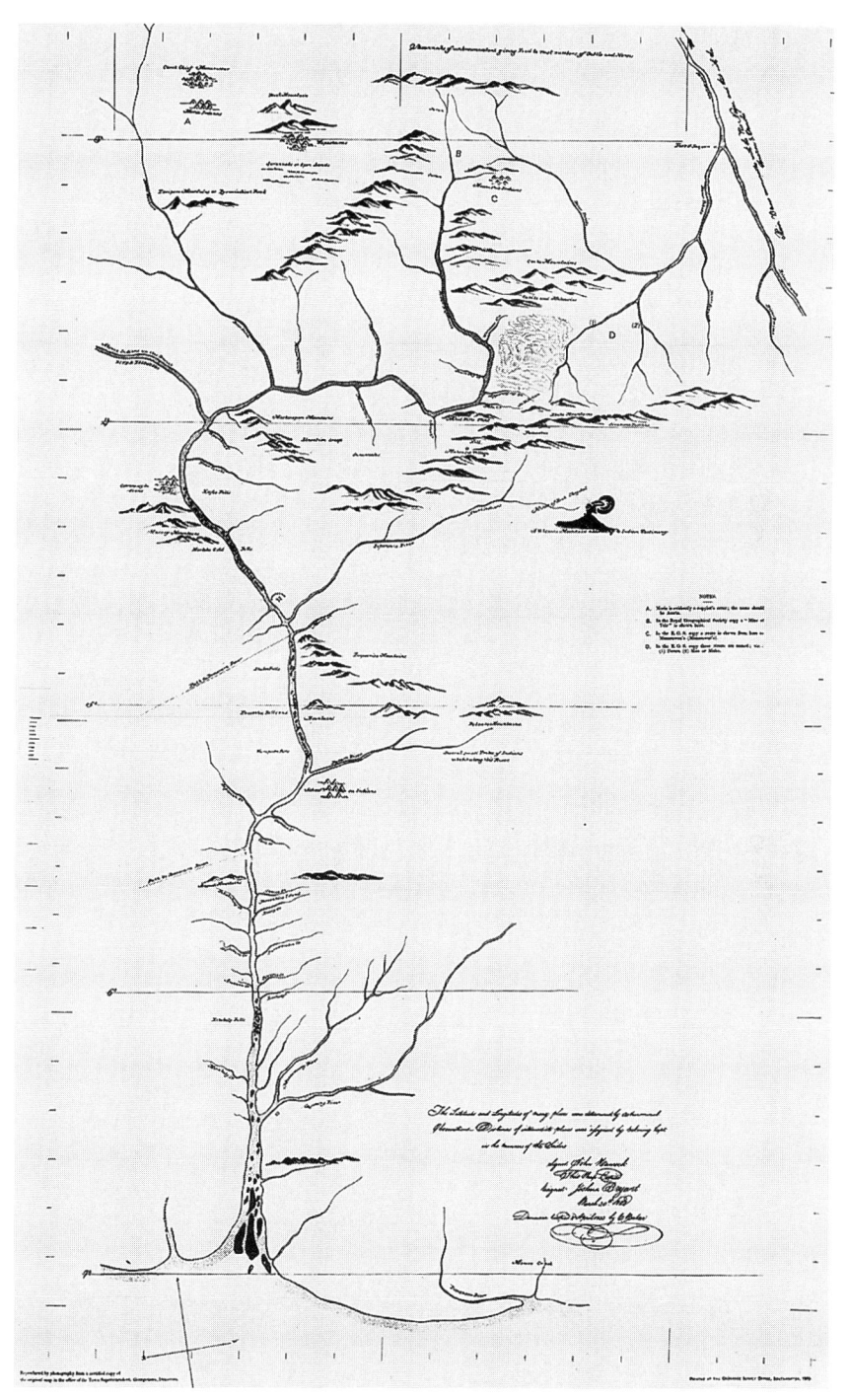

Figure 16. Hancock's sketch map: the cartography of colonial reconnaissance. Courtesy of the British Library.

haps, scorning it, he never bothered to secure a copy through his position as a sworn land surveyor. It is certain, though, that he discussed it with Schomburgk and described its shortcomings, which Hilhouse summed up in a memoir sent to the RGS: "[Hancock's] chart, as exhibited to the author, gives no more information than . . . others had before published, and will not bear the test of collation, appearing to be a mere private rough sketch or memorandum."[69]

Hancock's map, then, affords a clear instance of the kind of map Schomburgk was *not* to make. Yet Hancock's had been a traverse survey: a note in the lower right corner of the manuscript reads, "The Longitude and Latitude of many places were determined by astronomical Observations—distances of intermediate places were assigned by reckoning kept in the manner of the Sailors." Even so, it was a poor traverse and a bad map. It could not be collated into the geographical archive and was ruled an idiosyncratic depiction, a lightweight "sketch" of a personal passage. The official commission of its author was not sufficient to invest it with the reliability needed to make the map a geographical tool or a part of geographical knowledge. In part this assessment can be explained by the outdated cartographic stylistics employed by Hancock, who had no formal training as a surveyor or cartographer. He mixed pronounced elements depicted in elevation and entirely out of scale (the Amerindian encampments, the crystal mountain) with a base map drawn in plan. He demonstrated no hand for the cartographic conventions that signaled a reliable map. For instance, his rivers are shown straight and smooth, omitting the characteristic crenellated banks and irregular wavering that, while seldom conforming to hydrographic realities on a traverse survey at this scale, at least effected the symbolic mimesis that signaled the work of a cartographer and built confidence.[70]

But the most serious blow against Hancock's map was *not* that it would not collate—after all, there were no better maps of the region with which it could fail to agree—but rather that there was *no evidence of Hancock's fixed points,* no identified positional landmarks, no account of the instruments he used or the calculations he employed, no information concerning his method of keeping time, the ephemera he carried, the number of times he repeated his observations, what he observed, or what he was

69. BL, Hilhouse MS, 9.
70. A mapmaking handbook from 1846 reads: "Let it be understood that by legitimate map draughtsmen are meant those who have been regularly brought up in the profession of surveying, plotting and copying maps; it is maintained that persons so educated alone are legitimate map draughtsmen and that they alone are the proper persons to be entrusted with their constructions." The same text drew attention to the chief errors of amateurs, identifiably those of Hancock: "unsightly prominence of some portion of the map" and "unmethodical representation of natural objects." Wilme, *Handbook for Mapping,* "Introduction," 3.

standing on when he made a given observation. In short, though he claimed to have fixed points using his sextant, he had not provided any account of the points he fixed or the way he did so. Without these, Hilhouse relegated the map to the status of a "memorandum"—a relic of a single journey, not a cartographic codification of the place itself.

Two points emerge. First, the value of traverse surveys hung on the precision, accuracy, and number of their fixed points.[71] Second, because most traverse surveys could not be compared with other surveys, the assessment of the quality of those fixed points depended on a complex of factors related to the surveyor himself—his training, behavior, and instruments. That Schomburgk understood fixed points as the ultimate fruit of geographical exploration is evinced by his requesting that his collected travels be introduced by a reprint of Humboldt's *Essay on Some Important Astronomical Positions in Guiana*. At the conclusion of his last report published in the RGS *Journal* in 1844, Schomburgk chose to summarize almost a decade of geographical exploration in explicitly pointillist terms: "The general map of British Guayana, which I have conducted from my exploratory expeditions, is based upon the following data, viz., the determination of the latitude of 174 points, obtained by 4,824 altitudes of heavenly bodies, and the determination of hour angles for meridian distances, and the rate of chronometers for 223 different stations, ascertained by 5,801 altitudes of the sun or stars."[72]

Of colonial exploration Paul Carter writes: "To travel properly was to be able to fix one's position. In this way the Aristotelian dialectic between stasis and motion reasserts itself."[73] Motion was indeed the disrupting element of the traverse survey. The ranging glances of the explorer, the distances and directions of his daily work, brought unsettling instability too close to the stable stage of the map. Fixed points constituted the solution.[74] But only if they were reliable. Though they might serve as a "fix" for the inadequacies of the traverse survey, fixed points still depended on the geographical explorer, who had to use his narrative and his other tex-

---

71. The longitude and latitude elements of the coordinates were achieved with very different degrees of difficulty. Latitude could be derived from a relatively straightforward observation of the altitude of a heavenly body (most simply, the polestar). Longitude calculations were most reliable when the explorer could carry a chronometer (allowing direct comparison of local time with the time at some fixed meridian), though other techniques (for instance, observing the occultation of given stars by the moon and comparing the local time of this celestial event with tables that gave a corresponding Greenwich time) could produce, in the early 1840s, calculations accurate to within (roughly) four geographical miles.

72. *JRGS* 15 (1845): 102.

73. Carter, *Lie of the Land*, 125.

74. In 1682 only forty such fixed points had been established in the entire world, all of them in western Europe; by 1706 this number had increased to 109. By 1817, after the perfection of the reflecting sextant, the portable artificial horizon, and the altazimuth theodolite, more than 6,000 such points had been fixed. Skelton, "Map Compilation, Production and Research," 53–55.

tual productions to buttress their fixity and his qualities as a fixer. In the three sections that follow I discuss a variety of ways a geographical explorer could bolster the accuracy and reliability of the fixed points of his traverse survey. Good instruments were key, but so were allusions to an explorer's disciplined practice, as well as to his institutional and personal affiliations. All built confidence in his observations. In examining how a point could be expertly fixed, I use Schomburgk as an example, but the general ideas will apply to other surveyors engaged in similar work in the first half of the nineteenth century.[75]

FIXED POINTS I: THE HUMBOLDTIAN TRADITION

Shortly after meeting with Schomburgk in April, Hilhouse wrote a letter to the RGS reporting that he had given the new arrival a list of the instruments he would need to make a successful geographical expedition. Hilhouse added, "Dr. Hancock omitted these" and pointed out that, as a result of Hancock's inadequate instrumentation, "no object east of Joaquim has been fixed."[76] This letter provides an excellent example of the importance Hilhouse attached to proper instrumentation in executing a suitable traverse survey.

While Hilhouse expounded confidently on the subject of instrumentation and its significance, he knew that he was not acquainted with the highest instrumental geographical science of his day. In the *excusatio* with which he opened his "Memoir on the Warrow Land of British Guiana" he confessed as much, excusing himself for what he felt were his shortcomings on approaching such a geographical task, "unenlightened as I am by the science of *the great American traveller,* whose researches were confined to the south and west extremities of the province."[77] The traveler was Humboldt, whose instrumental, measurement-centered style of geographic travel represented the new standard in scientific geography in the period. Hilhouse appears to have been acutely aware that he was not *au fait* with what Susan Faye Cannon has called "the great new thing in professional science in the first half of the nineteenth century": Humboldtian science.[78] That said, Hilhouse must have felt that the latest thing had come to Georgetown when Schomburgk turned up, a veritable Prussian

---

75. Schomburgk's own surveying practices evolved, and it would be a mistake to present them as static, but in the sections below I am interested in illustrating the ubiquity of certain approaches to shoring up the inherent weaknesses of the traverse survey.
76. RGS, Correspondence, Hilhouse to RGS, 17 August 1835.
77. *JRGS* 4 (1834): 322. Emphasis mine.
78. Cannon, *Science in Culture,* 105.

toting his copy of Humboldt's *Personal Narrative* with him to the colony, and the *Equinoctal Plants* as well.[79]

The Humboldtian creed, as Cannon puts it, and as his works themselves clearly demonstrate, was measurement, and the promotion of accurate worldwide observations. That fixed points were at the core of the Humboldtian geographical program (at least as it was understood in Britain) is reflected in an 1812 essay on his voyage: "To Humboldt posterity will decree the superior honour of fixing with *astronomical* exactness, and by various observations, the degree, and even the minute and the second both of longitude and latitude, of each hill, volcano, lake, settlement, head and mouth of each mighty river."[80] Instruments were at the heart of the project, so much so that they became the symbol of the scientific traveler himself and inseparable from his scientific identity.[81] Such an emphasis on rigorous and repeated mensuration, and on the instrumentation to make it reliable, made a junior Humboldtian a very appealing envoy into the bush in the capacity of geographical explorer. An obsession with instruments, and the mapping of observations, distinguished Humboldtians.[82] As William Goetzmann has pointed out, "Humboldtian" surveyors, or at least those suffused with a heady dose of Humboldtian romance and at least a fair acquaintance with the most fundamental instruments in the repertoire, were recruited as boundary surveyors in the North American westward expansion.[83] Cannon goes further, pointing out that the "finicky" precision seeking and inordinate instrument fiddling demonstrated by true Humboldtians sometimes drove to distraction more practical-minded boundary surveyors, and also the funding agencies paying for Humboldtianism they had not bargained for.[84] Humboldtianism constituted an ideal instrumental fix for a solitary explorer headed for an interior with the obligation to create a map of his passage.

As much as Humboldtianism afforded an *instrumental* fix to the perennial problem faced by a geographical explorer fixing points on a traverse survey, it provided a *cultural* fix too, for among most of the scientifically literate and semiliterate alike Humboldt's very name conjured up an image of the "greatest scientific traveller who ever lived . . . the parent of a grand progeny of scientific travellers."[85] Not all, however. While Hum-

---

79. RGS, Correspondence, Schomburgk to RGS, n.d. [August 1833].
80. E. H. B., *Geographical, Commercial, and Political Essays,* 39. Emphasis in original.
81. Swijtink, "Alexander von Humboldt's Instrumental Life."
82. Cannon, *Science in Culture,* 95 and 96.
83. Goetzmann, *Exploration and Empire,* 191, 303–4, and Goetzmann, *Army Exploration,* 422.
84. Cannon, *Science in Culture,* 103.
85. The quotation is from Darwin, cited in Brock, "Humboldt and the British," 367.

boldt's name charmed much of the educated British readership and evoked precise measurements and trunks of instruments,[86] it carried with it some other baggage as well: sacks of verbiage. Humboldt liked to write, and prosing before nature stood as close to the core of the Humboldtian project as fine sextants and durable mountain barometers. This tendency to narrative excess alienated some. Barrow, to whom Schomburgk reported at the RGS, had publicly declared himself unravished by the prolix Prussian.[87] This would prove a problem for Schomburgk, whose Humboldtian narrative flights would meet with stern reproof from the editors of the Society's *Journal*.

Was Schomburgk the Humboldtian he was thought to be and that at first glance he appears? The wish list of instruments Schomburgk dispatched from Tortola to the RGS looks as if it was pulled directly from the *Personal Narrative*:[88]

> Chronometer
> Reflecting telescope
> Sextant
> Theodolite
> Azimuth compass
> Artificial horizon
> Pendulum
> Thermometer (for determining altitude by the boiling point of water)
> Barometer/electrometer/hygrometer[89]

It likely was. Moreover, Schomburgk's original RGS commission stated that Schomburgk's "distinct object" was to connect the "positions" he obtained in British Guiana with "those of M. Humboldt on the Upper Orinocco."[90] Schomburgk was thus on a Humboldtian quest in Guayana. Reaching Esmeralda—Humboldt's easternmost point—from the east meant more to Schomburgk than simply extending a web of fixed points across lowland South America. There was a sacred quality to the telos of his Guiana circulations:

> At length we came into view of a fine savannah extending to the foot of the mountain, which I knew, from Humboldt's description, to be that of Esmeralda. . . . I cannot describe with what feelings I hastened ashore; my object was realised, and my observations, commenced on the coast of Guayana, were now connected with those of Humboldt at Esmeralda.[91]

---

86. John Allen wrote of Humboldt's *Tableau Physique* in the *Edinburgh Review* 16 (1810): "No name stands higher than that of Humboldt among the lovers of geographical and physical science." Cited in Brock, "Humboldt and the British," 369 n. 20.
87. Brock, "Humboldt and the British," 369.
88. Cannon, *Science in Culture*, 75.
89. RGS, Correspondence, Schomburgk to RGS, n.d. [August 1833].
90. *JRGS* 6 (1836): 7, 10. The secretary of the Society called this Schomburgk's "great object."
91. *JRGS* 10 (1841): 243.

The arrival occasioned a passionate reflection on the hardship of the journey and the fortitude of his crew, though his own fortitude came to mind as well, accompanied by fond thoughts of his renowned countryman and inspiration:

> It is but due to that great traveller to acknowledge that at times when my own physical powers were almost failing me, and when surrounded by dangers and difficulties of no ordinary nature, his approbation of my previous exertions cheered me on, and encouraged me to that perseverance which was now crowned with success. The emaciated forms of my Indian companions and faithful guides told, more than volumes, what difficulties we had surmounted.[92]

Lest the religious quality of this moment be lost or the sense of a pilgrimage fulfilled be overlooked, Schomburgk cast his eye around the granite crags that hung over the sad little town and let his gaze settle on the cross perched on a cliff above. This "airy form" stood out "boldly in relief with the blue sky as the background . . . and reminds us that although nature and man appear in the savage state, there are still some in this wilderness who adore the Deity and acknowledge a Crucified Savior."[93]

The place of Humboldt in this advent scene is clearly more hagiographic than instrumental; he hovers over Esmeralda more as a devotional figure than as a scientific one. But the moment of arrival merely set the scene for the observation of the latitude that was the true consummation of the explorer's desires. By means of "eight observations by meridional altitude of the stars α Aurigae, α Columbae, α Argus, α and β Centauri, α and γ Ursa Majoris,"[94] Schomburgk was able to give the latitude of Esmeralda as 3°11′3″ N, more than a confirmation of the observation of the great man himself, who put it at 3°11′ N. The addition of three seconds of arc, and the placement of this comparison of observations at the end of a chapter of the published expedition journal, demonstrates how measurement could play a central role in the process of metalepsis introduced in the previous chapter. The meaning of Schomburgk's presence in Esmeralda is guaranteed by Humboldt's passage, but Schomburgk's additional significant figures place Humboldt's previous passage onto a Schomburgkian point. Schomburgk is a Humboldtian, but Humboldt has been remapped by Schomburgk's superior observation.

In part. In fact, several observations are missing. Where, for instance,

---

92. Ibid.
93. For an essay on the pervasive trope of the cross in the woods, and particularly its place in German romantic evocations of the forest, see Schama, *Landscape and Memory,* chap. 4, "The Verdant Cross."
94. *JRGS* 10 (1841): 243.

is a measurement of longitude? Where are the multiple observations of lunar distances that would allow Schomburgk to place Esmeralda at its precise coordinates? The answer lies in the expedition's journal entry on 4 December 1838, the day that Schomburgk's most ambitious circumambulation in the South American interior lost track of time and hence surrendered any pretensions to the certainty of its track as well:

> 4th— This morning was disastrous to our future astronomical observations; in winding up my watch the chain broke, and we were thus left without a time keeper, a misfortune without remedy, for, had I sent back to the colony, two months must have elapsed before a messenger could possibly have returned, and then the season for travelling would have been nearly over. I resolved, therefore, to go on and do as well as I could; but henceforth we had to estimate the time.[95]

And hence estimate their longitude as well. This passage merits mention in connection with Schomburgk's status as a Humboldtian interior explorer exactly because it shows that despite Schomburgk's attention to measurements (he was forever thrusting his thermometer into sandbanks, stagnant pools, and creeks, listing the stars he had observed at their meridional apogee, and ascertaining altitudes when he could) he was, in fact, not so terribly well equipped as a successful Humboldtian scientific traveler. He wrote: "In Guayana the traveller ought to be provided with instruments in triplicate, and they should be of the best construction." But, he added, "I was unfortunately so peculiarly situated, and was obliged to make so many sacrifices to procure other instruments, that I was not even provided with duplicates."[96] Even in his excuse, however, Schomburgk emphasized his "instrumentalism": he had brought only one timepiece because he needed so many other instruments. The RGS and the colonial government had failed to provide him with the instruments he continually requested. But there was no recourse, in the literal sense. In the snapping of the watch chain, Schomburgk was cut loose from Greenwich and cast adrift with respect to the east-west axis of the world. The same snap severed one of his main links to Humboldt and the Humboldtian science to which he aspired.

This "disaster" affected only the latter half of a single one of Schomburgk's multiple expeditions into the interior, but it reveals the tenuousness of his claim to be a true Humboldtian. He may have been a devotee of the great scientific traveler, may have styled himself after his countryman, but he was not what Cannon calls "a compleat Humboldtian." Nor does the evidence for this assertion hang exclusively on a weak winding chain. A "compleat Humboldtian," Cannon writes, "thought about what

---

95. *JRGS* 10 (1841): 216.
96. Ibid.

he collected" and deployed his collections of facts in the interests of powerful and general explanatory theories.[97] Humboldt disparaged cabinet mentalities and asserted that "observations are not really interesting except when we can dispose of their results in such a way as to lead to general ideas."[98] The most significant tool for this was the isoline map. These graphs were the manifestation of the synthetic and multivariable analyses that "high Humboldtians" sought.

Schomburgk made no contributions in these areas: he did not draw any isoline maps, nor does he discuss them or suggest that he is collecting data that would be best configured in such a way. Only on his last expeditions did he carry a pair of needles that enabled him to make observations of terrestrial magnetism; he seems not to have mapped these data.[99] Nor was Schomburgk a theoretician or speculator. His botanical knowledge, though considerable (and it grew during the course of his researches), never overstepped the authoritative figures for whom he collected and who wrote up his finds: Lindley, Hooker, and Bentham.[100] Nor did Schomburgk follow the professional model laid down by his inspiration. Schomburgk used his geographical work as a stepping-stone to a knighthood and diplomatic postings. He became, in the end, a colonial diplomat, not a metropolitan scientist.

Yet Schomburgk traded on Humboldtian rhetoric, and even shone, as I have suggested, in the scientific effulgence of his master. At some point between his departure from Tortola in 1835 and his setting off for Esmeralda in 1838, Schomburgk managed to enter into a correspondence with Humboldt, a correspondence that the younger man, as shown in the Esmeralda passage above, found inspirational in the extreme. The relationship blossomed, and Humboldt composed a very flattering preface for his protégé's *Travels* in 1841. Humboldt praised Schomburgk's collections of flora and fauna before addressing the main object of his expedition, "the astronomical connection of the British Guiana littoral with the most Easterly point of the upper Orinoco to which I had brought my instruments."[101] He went on to report, fudging somewhat the much-bemoaned demise of the timepiece, that this geographical "problem has been solved by Robert Schomburgk to the satisfaction of that celebrated Society [the RGS]."[102] The pilgrimage quality of that mission was not lost

97. Cannon, *Science in Culture,* 95.
98. Ibid.
99. They were Hansteen's needles, L(a) and L(b), listed in table 52 of the *Magnetic Survey of Great Britain.*
100. Bentham appears to have alluded to Schomburgk disparagingly in correspondence with Sir William Jackson Hooker, calling him a "species monger." I thank Richard Drayton for this citation.
101. Humboldt, "Preface," vi.
102. Ibid.

on Humboldt, and neither was his own place in setting up the objective. He wrote of his reasons for consenting to contribute the preface: "It was my duty to express publicly my heartfelt respect for a very talented traveller who imbued with one idea, the resolution to make his way from the valley of the Essequibo as far as Esmeralda, from East to West, has reached the appointed goal after five years' effort and suffering, the amount of which I can appreciate in part from my own experiences."[103] Humboldt understood Schomburgk's Humboldtian pilgrimage in its full spiritual intensity.

Humboldt also praised Schomburgk's "large, splendid" publication on the region, the *Twelve Views in the Interior of Guiana* (to which Humboldt had subscribed). Modeled on Humboldt's own *Views of the Cordilleras and Monuments of the Indigenous People of America* of 1810, the text featured lavish lithographs correlated with Schomburgk's fixed astronomical observations. For all this praise, however, Humboldt did not neglect to offer his apostle a backhanded compliment, one important in considering Schomburgk's status as a "compleat Humboldtian": Schomburgk, Humboldt declared, had "given himself his own scientific education," and what he did not learn in books "was supplied him by his life in the open air, by the sight of the starry skies in the tropical world, by direct contact with living nature."[104] This depiction of Schomburgk as the *géographe sauvage* simultaneously valorized his geographical contributions and, subtly, called them into question. For all his merits, he was an autodidact if not an amateur; he had earned through persistence the status of a Humboldtian disciple.

Schomburgk traded on his Humboldtianism by emphasizing and cultivating his links with Humboldt himself. Those auspices enhanced Schomburgk's scientific credibility by associating the traveler with the most esteemed instrumental science of the period, and hence weighted his fixed points with Humboldtian authority, despite his uneven record in using and maintaining his actual instruments. Schomburgk's skills—familiarity with navigational instruments, a passable acquaintance with botany—could wear the mantle of Humboldtianism, particularly when the practitioner had Prussian roots and could boast of intimacy with the man himself. In his address to the Society in 1843 William Hamilton, the president, announced that "none of our colonies have been more fortunate than British Guayana in having so excellent an explorer as Mr. Schomburgk," and the following year Roderick Murchison praised his "acquirements as an astronomical observer," calling him "one of those, in

103. Ibid., v.
104. Ibid., viii.

fact, formed in the school of Humboldt, whose researches and observations . . . make us fully acquainted with the regions they explore."[105] The mantle of Humboldt, as much as Humboldtian instrumentalism, proved a powerful way to enhance the value of a traverse survey.

FIXED POINTS 2: DISCIPLINE AND NAVIGATION

In the very last section of his writings on British Guiana for the RGS—his public epigram on a decade of explorations and exertions—the newly knighted Sir Robert Schomburgk elected to pause and reflect on Bunten's siphon barometer 430:

> I cannot conclude this memoir without drawing particular attention to Bunten's siphon barometer, whose advantages are incalculable for those who traverse wild and pathless regions. The barometer #430 of his construction has accompanied me during my last expeditions through forests and over mountains, and in my boat navigation of impetuous rivers; and on my return to London I found, on comparing it again with the barometer at the Royal Society, that it had not varied. On previous expeditions I have used Troughton's, Englefield's, and Newman's barometers, and though every precaution was taken with them, I never succeeded in bringing any one of these instruments safely back to the coast regions.[106]

This instrumental encomium testifies to Schomburgk's matured Humboldtian sensibilities. The litany of instrument makers' names points to the reliability of his observations and his own calibration.[107] In its reliability, durability, and invariance, the barometer becomes the explorer's most faithful companion, the "guarantor of his elevation" in several senses. In the emphasis placed on the instrument's having returned from the interior unchanged and unperturbed, the description of barometer 430 can be read as charged with metonymic overtones. It is as if the warhorse barometer has allowed Schomburgk to write about himself—the rigors, discipline, and constancy demanded through his journeys.

Good fixed points demanded more than a good instrument; they demanded a good man too. Lorraine Daston and Peter Galison have characterized the moralization of the production of "mechanically" objective images in the nineteenth century.[108] They point out that one meaning of objectivity in the period derived from a negative definition: freedom *from* the mediating and subjective presence of the observer. This meant that one route to the production of objective representations lay in censoring

---

105. *JRGS* 13 (1843): xcvi, and *JRGS* 14 (1844): xcvii.
106. *JRGS* 15 (1845): 104.
107. For a discussion of the calibration of the observer, see Schaffer's discussion of the "personal equation" in astronomical observations: Schaffer, "Astronomers Mark Time."
108. Daston and Galison, "Image of Objectivity."

the intrusions of the personal. Certain qualities of the observer—rigorous discipline, asceticism, self-denial—could guard against contamination by "unruly" and idiosyncratic elements.[109] Daston and Galison focus on the production of images for medical and scientific atlases, but their observations bear on the core dilemma of geographical representation in the period—objective maps from subjective sources—and more specifically on the challenge faced by a traverse-surveying geographical explorer like Schomburgk. In this section I examine how Schomburgk used his narratives and correspondence to alert readers to his discipline: both his own and the discipline he exercised over his crew. By being mechanically reliable with his instruments (and regulating his associates), the geographical explorer could bolster confidence in his fixed points.

Understanding this discipline demands attention to the military, and specifically naval, traditions of mensuration and observation. I mentioned above that the roots of the traverse survey lay in the army reconnoiter and the practices of naval navigation. This claim can be extended. The particular significance of the naval tradition lay not simply in its affording the equipment and protocols of calculation for fixing positions in an unfamiliar and "featureless" expanse. Even more important, naval practices were predicated on reciprocity between a map and a passage. The intimate exchange between nautical navigation and nautical surveying—on which the welfare of ships and shipping depended—created a framework in which the disciplined passage could produce charted territory. The computational character of naval charts made this reciprocity a testable proposition. Interior geographical explorers drew on this framework and its traditions of discipline to substantiate the reliability of their mapped passages as much as they drew on the sextants, chronometers, and almanacs of seagoing voyagers in order to effect those passages. This section, then, proceeds in two parts. First, I examine these naval navigational and surveying traditions with an eye toward showing how they established reciprocity between map and passage, a reciprocity contingent on military discipline.[110] Second, I look at how the geographical explorer emphasized the various disciplines of his passage.

Maritime navigation on the open sea developed as an attempt to keep track of the question Where am I? in the absence of any visible markers.[111] A nautical navigator uses positional fixes to maintain a "positional con-

---

109. Ibid., 118.
110. This reciprocity is mentioned in Sandilands, "History of Hydrographic Surveying," 113.
111. This is the conclusion of the "computational account" of navigation, *Cognition in the Wild*, by cognitive anthropologist Edwin Hutchins. Conrad writes an elegant account of the sailor's ability to "keep track" of himself: "The seaman takes his Departure by means of cross-bearings which fix the place of the first tiny pencil cross on the white expanse of the track chart, where the ship's position shall be marked by

sciousness" by means of reckoning on some spatial analogue (like a chart).[112] A look at the handbooks of navigation and naval surveying in the early nineteenth century confirms the inextricability of good navigation and good chart making.[113] Just as a good track chart of a passage along a coast constituted a chart of that coast itself, the proper practice of navigation in the first half of the nineteenth century implied careful attention to maintaining and improving charts. Nautical astronomy was inseparable from proper attention to cartographic detail. Raper's standard *Practice of Navigation* of 1840, in addition to detailing the practices of departure taking and the protocols for calculating time, longitude, and latitude from shipboard observations, also contained extensive exhortations on the proper maintenance of the ship's journal and its chart, as well as a section on a ship's "obligation" to stop and establish the position of any shoals or headlands that did not conform to the chart being used.[114] Extending and improving charts was an extension of good navigational practices.

Vice Admiral Beechey reinforced this point in his essay on hydrography in the 1849 *Admiralty Manual for Scientific Enquiry,* where he detailed the obligations of a ship approaching a coast. They were "that the general feature and aspect of every country should be noted from the moment hills rise above the horizon; that all remarkable objects which may be recognized and by which the position of any port or other locality may be known, either at a distance when the weather is clear, or in close when

just such another tiny pencil cross for every day of her passage.... and there may be sixty, eighty, any number of these crosses on the ship's track from land to land." Conrad, *Mirror of the Sea,* 2. The classic history of navigation (through the eighteenth century) is Taylor, *Haven-Finding Art.*

112. Hutchins argues that what is particular about navigational charts in the Western tradition is that there is more than just a "correspondence" between chart space and real space. "A navigation chart," he writes, "is a carefully crafted computational device." Hutchins, *Cognition in the Wild,* 61. It is not only a coordinate space but, like a graph in analytic geometry, a kind of analog computer, allowing a properly equipped user to "read" and "write" one-dimensional constraints on a single coordinated graphical field. (For a "computational account" of these constraints, see ibid., 50–52.) Charts constitute both a record of the ship's course in relation to the world *and* instruments for determining that course: "The chart is the positional consciousness of the ship: the navigation fix is the ship's internal representation of its own location." Ibid., 61.

113. For accounts of the Admiralty surveys, see Day, *Admiralty Hydrographic Service;* Sandilands, "History of Hydrographic Surveying"; and Ritchie, *Admiralty Chart.* From 1810 to 1830 British expenditures on such surveys, which involved multiple ships and the coordinated work among their crews to make running triangulated surveys along coasts, rose about 2,400 percent. In 1828 alone more than twelve such surveying expeditions were working on coasts from Mauritania (Boeteler) to the Bering Straits (Beechey); by 1829 the Admiralty's charts were formally available to merchant shipping concerns. See Day, *Admiralty Hydrographic Service,* 43. A more general account of naval surveying can be found in "Surveyors of Sea and Shore," chapter 10 in Wilford, *Mapmakers,* 144ff. Although the book is uneven, Wilford rightly sees Cook's land-surveying experience in Holland as a key to his hydrographic style. For an institutional history of North American naval surveys, see Manning, *U.S. Coast Survey.*

114. Raper, *Practice of Navigation and Nautical Astronomy,* "Shaping the Course." This book went through twenty-one editions and remained in print into the twentieth century.

mist or haze prevails, should be described as graphically as possible." Nor could any mere description "equal a tolerably faithful sketch accompanied by bearings." "On the discovery of unknown lands or dangers," he warned, "the first endeavor after the vessel is placed in safety should be to fix the position of the place as accurately as the means of observation admit, and *not to quit the place until the danger is satisfactorily placed upon the chart.*"[115] For more detail on how to do so he referred his readers to Edward Belcher's *Nautical Surveying* for full instructions on how, if time allowed, to run a triangulated survey of the shore. Beechey closed with a brief abstract on how to conduct a quick survey of a port that could be linked to an astronomical fix, as well as how to conduct an "itinerant survey" of a coast by measuring distances using canon shots (timing the gap between the flash and the sound gave the length of the baseline).[116]

All these practices were laid down in greater detail by Belcher, who went beyond insisting on the obligation to chart as one made use of a chart, all the way to suggesting that a kind of "transport" of positional ecstasy could be evoked in men by affording them access to instruments. "What have our writers on navigation been about," he asked in the 1835 edition of his book, "to have omitted to excite their readers?" A veritable mania of positional curiosity could be whipped up by a captain capable of communicating the niceties of converting the humdrum business of navigation into the production of accurate charts. "Instantly," he promised, "in ships employed in surveying, the sextant is in request; even the pleasures of the table are forgotten." Lunar tables quickly eclipsed the mess table: "Astronomical pursuits, surveying, etc., have a peculiar attraction. Let but one moderate draught be taken, fairly tasted, and a species of intoxication follows, a 'scientific mania' ensues. Example only is wanting, and if that happen to be the principal (the Captain or Lieutenant) the contagion rapidly spreads, it becomes the fashion."[117] This passage attests to the reciprocity of nautical surveying and navigational practices. It also demonstrates that maritime culture instilled a fascination for reconciling positional observations with the representational fields necessary to transform such observations into records and calculating tools. Navigation and coastal surveying were very much understood as a complex of projects, in which passages could be translated into maps and vice versa, provided practitioners adhered to the disciplined protocols of the community.[118]

---

115. This quotation and those above are from Herschel, *Manual of Scientific Enquiry*, "Discovery of Land," 86. Emphasis mine.

116. Ibid., 84.

117. This quotation and those above are from Belcher, *Treatise on Nautical Surveying*, 3.

118. See Schaffer, "Accurate Measurement." I will return below to the role of discipline, particularly bodily discipline, in this process.

To this end, the "transports" of enthusiasm were not to interfere with the bodily discipline necessary to ensure that such observations were objective and reliable. The naval discipline that characterized shipboard life had to be emphasized and extended where navigational observations were concerned.[119] Simon Schaffer, writing on precise measurement in midcentury Britain, identifies the military character of bodily discipline, which allowed an array of qualities of individuals—regularity of step, "a disciplined forearm," steady, temperate hands—to convert the experience of the world into quantities.[120] Instructions to naval navigators and surveyors were just as intimate, dictating how a measuring officer was to stand, hold his instrument, and exercise his calculations. Commander Belcher drew cadets' attention to the correct way to handle a watch, warning them that altering its position could change its rate; placement of the calibrating hand on a sextant was just as important.[121] Such physical discipline constituted its own means of substantiating the accuracy of naval observations. From the rigorous lockstep protocols for calculations to the physically explicit instructions on the position of the body, good observations were contingent on good discipline.

How did these naval traditions bear on the traverse survey? Geographical exploration and navigational hydrography shared more than just the swampy littoral as an overlapping zone of interest. This discussion of navigational practices shows that the traverse surveyor and the ship's navigator at sea shared a great deal. In the first place, the navigational preoccupation with maintaining positional consciousness, and with the maintenance of the representational field on which position could be calculated (and a record of that position inscribed), was an exact analogue of the requirements of the traverse surveyor. Naval navigation afforded a tradition in which the reciprocity of chart and passage—that correct passages made and remade charts, that correct charts made passages possible—had been firmly established. The computational quality of the nautical chart (its multiple embedded positional constraints) transformed a passage into a testable proposition inscribed in a graphic form. This "testability"—the computations that went into making the chart and those that could be made from it—afforded a powerful framework in

119. On naval discipline, see Dening, *Mr. Bligh's Bad Language*. For a discussion of the integration of scientists into maritime culture, see Rozwadowski, "Small World."

120. A citation from Garnet Wolseley's *The Soldier's Pocket-Book* reveals the intimacies of military mensuration in such injunctions as "everyone should know the exact length of his ordinary pace, and be able to pace yards accurately; he should know the exact length of his foot, hand, cubit and sword, and arms from the tips of his fingers of left hand to right ear; he should know the height of his knee, waist and eye." Cited in Schaffer, "Accurate Measurement," 135.

121. Belcher, *Treatise on Nautical Surveying*, 5–6.

which to assert the reliability of the positional fixes made on traverse surveys. Moreover, the techniques, instruments, tables, and protocols for calculating naval navigation were ideally suited for adapting to the new needs of interior explorers in very direct ways. By enabling the user to "fix" points of a trajectory along the blank "track chart" of terra incognita, navigational techniques constituted a superior way for an interior explorer to "find himself" after he had succeeded in losing himself in unfamiliar terrain.

For evidence of the degree to which the interior expedition carried inshore the equipment and techniques of naval navigation, we need look no further than one of the first letters that Schomburgk sent from Tortola to the secretary of the newly founded RGS, listing the kit he would need to conduct his "expedition of discovery into the interior of British Guiana." Prominent in this list was a "nautical almanac," the book that would allow him to calculate longitude by means of a list of the angular distance between the moon and sun or moon and given stars for every third hour Greenwich mean time.[122] One author instructing those engaged in traverse surveys lamented that almanacs were not printed "divested of all information that is purely nautical," since pages on currents and tides did nothing but burden those who dragged navigational guides into continental interiors.[123]

How Schomburgk came to be acquainted with this set of specialized skills is difficult to determine, but by the time his paper on Anegada was read to the Society, he had already been three years in the West Indies, where captains, sailors, and the details of navigation—the lifeblood of any archipelago—circulated in every settlement.[124] During this time he managed to develop sufficient aptitude for navigational concerns that by 1832 he was able to construct a navigational chart received favorably by the RGS, many of whose members had seen long terms of duty in the Admiralty service. Schomburgk's remarks on Anegada demonstrate that he was well acquainted with maritime navigation. He was able to query captains on making landfalls; he discussed currents, soundings, and details of nautical life with confidence.

Schomburgk's point-fixing observations were characterized by a military commitment to duty and a strenuous exercise of discipline. He wrote

---

122. RGS, Correspondence, Schomburgk to RGS, n.d. [August 1833]. For an account of its use, see Jackson, *Course of Military Surveying*, 268.

123. Jackson, *Course of Military Surveying*, 287. Raper's *Navigation*, the *Nautical Almanac*, and an assortment of cheaper almanacs were included among the lists of "necessary instruments" recommended in 1856 by the author of *The Art of Travel*. The cheaper versions, like the *Seaman's Almanac* and White's *Ephemeris*, were all useful, he wrote, "to select and cut tables out of." Galton, *Art of Travel*, 277.

124. Rivière has informed me that Schomburgk mentions learning the elements of naval surveying from the Danish military commander and harbor master of Saint Thomas.

regularly of the hardships of keeping himself awake for the better part of night after night, waiting for clearings in the clouds through which he could catch a fleeting glimpse of the meridian transit of a bright star. In desperation at the censure by his sponsoring RGS officials, Schomburgk wrote an impassioned testament to his geographical discipline: "My devotion to my geographical observations have [sic] made me a perfect slave!"[125] Only such a "perfect slave" could serve as the amanuensis for nature at the colonial periphery, a transparent agent for the transcription and transmission of positions.

Instruments demanded particular care. By binding the fragility of the native body to the fragile brass and glass, the explorer bought corporeal insurance for his instruments. Francis Galton's *Art of Travel*, written by an honorary secretary of the RGS, advised: "Entrust instruments and fragile articles to some respectable old savage, whose infirmities compel him to walk steadily."[126] Care in transporting such instruments and their uneventful passage through difficult terrain—allowing them to be calibrated on the explorers' return—constituted strong evidence of a disciplined and orderly traverse survey. Schomburgk carefully noted how he configured the men in his boats so as to "convey my instruments more safely," even though the shuffling of supplies meant there was room for only two or three persons in each vessel.[127] On his work in the 1840s in the service of the Crown, his precious chronometers, Frodsham 389 and Arnold 6062, which bore Greenwich through the jungle, never left the person of the most trusted "canoe man," to whom they were attached by means of a "tin canister . . . slung by a strap across his shoulder."[128]

What was expected of a superior traverse surveyor was the reflex of positioning himself and then determining the orientation of all objects he could see. An explorer wishing to contribute to the archive of fixed points "cannot better occupy himself than by rating his own chronometer, determining daily from such position the true bearing of some distant object, etc."[129] Self-discipline was as important as instrumentation. Captain J. E. Alexander completed an expedition into the interior of British Guiana in 1831, and with his map and narrative he submitted to the Society's *Journal* a note on the monastic bodily discipline of a successful explorer: "My habits are to rise early—to wash and sponge immediately from head to foot—to take some active exercise always before break-

---

125. RGS, Correspondence, Schomburgk to RGS, 20 August 1836.
126. Galton, *Art of Travel*, 278.
127. *JRGS* 10 (1841): 21.
128. Ibid., 19. One might wonder why they were not spread out.
129. Belcher, *Treatise on Nautical Surveying*, 29.

fast . . . avoiding acid or even green food; and at night, to rub my whole body with a very coarse hair-glove or flesh-brush, thus clearing the pores, and equalizing the whole circulation."[130]

The geographical explorer had an obligation not only to discipline himself, in order to dramatize the ascetic purity of his observational mission, but also to maintain a firm discipline over others, particularly his crew and the indigenous people he met. While this discipline was distinct from the self-discipline necessary for trustworthy observations, the two went together. The successful passage of the explorer depended on both: only a disciplined explorer who cared for his instruments produced reliable fixed points; only disciplined, orderly passages could afford the meticulous observations necessary to create a reliable map. By maintaining discipline, the explorer ensured that he could bring his fixed points back.

Hilhouse doubtless gave Schomburgk abundant advice on the payment of his boatmen and the correct way to maintain advantageous trading relations with the Amerindians on whom he would rely for food and directions, but a main reason for Schomburgk's first expeditionary fiascoes lay in the breakdown of such relations. He was forced to acknowledge as much in his report to the Society: "Our Indians having procured themselves, during the period of the intercourse we had with them, all the necessities they stood in need of, they relapsed into their old indolence: neither knives, nor combs, nor scissors—for which they would have sold their birthright when we first arrived—could now induce them to leave their hammocks."[131] Though when they felt like it they left their hammocks readily enough: "desertions" by his so-called crews plagued Schomburgk's early expeditions, prevented him from getting above the falls of the Corentyne, and made him look foolish several times, particularly when the runaways took the boat with them.[132] Nor did he have better luck securing provisions. The people of Tomatai promised to bake enough cassava bread to provision the expedition for a month. At the end of the agreed period, "They advanced me numerous excuses for not having complied with the promise, and desired me to wait three days longer, at the expiration of which eight or ten cakes were brought, a quantity which was not sufficient for a single day's subsistence, and neither threats nor promises could induce them to sell us any more."[133]

Schomburgk's willingness to punish crews and, interestingly, his pursuit of the legal power to do so, both evolved rapidly in the course of his

---

130. *JRGS* 2 (1832): 72.
131. *JRGS* 6 (1836): 260.
132. As several Arawaks did on the Corentyne. *JRGS* 7 (1837): 291.
133. Ibid.

work in the interior. There was a distinctively martial flavor to the narration of the third expedition: "Our column put into marching order, Peterson at the head, carrying the British Union flag, under which we had been marching for the last three years through hitherto unknown parts of British Guayana."[134] By December 1838 Schomburgk was prepared to hold his Amerindian traders to their word by force, as when Schomburgk responded to confusion over the price of a corial by requisitioning it at what he thought a fair price. He wrote: "I was prevented using my firearms only by the fear that false accounts might be circulated among the Indians as to our real purposes."[135]

This trajectory toward a more commanding expeditionary discipline was accelerated by the Crown commission that brought Schomburgk back to Georgetown in 1841 in the capacity of boundary commissioner. This time he wasted no time in using his new clout with the governor to ensure that he was appointed a magistrate, a position that gave him considerable legal power over his crew. It allowed him to "punish by fine, forced labour, reduction of daily allowance of rum, tobacco or ration allowance, any subordinate disobeying his own written instructions or orders of any senior officers of the expedition."[136] Schomburgk became a mobile punitive force. Moreover, he had a contract written up and read out to the crew he enlisted, and he obliged them to "sign" it with their marks. This practice was emulated by later travelers, since it gave them a legal basis for punishing uncooperative crew members.[137] The particular significance of the contract in Schomburgk's case lay in his jurisdiction as a magistrate. Schomburgk's own copy of the *Local Guide* for British Guiana states that

> 1) The stipendiary magistrate shall have an exclusive jurisdiction for the enforcement of all contracts of service, and for imposing all penalties for the breach, neglect, or non-performance thereof.
> 2) This jurisdiction shall be exercised in a summary manner.[138]

The magistrates answered directly to the governor, and their task was explicitly the enforcing of contracts for service. A sample contract, which Schomburgk may well have used as the structure for the contracts he drew up for his crew, was included in the guide. In his new capacity as magistrate, Schomburgk embodied many of the same functions served by the colonial officials known as superintendents of rivers and creeks: he could

---

134. *JRGS* 10 (1841): 195.
135. Ibid., 216.
136. M. R. Schomburgk, *Travels,* 1:278.
137. Webber, *British Guiana,* appendix.
138. BL, Schomburgk Autograph, is Schomburgk's copy of British Guiana, *Local Guide of British Guiana.* See also below, chapter 4, note 26. The quotation is from ibid., cap. 4, secs. 1 and 2.

apprehend and punish; he was mobile surveillance at the same time that he was a land surveyor; he enforced the same colonial sovereignty that his observations installed.

Schomburgk used this new authority without hesitation. Early in the Pirara expedition, a crew member was found with an illicit bottle of rum. Though he claimed he had only attempted to salvage a leaky cask, he was sentenced to three days' hard labor "to serve as an example."[139] When several Amerindians who had been recruited as porters refused to rise from their hammocks, Schomburgk took matters into his own hands, slashing their hammock ropes with his cutlass. His brother recorded the incident and reported that the humiliated Macushis "obeyed his orders and carried out their duties without a murmur."[140] The next morning, however, they had vanished, forfeiting their payment and heading back to their villages on foot.

In some cases Amerindian defections were not crises, but in others they could be. There were many situations in which the departure of the crew would likely have meant the death of the expedition members, a fact that reinforced the dependence of the explorer, whatever his juridical authority, on the Amerindians who guided him.[141] When Schomburgk could not appeal to his power as a magistrate, he could try to invoke Amerindian concepts to maintain order. Near the settlement of Watu-Ticaba, in the middle of a difficult stretch of rapids, four members of the Macushi crew disappeared, disgruntled by the reduced rations. Schomburgk, gathering those who remained, invoked the powerful Amerindian concept of the Kanaima, or revenge murderer, as he harangued them to stay with the expedition. He asserted that those who ran away, leaving the expedition to perish, were Kanaima, the assassins—part human, part spirit—that were thought to haunt the bush and kill by the most grotesque and duplicitous means.[142]

With the non-Amerindian members of the expedition, Schomburgk could enforce his magisterial powers without any intricate cross-cultural translation and within the context of traditional military discipline. One of the members of the surveying party received a dressing-down from Schomburgk one afternoon for having neglected to wear his uniform jacket on a Sunday, a breach of expedition etiquette. Later the same day Schomburgk wheeled on the men and threatened, in the words of one diary keeper on the expedition, "that if they disobeyed his orders in the

---

139. Goodall, "Diary December to June," 64.
140. M. R. Schomburgk, *Travels,* 2:200.
141. For more detail, see Burnett, "Science and Colonial Exploration."
142. M. R. Schomburgk, *Travels,* 2:296. On Kanaima, see Roth, "Amerindian Kanaima or Avenger," and Anthon, "Kanaima."

slightest degree, he would punish them to the extent of his powers."[143] He called them out and read his commission aloud to the assembled surveying party at Pirara that Sunday afternoon, wearing the military regalia he adopted in his Crown capacity. Though privately mocked by the actual members of the military in attendance, Schomburgk was acting out a version of the Sunday reading of the Articles of War before a flogging.[144] Schomburgk's truculence had different meanings in different situations, but when he narrated scenes in which he subjected his crews to various disciplinary measures he was narrating an orderly expedition. Doing so ensured that his traverse surveys were understood to be methodical affairs where Schomburgk himself controlled the goings-on, maintaining a suitably martial aspect. A disciplined environment suggested disciplined and reliable observations, conducted within the naval tradition on which they drew.

In this section I have called attention to the importance of this naval tradition as a point of reference for a traverse surveyor aiming to reinforce the quality and reliability of his work. Nautical instruments were key, as was the reciprocal relationship between maps and passages, a hallmark of nautical navigation in the period. The naval tradition also afforded a paradigm for a variety of forms of discipline—of inferiors by superiors, of self by self—that underscored the moral qualities of the observer. Since the fixed points of the traverse survey depended on that observer and the sublation of unruly elements—of self, of context—these invocations of order reinforced the fixity of those points. Murchison announced exactly that in his address to the Society in 1844, where before praising Schomburgk's astronomical observations, he invoked his qualities as an observer: "Conciliatory in his manners, yet firm, cool in judgement and prompt in action; inured to privation and fatigue, and undaunted by difficulty and danger, zealous and persevering—such is his moral character."[145]

FIXED POINTS 3: LANDMARK, IMAGE, ENCLOSURE

The man who neglected his Sunday uniform was Edward A. Goodall, the draftsman who accompanied Schomburgk's surveying expeditions from 1841 to 1844.[146] Shortly after his arrival in the colony Goodall had a glimpse of just how disciplined an expedition Schomburgk sought. Enthusiastic about the tropical scenery, Goodall spent mornings sketching

---

143. Goodall, "Diary July to December," 41.
144. Dening, *Mr. Bligh's Bad Language,* 64.
145. *JRGS* 14 (1844): xcvii.
146. For a brief biography of this son of the more celebrated London engraver of the same name, see Menezes, "Sketches of Amerindian Tribes," 11–13.

scenes in Georgetown and afternoons reading Waterton's *Wanderings*. He had even gotten some advice from a local plantation owner about the subjects likely to sell to the art-starved plantocracy and had begun to warm to the idea of making a bit of money. Inspired, he rose early to do some studies, one of which he was just finishing when Schomburgk approached and admired the work. In fact, Goodall recorded in his diary, his new superior "liked it so much that he put his seal upon it, and said that . . . he must get a stamp with 'Guiana Expedition' engraved, with which he intends stamping all my sketches so that I will not have a single sketch that I can call my own."[147] Schomburgk wanted control over the visual production of the expedition.

The incident draws attention to the importance of visual images in the work of a traverse surveyor. Schomburgk's enthusiasm for establishing proprietary control over Goodall's sketches was not unrelated to his own artistic weaknesses. Hilhouse had made a point of informing the RGS that its envoy showed disappointing ineptitude with a pencil.[148] And this was, in Hilhouse's eyes, a severe handicap for a geographical explorer in British Guiana, as well as a bad omen. He warned in an article in the Society's *Journal:*

> Another qualification is indispensable [for the explorer]—a ready pencil in watercolours; this saves the annoyance and expense of a companion, and the pencil is a faithful and credible witness, without which the pen is too often held as the mere tool of traveller's stories. It is my firm belief that no man can narrate an intelligible and faithful description of any object in nature, except that he can also make a drawing of it.[149]

That the RGS chose to publish such advice is evidence of the importance of the visual in the work of the geographical explorer. The sketch was not merely an illustration accompanying the narrative or an ornament for the map's cartouche. It served as an authenticating tool, a corroborating witness. As Anne Godlewska has pointed out concerning the *Description de l'Égypte,* the celebrated Napoleonic effort to capture Egypt on paper: "Sketches and maps were not designed to illustrate but to share the load of description and analysis."[150] In the context of a large and well-organized colonial reconnaissance project (like Napoleon's multifaceted Egypt expedition), the image is part of the carefully orchestrated integration of representations that constituted the imperial construction of Egypt, a French invention of a possessable place. In the context of soli-

---

147. Goodall, "Diary July to December," 41.
148. RGS, Correspondence, Hilhouse to RGS, 12 April 1836.
149. *JRGS* 2 (1832): 327.
150. Godlewska, "Map, Text and Image," 11.

tary traverse surveys like those of Hilhouse and Schomburgk, the interlocking representational components—image and narrative—were even more than an "enormous discourse built around topographic mapping" that turned a map into an "integrated and multi-media text" capable of transporting the image of a colony "perfectly packaged for European consumption."[151] They were also part of the elaborate attempts to shore up the reliability of the explorer's traverse, and particularly the significance of his fixed points.

Here is the crucial issue: fixed points alone did not make a place, regardless of how precisely they had been fixed. After all, those coordinates preexisted the passage of the explorer; they had been predicted, postulated, even *guaranteed* by the space-encompassing grid of longitude and latitude. How was fixing them anything more than confirming what was already known? The only way that points on the grid could be used to construct a map of the place itself was through the representation of the *character* of those points, the character of the *ground* at the point identified. Mapmaking demanded more than just pairs of astronomically defined coordinates. What was needed was an account of what existed at those points. Only through such an account could the map be reconciled with the ground; only then could the map become a testable proposition.

Seen in this context, the images prepared by the traverse surveyor take on new significance: they were central to the construction of the landmarks that anchored fixed points to the ground. Those landmarks were the critical intersection of the route and the graticule, the view from within and the view from above. To serve this function the points had to be clear both on the chart *and* on the ground; the map could be reconciled with the place only if the fixed points on the chart could be linked to *visible points on the ground*. The value of the geographical work of a traverse surveyor hung, then, on landmarks—the intersection of points on the representational field of the chart and points in the fields of the interior.

But what was a landmark in the trackless savanna, the relentless green of a jungle? The interior of British Guiana presented no fixed settlements. Amerindian encampments were transient. To colonial eyes inhabitants burned their settlements to the ground and vanished on whim, dispersing back into the forest only to reappear elsewhere. Nor were there architectural ruins of sufficient magnitude to provide some anchor against the perpetual cycles of flood and drought that left the landscape in continual flux. Confronted with the need to map this simultaneously dynamic and problematically homogeneous landscape, traverse surveying explorers

---

151. Ibid., 25.

like Schomburgk had to make use of images and narratives to layer given sites with significance, to represent them to others, and in doing so to make landmarks. These landmarks constituted the features of the place, and they were the essential links between the meandering traverse of the explorer and the rigid graticule of longitude and latitude. Understanding the work of nineteenth-century geographical exploration hinges on understanding the meaning and making of landmarks.

Naval navigators could leave no marks on the surface of the sea. The task of landmarking belonged to a different tradition, that of the civilian land surveyor. Unlike navigational astronomy, the basic tradition of "metes and bounds" land surveying was rooted in the practice of erecting landmarks and, crucially, *linking* them to make boundaries.[152] If navigational astronomy was about routes and trajectories, a practice that generated lines out of points, surveying was about enclosures, about generating enclosed territories out of lines—lines walked, drawn, and sighted. Land surveying involved inscribing positions not only onto diagrams and plans, but onto objects in the world as well. The surveyor had to see suitable landmarks, make permanent inscriptions on trees and stones, and erect cairns and obelisks in order to reconcile places in the world with his plot of a property. If navigation was designed to find an individual who was lost and without landmarks, surveying was about making landmarks. It was a pastoral project, explicitly possessive and inextricable from the European tradition of land tenure.[153] It was the art of establishing and representing ownership, not orientation. It was about locating the place you wanted, not the place you were.

In the colonial context these traditions of enclosure, landmarking, and land tenure were grafted onto the navigational practices of the traverse surveyor, whose passages produced landmarks, linked them to form boundaries, and generated maps on which claims of colonial possession were made. Hilhouse and Schomburgk were trained as land surveyors first, and Schomburgk demonstrated that he considered this training significant to the traverse survey in a colonial interior when he wrote the RGS that in addition to his properly naval kit—chronometer, sextant, nautical almanacs—he would need to bring his chain as well.[154] The tradition of land surveying provided the context in which traverse surveyors selected and marked the sites that became landmarks on their maps.

152. Kain and Baigent, *Cadastral Map*, 273–75 and passim.
153. The most recent work on the history of cadastral mapping is Price, *Dividing the Land*. For a close examination of such surveys in British plantation colonies in the Caribbean, see Higman, *Jamaica Surveyed*.
154. RGS, Correspondence, Schomburgk to RGS, n.d. [August 1833].

Plate 1. A view of Pirara: envisioning El Dorado. Courtesy of the Yale Center for British Art, Paul Mellon Collection.

Plate 2. A flower of destiny: the *Victoria regia*. Reproduced from volume 16 of the *Floricultural Cabinet* of 1848; courtesy of the New York Public Library.

Plate 3. The frontispiece of the *Twelve Views:* a welcome at the fertile gate. Courtesy of the Yale Center for British Art, Paul Mellon Collection.

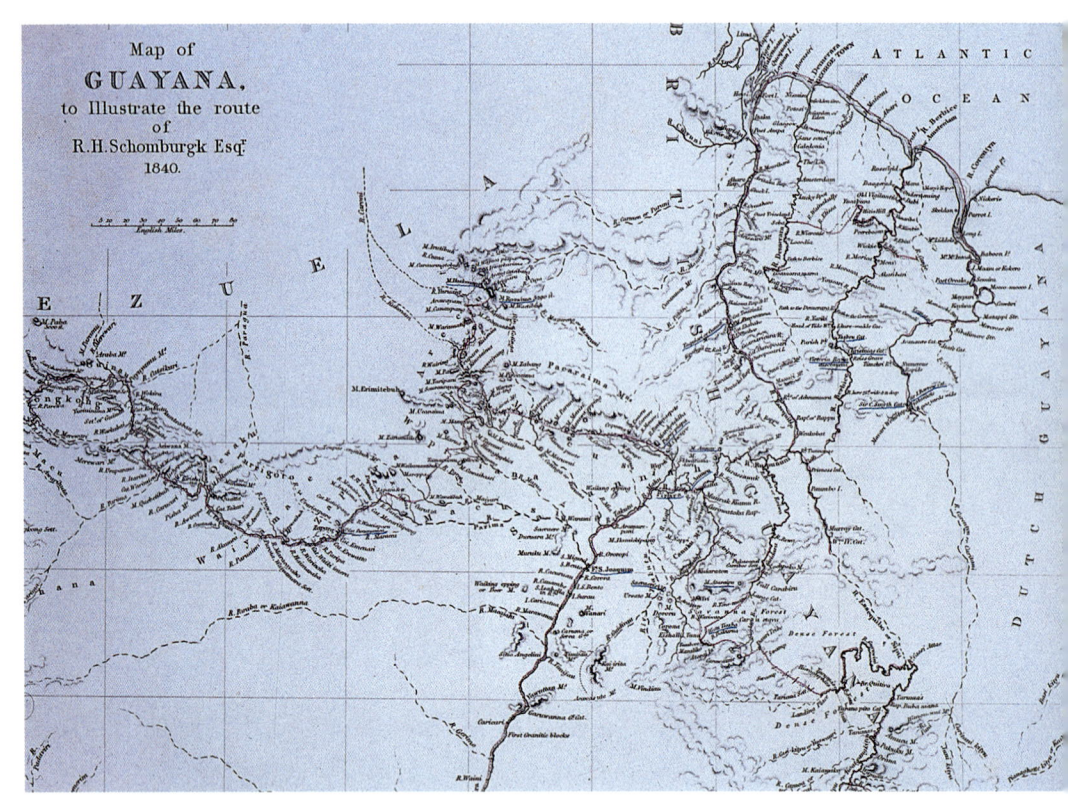

Plate 4. The key map of the *Twelve Views:* an overview of all the landmark sites. Courtesy of the Yale Center for British Art, Paul Mellon Collection.

Plate 5. Comuti or Taquiari Rock: the pedestal. Courtesy of the Yale Center for British Art, Paul Mellon Collection.

Plate 6. Ataraipu, or Devil's Rock: the pyramid, charged with a threatening mythology. Courtesy of the Yale Center for British Art, Paul Mellon Collection.

Plate 7. Puré-Piapa: the column. Courtesy of the Yale Center for British Art, Paul Mellon Collection.

Plate 8. Roraima: the Colosseum and the promise of a lost world. Courtesy of the Yale Center for British Art, Paul Mellon Collection.

Plate 9. Esmeralda: an emerald myth, and a site weighted with geographical significance. Courtesy of the Yale Center for British Art, Paul Mellon Collection.

Plate 10. Fort São Gabriel. Courtesy of the Yale Center for British Art, Paul Mellon Collection.

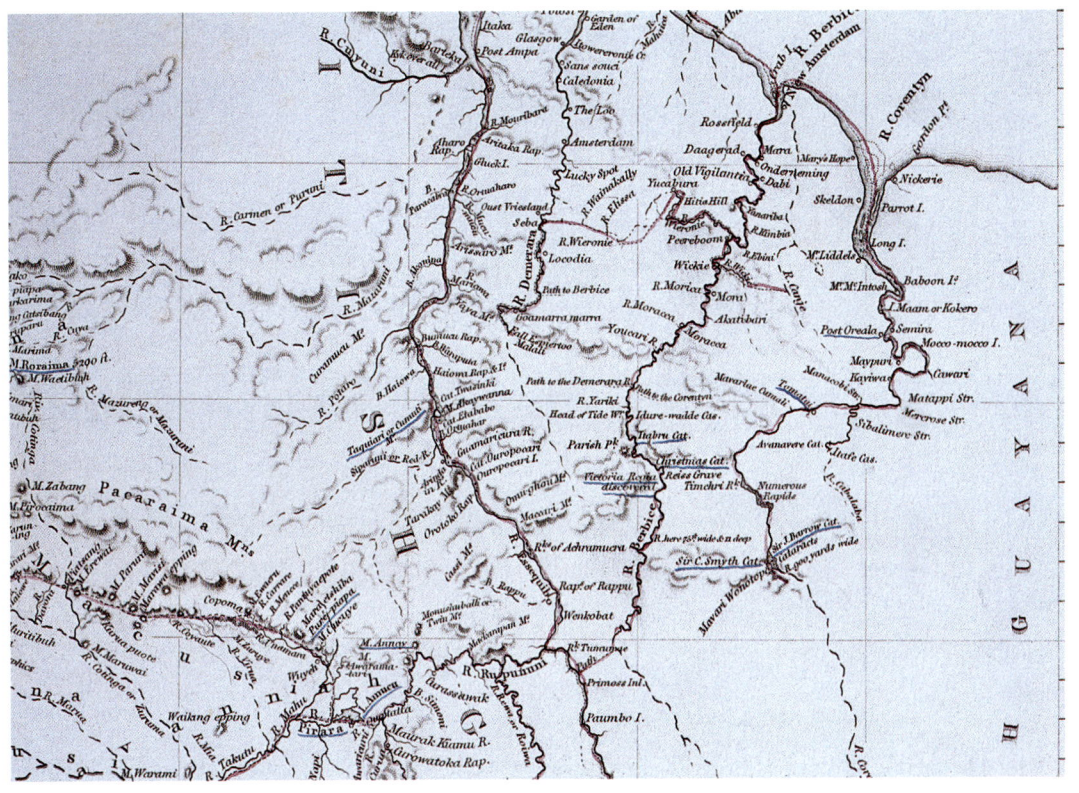

Plate 11. A detail from the key map of the *Twelve Views.* Courtesy of the Yale Center for British Art, Paul Mellon Collection.

Plate 12. British Guiana commemorative stamps: landmarks as icons. Collection of the author.

Plate 13. Kaieteur Falls: "far grander than Niagara!" Reproduced from Brown's 1876 *Canoe and Camp Life;* courtesy of the New York Public Library.

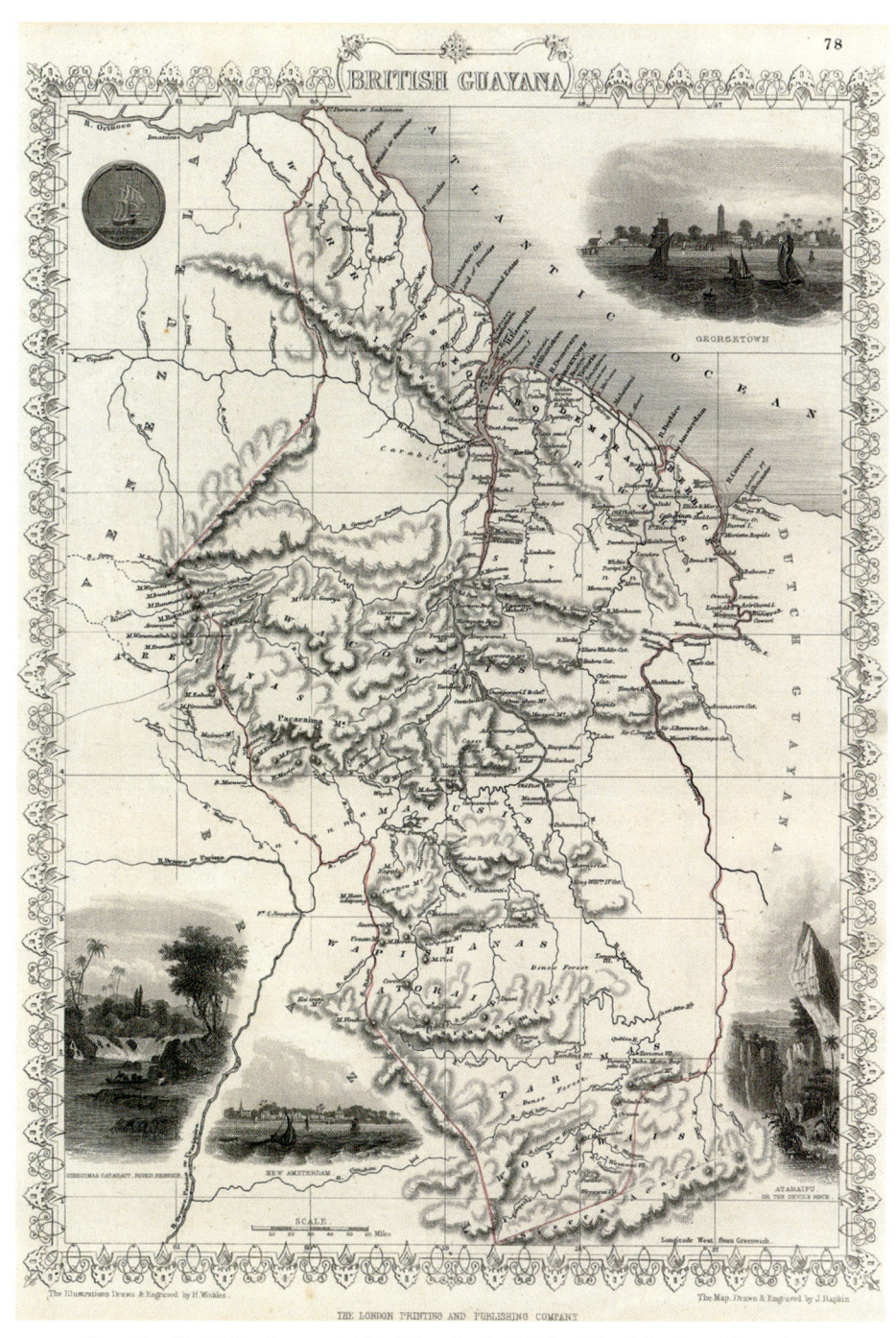

Plate 14. The Tallis atlas map: British Guiana for the British public. Collection of the author.

Plate 15. Bentley's Christmas Cataracts: a site of disaster. Courtesy of the Yale Center for British Art, Paul Mellon Collection.

Plate 16. Kundanama junction: playing out the fear of the river. Courtesy of the Yale Center for British Art, Paul Mellon Collection.

Plate 17.   Detail from Bentley's image of the Christmas Cataracts: pondering the fall. Courtesy of the Yale Center for British Art, Paul Mellon Collection.

Plate 18.   Making the boundary: a sketch by Goodall of a surveying scene near the Takutu. Courtesy of the British Library.

A good land survey depended on the surveyor's capacity to see the territory well, to see it from the right places, with sufficient accuracy and detail of observation. Good surveying also demanded attention to *closure*. It was not enough to see an object only once, to fold it into the web of the survey by a single sight line. Much better was to enclose the terrain, returning to see each point again from a new orientation so as further to confirm its setting with respect to the others. As one land surveying textbook explained, "The oftener you close your work . . . the less liable you will be to error."[155] Traverse surveyors in Guiana went to great lengths to enclose their work by going up one river, marking a tree or erecting a pillar, returning to the coast to ascend a *different* river, and then, on reaching the correct latitude, traversing the land in between in order to reach their previous mark, an event often accompanied by considerable celebration. These traverse enclosures ultimately enclosed the plot of the colony itself, through Schomburgk's boundary surveying expeditions in the early 1840s (the subject of chapter 6).

Schomburgk understood from early on the particular importance of the landmark in the work of the traverse survey. He also demonstrated that he was aware of the power of words and images to create a landmark that would give meaning to a fixed point. In his disappointing first expedition up the Essequibo, he went to some lengths to conjure a landmark out of the waterfall that stopped his progress. His (Humboldtian) prose gathered force in evoking the turnaround point of the journey:

> The great cataract of the Essequibo was before us. Numerous conical hills of granitic structure, and about 300 feet in height, covered with luxuriant verdure, narrow the river down to within fifty yards, where the whole body of water dashes down a precipice of fourteen feet, then foams over a rugged bed of rocks for about twenty yards, and again precipitates itself, ten feet, to the basin below,—the rich vegetation luxuriating in all the fertility of a tropical clime,—the masses of granite projecting into the river, and hemming it in to narrow limits; and the foaming waters in the background bearing away everything opposed to its progress, combined to form the most beautiful and picturesque scene we had witnessed during the course of our expedition.[156]

And lest his readers' attention be drawn to the fact that Schomburgk had ended up taking meteorological observations at a small village on the *terra cognitissima* of the Rupununi instead of attaining the sources of the eastern Essequibo, lest it be thought for a moment that he had not in fact reached any terra incognita on the more than £700 he had laid out for the

---

155. Tate, *Treatise on Practical Geometry*, problem 10:4.
156. *JRGS* 6 (1836): 267. For an analysis of Humboldt's (similarly active) prose descriptions of nature, see Pratt, *Imperial Eyes,* 123.

boats, crew, and equipment, Schomburgk provided an indigenous testimonial of territorial virginity and a climactic onomastic offering in the same narrative: "As all the Indians of our party agreed in declaring that no white man had ever before reached this fall, and as from every inquiry I made I could obtain no native name for it, I considered myself justified in naming it King William's Cataract, in honour of his Majesty, the Patron of the Royal Geographical Society."[157]

The naming ritual linked the point to a toponym. The fall itself Schomburgk fixed by means of a "satisfactory meridian altitude of Canopus" (giving latitude 3°14½′ N) and by deduced reckoning from his base camp at Annai (giving longitude 57°43′ W).[158] In his narrative he linked the place to some Amerindian traditions, pointing out that it was said to have been the dwelling of "Mahanarva, the last Cacique of the Indians," and that upriver was said to be the "abode of an evil spirit in the form of a serpent &c. &c." He spent the evening exercising his pencil and managed to produce a sketch of the site. The original is lost, but it was good enough that the RGS secretary had it put on stone and printed with Schomburgk's account in the *Journal* (fig. 17). The caption of the view gives the cataract's royal sobriquet and its position; the cataract was placed by name on the accompanying map. Schomburgk prepared a cache (a bottle with an inscription of his own name, that of the king, and the date), which he entombed at the base of the fall. Schomburgk had established his first landmark in the interior of Guiana.

It was a tricky business. Fixing the point demanded technical calculations, but constructing a landmark involved literary style, and this could be even more troublesome. Schomburgk discovered as much when his flights of romantic prose came under the correcting pen of the RGS editors. In his manuscript on the cataract Schomburgk had in fact composed several pages about the fall, and he had detailed the christening with considerably more local color. He had not written of the Amerindian tradition as "a serpent &c. &c." In fact, he had gone on at length about the serpent, its battle with a greater serpent, its subsequent metamorphosis, and particularly its power to control the qualities of the river. All this was cut.[159] In probing for the right language for a traverse survey he had actually requested that his reports be "submitted to the correcting pen . . . [as] I am sadly afraid that I have sinned against the style in some places."[160] But he never heard directly the withering critique of his narratives passed on by Sir John Barrow, president of the Society, who wrote in an internal

157. Ibid.
158. Ibid., 268.
159. RGS, Journal MS, Schomburgk, file 1.
160. RGS, Correspondence, Schomburgk to RGS, 15 September 1833.

Figure 17. King William IV Cataract: conjuring a landmark (out of failure). Reproduced from the Royal Geographical Society's *Journal*, 1836.

memo that Schomburgk was "too fond of writing to accomplish much" in the service of scientific geography.[161]

As Edney has written of the geographical narrative in the period, "It was impossible to write a purely geographical narrative without being influenced by at least some of the features of the dominant literary form."[162] So while authors of these narratives could try to distance themselves from more popular modes of travel writing—by their constant attention to instruments and observations, by the emphasis on identifying colonial prospects and issues of imperial security in lieu of personal details—their

---

161. RGS, Correspondence, Schomburgk file, Barrow's memorandum of 13 October 1836. The British critique of Humboldt's "wordy" style is laid out by Brock ("Humboldt and the British," 368–69). Barrow, Brock claims, "refused to be 'ravished' by the 'whistling' of Humboldt's name as 'the first of all travellers . . . a rare union of all that Plato, Thales and Pythagoras taught among the ancients.'" Brock goes on to summarize Barrow's critique of the *Personal Narrative:* "The title was wrong, for it was not a *personal* account, but full of quotations from other writers, nor was it a *narrative* since 'the most trite and trivial matters' were 'meretriciously tricked out in the garb of philosophy and mathematics.' Not only was Humboldt at fault for failing to write in the (actually already outmoded) tradition of travel writing, he was also at fault for continually breaking up the story of his experiences by long-winded dissertations. In other words, by mapping things together, by artificially putting things together and not necessarily describing them in the way they occurred . . . Humboldt was taking the *personal* out of the story." Brock, "Humboldt and the British," 369.

162. Edney, *Mapping an Empire*, 65.

texts inevitably ran the risk of appearing unduly personal or inappropriately artificed. The same could be said for the accompanying images. A view, after all, was a *view*. Not only were such images highly *particular* perspectives on a given site, they could hardly avoid being influenced by the conventions of European landscape representation.[163] The tension was ineluctable: details were needed to substantiate the map, to demonstrate method, to make the position of a single individual into a landmark that was fixed and testable; yet the distance between detail-rich prose and merely anecdotal travel writing was disconcertingly narrow. Some version of this tension is consistently at the core of the representational challenges of exploration in general, but the traverse survey in the nineteenth century faced a particularly paradoxical situation.[164] Linking fixed points to constructed landmarks demanded narratives and images, but these had the side effect of insinuating rhetoric and subjectivity into the objectifying rigor of the traverse survey.

Without focusing on the particular challenges of the traverse survey, Barbara Stafford has traced the evolution of a "scientific aesthetic" over the period 1760 to 1840, and she locates the development of a "plain" style of landscape representation exactly at the frontiers of European expansion. Travel and discovery demanded—and, she claims, produced—a fact-centered, unadorned, neutral, *scientific* style of image making. Such focused and empirical depictions were intended to supplement what were seen as the inherent shortcomings of language as a mode of characterization. Is this quite right? David Livingstone, by contrast, considers Humboldt an exemplary instance of the early nineteenth-century attempt to harness word and image *together* in the service of geographical description,[165] and more recently Michael Dettelbach has characterized Humboldt as "torn by how to satisfy the demand of both 'precision' and 'picturesqueness' in the depiction of vegetation and its geographic progress."[166] The idea of the picturesque in such close juxtaposition with concern for the empirical depiction of a place does not accord at all well with Stafford's thesis. For her the "scientific mode" of depiction distanced itself from the pictorial conventions and a priori concepts of place that characterized the picturesque. As a number of scholars have argued, there are problems with the radical disjunction Stafford seems to insist on be-

---

163. I will return to this theme in chapters 4 and 5.
164. Mary Fuller, for instance, claims that the "newness" of the New World forced its explorers into the multiple entanglements of rhetoric as they attempted to substantiate their new knowledge. Fuller, "Ralegh's Fugitive Gold," 221.
165. Livingstone, *Geographical Tradition*, 133.
166. Dettelbach, "Global Physics and Aesthetic Empire," 271.

tween the picturesque and the scientific modes of image making.[167] Applied to Schomburgk's work, as I will show, her distinction would be a misleading analytic tool.

For the traverse surveyor working in a distant continental interior, the need to construct landmarks—crucial to anchor his astronomically fixed positions—led to the deployment of a barrage of tools, both narrative and pictorial. "Plain" description and "plain" representation were supplemented by a welter of aesthetic conventions, historical associations, and a raft of other tropes and devices. Tensions were inevitable. The next two chapters examine all this in more detail.

167. It seems to suggest that she would deny the "ideology that permeates topographic views." Edney suggests that the "geographical gaze" subsumed elements of "plain" and "picturesque" styles. I return to this idea in the next chapter. See Edney, *Mapping an Empire,* 55. For another recent discussion of the relation between scientific and picturesque images in the period, see Klonk, *Science and the Perception of Nature,* chap. 3, "From Picturesque Travel to Scientific Observation."

CHAPTER FOUR

Marks on the Land: Landmarks, Aesthetics, and the Image of the Colony

VIEWS IN CONTEXT: WRITING THE JOURNEY

In his textual products, as in his instrumental poverty, Schomburgk was no Humboldt. The expedition that linked the disciple to the master in the shadow of the cross at Esmeralda (and let Schomburgk too float his corial down the Casiquiare canal while thinking about the hydrography of El Dorado) was not the graphomaniacal "thirty-volume voyage" of Humboldt.[1] But Schomburgk got a great deal into print. From 1839 through 1841 he published:

> "Report of the Third Expedition into the Interior of Guayana," nearly 100 pages in the *Journal* of the RGS;
> *A Description of British Guiana, Geographical and Statistical,* 155 pages on the resources and capabilities of the country;
> *The Natural History of the Fishes of Guiana,* two volumes with colored plates in Jardine's Naturalist's Library series;
> *Reisen in Guiana und am Orinoko,* more than 200 pages, with map and colored plates;
> *Twelve Views in the Interior of Guiana,* a folio travel book published by Ackermann;
> half a dozen articles on botanical and other specimens in a variety of journals, including *Hooker's Journal of Botany, Transactions of the Linnean Society, Transactions of the Botanical Society of London,* and others.

While there is a fair bit of repetition in these—the German *Reisen* is in large part a translation of his reports to the RGS from 1836 to 1841, and the text of the *Twelve Views* and the "memoir" accompanying *The Natural History of the Fishes of Guiana* both borrow heavily from those same reports—Schomburgk clearly wasted little time after his return to England in the autumn of 1839. In part this was a financial necessity. He had not secured as many botanical specimens as he had hoped, so when William Lizar, the Edinburgh printer, engraver, and publisher of the Naturalist's Library, offered £60 for descriptive text and illustrations of sixty-six fishes he had sketched in the course of his travels, the explorer quickly agreed.[2] Lizar, whose popular series (featuring hand-colored plates) sold at the very competitive price of six shillings per volume, took advantage

---

1. Pratt, *Imperial Eyes,* 115.
2. Sheets-Pyenson, "War and Peace in Natural History Publishing," 65 and 72.

of every innovation in printing technology to keep his editions profitable at runs that varied from 2,000 to 4,000 copies. He also stayed on the lookout for naturalists not yet commanding top prices for their work. In this enterprise he had the advantage of having married well: his brother-in-law was fellow Scotsman William Jardine, accomplished in ornithology, botany, and ichthyology.

It is safe to assume that Schomburgk knew Jardine's two-part work on hummingbirds, which had been published in 1833 and became the most popular books in the series: more than 7,000 copies sold in two years.[3] Schomburgk had several interesting hummingbird specimens among more than one hundred bird skins he had preserved and brought back with him, and he likely thought of Jardine as both a taxonomic consultant and a potential customer for some of his skins. A correspondence began, and Jardine suggested that Schomburgk submit his collected notes and fish sketches to Lizar for publication.[4]

The money would have been appealing and the publicity still more so. It is to the Naturalist's Library that historians are indebted for one of the few engravings of Schomburgk that exist: an image of a boyish explorer with tousled hair (fig. 18). As Susan Sheets-Pyenson shows, the Library was available to "all but the lowest levels of the Victorian middle classes,"[5] and the volumes, packaged in size and feel like the popular *Waverley* novels, were published in Dublin, London, and Edinburgh. Nor was Lizar an excising editor. In fact he sought to keep the volumes thick, even tweaking the margins and adding introductions, prologues, and announcements. "Thin-looking" volumes, he fretted, did not call forth six shillings with enthusiasm.[6] This padded format would have suited Schomburgk's needs. He had ample quantities of text in the reports he had prepared for the RGS, and he cobbled together bits of this narrative to compose a synopsis of his journeys that would be read by a considerably broader segment of the population than the readership of the RGS *Journal*.[7] Schomburgk's travel accounts would have reached their widest audience in early Victorian society through the medium of the Naturalist's Library, where members of the respectable middle class would have learned of Brazilian night raids on helpless Amerindian women and children.

3. Ibid., 55.
4. Schomburgk must have remained on good terms with Jardine. In 1842 Schomburgk dedicated to him a species of *Hypostoma*, the most interesting fish he caught on the Takutu expedition.
5. Sheets-Pyenson, "War and Peace in Natural History Publishing," 51.
6. Ibid., 66.
7. On his departure in 1841, Schomburgk left the task of compiling the memoir for the second volume of *Fishes of Guiana* to the wanderer of Nubia, J. L. Burkhardt, who again abstracted the *JRGS* accounts to compose the memoir.

Figure 18. The young geographical explorer, Robert H. Schomburgk: defining boundaries, crossing them. Reproduced from the first volume of Schomburgk's *Fishes of Guiana;* courtesy of the New York Public Library.

However, it is not in this form that his work would have reached the most exclusive readership, those members of the ruling class most intimately involved in the future of the colony. Schomburgk reached them with his sumptuous *Twelve Views in the Interior of Guiana,* which sold, on subscription, for the stiff sum of £4 4s. for the colored edition on large paper and £2 12s. 6d. for the two-tone lithographed copy uncolored.[8] R. V. Tooley lists the work in his bibliography of the most important English books with color illustrations before 1860,[9] and Anthony Dyson cites it as an exemplary instance of the multiple media increasingly used by publishers like Ackermann and Company in the late 1830s and 1840s to produce exceptional works of foreign travel.[10] Intaglio, woodblock, and the relatively recent innovation of color lithography were all combined to produce what John Ford has called "a major publishing achieve-

---

8. Note that the publication of his multiple reports in the *JRGS* would have reached a large and influential readership. Stoddart's list of the original membership draws attention to the high social standing of the members. Stoddart, *On Geography and Its History,* 60.

9. Tooley, *English Books with Coloured Plates,* no. 447. See also Twyman, *Lithography.*

10. Dyson, *Pictures to Print,* 32.

ment which has often gone unremarked, possibly because these lovely tinted lithographs have been less popular than the bright and showy aquatints" sought by collectors.[11]

Ackermann's capacity to coordinate the production of such volumes—which in the case of the *Twelve Views* demanded organizing the preparation of stones by at least three artists at two separate lithographic printers, in addition to procuring map plates, wood engravings, and text from other sources—likely came to Schomburgk's attention through the RGS, one of whose fellows, Commander William Allen, had just seen his *Picturesque Views on the River Niger* come to completion in similar format in 1840. This text, offered for the more modest sum of £1 5s. uncolored, had proved a successful enterprise, attracting a strong list of subscribers. Schomburgk purchased one.[12] The RGS took considerable interest in Allen's survey work on the river and his improvements on the 1830 Niger map by Richard and John Lander.[13] Allen secured the plate for that map of the Niger survey from John Murray in order to include it in his volume. It was a notorious map. The human cost of the surveys led the secretary of the RGS to declare that it had been "drawn and coloured with drops of blood."[14]

The juxtaposition of picturesque views and map of blood went to the heart of the matter. The notion of the Niger as "picturesque" would have given pause to a browser in Ackermann's bookshop at 96 Strand, considering West Africa's unshakable reputation as the "white man's grave."[15] In spite of this, Allen's publication marked the attempt to capture in an elegant book nearly a decade of costly and indifferently successful Niger explorations.[16] Its publication in 1840 immediately preceded the departure of the large government expedition to the Niger in 1841. This expedition, staffed with scientists supplied by the African Civilization Society and equipped with government ships, departed in the midst of a fanfare created in part by Sir Thomas Buxton. Drawing (and feeding) on what Philip Curtin calls the "new concern with the plight of aborigines in British Settlements,"[17] Buxton simultaneously harangued Parliament for more commitment to West Africa and roused a swell of public support for the Niger project, support Curtin calls "almost as important [for the future British image of Africa] as the expedition itself."[18] Allen's *Picturesque Views on the River Niger* would have come to light at the very

11. Ford, *Ackermann*, 123.
12. Allen, *Picturesque Views on the River Niger*.
13. Skelton, *Explorers' Maps*, 282.
14. Day, *Admiralty Hydrographic Service*, 26.
15. Curtin, *Image of Africa*, chaps. 3 and 7.
16. Ibid., 194, 321.
17. Ibid., 299.
18. Ibid., 302.

peak of this enthusiasm for the region, and by recasting the "white man's grave" as a bucolic riverscape, he contributed to the optimism that surrounded the Niger expedition.[19] The deaths of 48 of the 159 Europeans on the expedition in the first two months cooled public enthusiasm considerably.

The circumstances surrounding the publication of the *Twelve Views* share a good deal with those of Allen's book, even if interest in Guiana never reached the missionary fervor of the Niger expedition.[20] The younger Ackermann brothers treated the volumes as parallel projects and drew on two of the best lithographic houses in London, Hulmandel and Gauci, for the work on stone. In both Allen's and Schomburgk's cases, an explorer placed his cartographic work in the context of a set of landscape views and linked the images to the map by means of a journey narrative that seized on the enchanting character of the place as well as its history and prospects. In both cases the resulting work was published to the highest standards of the printer's trade—using new multiple-stone color lithography—and sold on subscription to an elite group of political figures, members of the aristocracy, colonists, and merchants. In both cases the publication coincided with a surge in interest in the area, interest that in both cases led to commissioned government expeditions.[21]

In neither case would it be wise (or possible) to construe each aspect of government involvement in the region as the direct effect of an illustrated travel book, but particularly in Schomburgk's case (where so little was in fact known in Britain about the region), the book would have played a significant role in making the place visible to a powerful community. The *Twelve Views* was a synthetic text—like the *Description de l'Égypte* on a (much, much) smaller scale—that combined views and narratives in a cartographic framework, and thus it constituted the sort of geographical construction of Guiana that made the place visible. By allowing a distant place to be seen as part of British landscape traditions, the *Twelve Views*

---

19. Twyman, *Lithography*, 196.

20. It is likely that on his return Schomburgk would have approached Ackermann and Company after learning of Allen's forthcoming work through Captain John Washington, the secretary of the Society. John Murray (also a member), who had produced the maps for Allen's volume, may also have been the link. See Herbert, "Royal Geographical Society's Membership," 69. Ford (*Ackermann*, 84–88) points out that Ackermann had long-standing links to South America. The founder, Rudolph (a native of Saxony who retained his German accent throughout his life), had been engaged in 1822 to print bond certificates for a Colombian loan. He invested as well, and later added Mexican and Brazilian holdings to his portfolio. He lost heavily in the South American defaults of the late 1820s. His commercial interests followed his investments, however, and by 1830 Ackermann and Company had more than a hundred titles in Spanish.

21. In fact the preparations for the two expeditions overlapped, and Richard Schomburgk wrote of mingling in London with the Germans of the Niger expedition, F. R. Vogel and C. G. Rotscher, who hailed from Schomburgk's hometown. M. R. Schomburgk, *Travels*, 1:7.

would have offered its readers a Guiana that was pleasingly British while at the same time piquant with exotic touches. Schomburgk's *Twelve Views* constructed an image of British Guiana among British ruling elites during a crucial period in British colonial decision making.[22] It was explicitly written in hopes of persuading British elites that the colony was not "merely a flat, traversed by dikes and devoid of picturesque effects" but rather had been stamped by the "mighty hand" of nature with a "portion of her sublimest effects."[23]

As a text dramatizing the romantic tropical scenery of the colony, the *Twelve Views* can best be understood as the "visibility enhancing" companion volume to Schomburgk's other major publication on the colony, *A Description of British Guiana, Geographical and Statistical.*[24] This work, also completed on Schomburgk's return to London and published in 1840, falls squarely into an established tradition of statistical geographical works, texts that took on particular importance in view of British "enlightened" and developmental imperialism in the period.[25] The *Description* laid out the physical aspect, vegetation, inhabitants, climate, and political divisions of the colony before exploring in greater detail its "Resources and Capabilities" as well as "Future Prospects." Underutilized resources and languishing fertility constituted a largely explicit argument for greater British colonial activity in the region, and Schomburgk included proposals for importing (nonslave) labor as well as schemes for encouraging agriculture and exploiting natural riches. Various tables presented in the text presented population figures, exports, tallies of criminal convictions, and the like, mostly borrowed directly from a copy of the *Local Guide of British Guiana.*[26]

The *Twelve Views* may have afforded a certain colonial visibility, but only for those who could afford it. Analysis of the subscriber list reveals

---

22. Curtin, *Image of Africa,* ix.

23. Schomburgk, *Twelve Views,* "Preface," b recto.

24. Schomburgk, *Description of British Guiana.* See his similar work on Barbados, published in 1848: Schomburgk, *History of Barbados.* For a discussion of the intersection of statistical and cartographic knowledge in the process of state formation, see Revel, "Knowledge of the Territory."

25. Livingstone, *Geographical Tradition,* 103. Examples include Boyle, *General Heads,* and Playfair, *Statistical Breviary.*

26. BL, Schomburgk Autograph (British Guiana, *Local Guide of British Guiana;* see also above, chapter 3, note 138), is inscribed inside the title page with Schomburgk's name and presumably belonged to him. The statistical pages are heavily marked, though this text would have been published too late to be of any use in preparing the *Description of British Guiana.* Schomburgk would, however, have had access to Baker, "Local Guide"; British Guiana, *Demerara and Essequibo Vade-Mecum;* and Henery, *Almanack and Local Guide.* Much remains to be said about the role of statistical texts in the workings of imperial governance, particularly in light of the recent treatment of statistics in Britain in Mary Poovey's *History of the Modern Fact.* See particularly chap. 7, "Figures of Arithmetic, Figures of Speech: The Problem of Induction in the 1830s."

an elite circle of 360 individuals who subscribed for copies in advance of the publication, most for the hand-colored edition. Of the 360, 209 resided in Guiana and most of these were plantation proprietors, but a considerable number were colonial officers, including the governor (three colored copies), the attorney general, the government secretary, the inspector of police, the highest-ranking member of the judiciary (the chief justice), and a number of other judges and justices of the peace, one sheriff, the customs officer, and assorted other bureaucrats and officials. Among the subscribers residing in Europe, "Her Majesty's Royal Library" took three colored volumes, as did the king and queen of Prussia and the queen dowager of Great Britain, Adelaide. The duke of Sussex and the duke of Devonshire took copies. To the latter, William Spencer Cavendish, a patron of the botanical sciences, Schomburgk dedicated the *Twelve Views*. In all, eighteen titled members of the nobility subscribed, along with fourteen colonial officers, among them the British consul at Madeira (shortly to begin encouraging emigration from the populous island to the labor-starved colony), the chief justice of Saint Lucia, and several former colonial governors. Lord Stanley took a colored edition, shortly before he became colonial secretary in Peel's government in 1841. He would be prime minister on and off through the 1850s and 1860s when British Guiana goldfields were discovered and contested with Venezuela. Three members of Parliament subscribed, as did six clergymen (including a colonial bishop and the dean of Winchester). The army and navy were well represented, totaling eight officers, including two lieutenant colonels and one captain in the Royal Navy. Among scholars and scholarly institutions the RGS was not alone: the University of Edinburgh took a copy, as did Sir William Hooker at the University of Glasgow. King's College London did not subscribe, but Professor Thomas Bell did. In all, ten scholars and medical men purchased copies.

What emerges from a review of the readership of the *Twelve Views* is first and foremost the book's importance in the colony itself. The sale of 209 copies of such an extravagant volume within the colony means, remarkably, that 5 percent of the white population owned a copy.[27] Since nearly all the subscribers were married men, the figure of books per white household could nearly double that figure. Added to the shelves of libraries at such a strikingly high percentage of the houses of landowning elites, the *Twelve Views* represented the most elaborate source of information about the interior. The territory of Guiana took shape in the minds of colonists who paged through the *Twelve Views* in George-

---

27. Based on an estimate of four thousand whites in 1839. Schomburgk, *Description of British Guiana,* 48.

town.[28] At the same time, the presence of such a volume in the hands of the (future) fourteenth earl of Derby, Lord Stanley, not to mention in the hands of Queen Victoria herself, suggests the way its "image of Guiana" shaped attitudes not only in the colony itself but also in the institutions and among the individuals with custodial responsibility for the British Empire.

VIEWS AND STRATEGIES OF IMPERIAL VISIBILITY

Empires are hard to see.[29] The better they are the wider they extend, and the more difficult to encompass they become.[30] Arriving in Georgetown in 1835, Schomburgk faced a British colonial possession not only celebrated (ironically) for its obscurity—little known, nebulous—but also on the verge of collapse.[31] His advertisements for subscriptions to the *Twelve Views* were published in the colony in the immediate wake of full emancipation and during a period of economic turmoil there. The prospect the book promised—the opportunity to consume the colony at a glance[32]—likely fueled its local popularity, a popularity that outstripped all expectations.[33] Nor can the book's appeal be separated from the paucity of visual images depicting the colony. If empires pose visual problems, then British Guiana as part of the British Empire had a very serious problem indeed. No printed book with colored plates depicting its topography existed before 1840.[34]

When Curtin writes of the British image of Africa between 1780 and 1850, he is concerned less with the visible Africa per se—as in what Europeans thought the physical landscape of Africa *looked like*—than with the set of shared assumptions concerning the character of the place and its inhabitants. But how deeply visual imagery informed these attitudes

---

28. Schomburgk's work also affected colonial involvement in scientific investigations, particularly those concerned with positional astronomy and its application to geography: the "observatory" that was erected in Georgetown to establish baseline observations for the Crown-sponsored surveys in the early 1840s became (along with a set of instruments left by Schomburgk) the nucleus of the Astronomical and Meteorological Society of British Guiana, founded in May 1844. See the Society's manuscript minutes in the NAG. Their "Prospectus" discusses the importance of "a general familiarity with the use of ordinary Astronomical Instruments" given the "partially explored condition of this Colony."

29. For discussions, see Dettelbach, "Global Physics and Aesthetic Empire," 272–77; Ryan, *Cartographic Eye*, 100; Edney, *Mapping an Empire*, 24–25; and Mundy, *Mapping of New Spain*, chap. 1.

30. Dettelbach, "Global Physics and Aesthetic Empire," 273.

31. For the economic situation, see Adamson, *Sugar without Slaves*. British Guiana's obscurity is captured in its uncharitable West Indian moniker, the "land of mud." Recall parliamentary confusion as to whether Guiana was an island: see above, chapter 2, note 1.

32. McClintock, *Imperial Leather*, "Progress Consumed at a Glance," 36–44.

33. Schomburgk, *Twelve Views*, "Preface," b recto.

34. The only exception might be Alexander, *Transatlantic Sketches*. This was available with several engravings of Guiana that could be purchased colored. CRL, RBR has manuscripts of these images.

cannot be overlooked. The work of Cosgrove, Daniels, and others has examined the ideological content of landscape representation,[35] and Smith, Stafford, Miller, Ryan, Edney, and others have extended this attention to visual culture to geographical exploration and empire. Efforts to understand the place of the topographical view in the geographical construction of distant colonies have consistently drawn attention to the tension identified in the previous chapter: views offered detail that, under the right conditions, could reinforce descriptive narratives and supplement maps; but at the same time, the view represented a singular perspective on a single place, meaning the image lacked the totalizing and perspectiveless power of the map.[36]

In trying to make sense of topographical views as tools for geographical knowledge in the nineteenth century, David Buisseret suggests that they can be thought of as "graphic representations of place," somewhere "between" maps and mere pictures.[37] Matthew Edney makes a more refined argument along the same lines. He shows that, charged with the same empirical confidence that positioned the surveyor outside the very land in which he found himself, the topographical view stood comfortably in the company of geographical measurements and cartographic depictions.[38] There is then, among several commentators, the idea that topographical views of landscape belong, conceptually, "between" pictures and maps. This argument is grounded in the claim that the topographical view, while preserving elements of the perspective of the viewer, was already distanced from that viewer, just as the map, composed out of multiple perspectives, was thought to distance perspective altogether and become a "view from nowhere." To show the maplike character of the topographical view, Edney cites several instances of geographical narratives illustrated by views that cannot be fixed to any particular location but must be understood to be "*typical* or *characteristic*" instances of Indian landscape.[39]

---

35. This work includes Daniels, *Fields of Vision;* Barnes and Duncan, *Writing Worlds;* and Cosgrove and Daniels, *Iconography of Landscape*. These works are continuing a deeper scholarly tradition that would include Smith, *European Vision;* Glacken, *Traces on the Rhodian Shore;* Jackson, *Interpretation of Ordinary Landscapes;* and Tuan, *Space and Place*. I have also found the following recent works helpful: Jackson, *Sense of Place;* Walter, *Placeways;* Stilgoe, *Alongshore;* and Schama, *Landscape and Memory.*

36. From here on I use "view" to mean a graphic depiction of place in the European tradition of topographical drawing. Distinguishing this tradition rigorously from beaux arts landscape images is impossible, but the broad distinction holds. See Bicknell, *Beauty, Horror, Immensity,* x–xi, and Einberg, *Origins of Landscape Painting,* 5.

37. Buisseret, *From Sea Charts to Satellite Images,* chap. 5. For a discussion of the exchanges between surveying and the work of British landscape painters like John Constable and J. M. W. Turner, see Daniels, "Re-visioning Britain."

38. Edney, *Mapping an Empire,* 72.

39. Ibid., 71.

That such general (and hence already somewhat "aperspectival") views were a part of the geographical discourse of the early nineteenth century cannot be denied. We have already seen an instance of such an image in the previous chapter: Hilhouse's image of the "bush" in Guiana. The frontispiece of Schomburgk's *Twelve Views* offers another good example (plate 3).[40] Given that the image depicts the gargantuan *Victoria regia*, it might be supposed that the river represented is the Berbice, the site of the famous flower's "discovery." But the upper reaches of the Berbice, from Schomburgk's own descriptions, lack picturesque elevations of any sort. If it is the Berbice, then, a very handsome peak has been imported to anchor the horizon and provide a suitable object to be veiled by the scrim of mist that enchants the vanishing point. The nearer hills are covered in meadow and wooded vales, and a pleasing foreground frame of jungle (in full flower) has been pulled aside into an elegant coulisse by elaborately adorned natives. This view, then, is not so much a picture of a place as a picture of ideas about a place: fertility, ease of navigation, natural beauty, congenial indigenes, salubrious climate, and picturesque promontories.

The frontispiece, however, is an exception. The other topographical scenes of the *Twelve Views* do not depict nonplaces at all. In fact, the location of each of the other views is given to the minute of arc for both longitude and latitude, as the position of King William IV Cataract was in Schomburgk's RGS lithograph of 1836. Moreover, in the *Twelve Views* each view is itself precisely mapped by means of the color-coded map that begins the volume (plate 4).[41] How can this fixation with the astronomically fixed view be understood? As I suggested in the previous chapter, the value of the traverse surveyor's fixed points depended on attaching those points to the ground itself. This gives us the framework for understanding why the scenes of the *Twelve Views* are captioned by their precise geographical coordinates. Understood in the context of the traverse survey, the *Twelve Views*' presentation of the key map and coordinate-fixed views must be understood as part of a powerful representational strategy. It is a strategy described by Edward Tufte in his *Envisioning Information*. When detailed images (like views) are coordinated within a structural array (like a key map), Tufte calls the result a "micro/macro" graphic. His description fits the structure of texts like *Twelve Views*. What Tufte specifies is that the maplike nature of a topographical view, as well as its value as a cartographic instrument, cannot be judged without reference to context. Detailed information, situated in the appropriate context, reads differ-

---

40. The image was included, Schomburgk pointed out in the preface, as a "gift" to subscribers beyond the promises of the prospectus. Schomburgk, *Twelve Views*, "Frontispiece," *b* recto.

41. Godlewska identifies the importance of mapped views in the *Description de l'Égypte*, but she is referring to individual views accompanied by local-scale maps. Godlewska, "Map, Text and Image," 12.

ently; detail that might yield a representational morass can clarify when set in the appropriate framework. Micro/macro graphics are powerful because "detail cumulates into larger coherent structures. . . . Simplicity of reading derives from the context of detailed and complex information, properly arranged. A most unconventional design strategy is revealed: *to clarify, add detail.*"[42]

By situating picturesque views within a geographical field by means of the key map and coordinates, Schomburgk and those other early nineteenth-century geographical explorers who relied on similar techniques can be seen as deploying the outdated structure of the chorographic/geographic distinction in a new way.[43] What has been depicted in these landscape views is a chorographic map not of a city or town, but most often of *a place from which seeing itself was possible*. The landmark objects in Schomburgk's *Twelve Views* are most often promontories or elevations, points that offer a view of the place itself. What the key map provides, then, is an overview of key points of overview, affording readers the opportunity to "stand above" the territory at two distinct scales. At the same time, the map buttresses the objectivity of the views, which, despite their subjectivity and conventionality, have definite places in the cartographic array. Nor is the reinforcement only one-way: by depicting the "places" at "points," the views emphasize that the map is not merely a field of coordinates, but rather a field of coordinates each of which stands for a place. In such a structure the plain and picturesque gazes of early nineteenth-century geography are deployed in such a way as to fortify their respective representational powers.[44]

The key link lies in the landmark, the object at the intersection of the multiple gazes of the traverse surveyor, who must look up to the stars to position it in the graticule of longitude and latitude and must look out from it to have a panoramic prospect of the place. So constructed, the landmark can become a "nodal point" in the construction of the colony, a position of relative stability in multiple fields—geographical, social, and political.[45] By differentiating space and organizing it hierarchically, such nodes foster the regime of colonial control, install regions and boundaries, and give permanence to individual experience. They allow individuals to situate themselves with respect to privileged positions and constitute a powerful technique for "capturing" space.[46]

42. Tufte, *Envisioning Information*, 37.
43. For a discussion of the integration of chorographic images by means of a key map in the late sixteenth century, see Mundy, *Mapping of New Spain*, 7.
44. Edney, *Mapping an Empire*, 63.
45. I borrow the idea of the "nodal point" from Noyes. See *Colonial Space*, 107.
46. Ibid., 108.

Landmarks are such nodes, and the *Twelve Views* is a text that produces landmarks, layers them with meanings, and assigns them values in aesthetic, political, scientific, and social as well as cartographic frames of reference. These points contribute to the "saturation" of colonial space by making possible legacies, pilgrimages, and the cycles of metalepsis that I have argued root colonial possession (chapter 2).[47] The landmarks of choice were promontories, since they became not only places to see but places that enhanced seeing. By affording multiple *overviews,* such an array of promontories intimates a fantasy of total *oversight,* of the complete surveillance of the territory of the colony by the colonial eye.[48] Roland Barthes claims the Eiffel Tower achieved its singularity because it "transgresses this habitual divorce of seeing and being seen" and in doing so allows the structure of the city to be seen.[49] Laid out before the viewer, the city becomes for the first time an object. Similarly, Comuti (Taquiari) Rock (plate 5) makes Guiana a place by affording the first glimpse of the alluvial plain from above.

But the significance of promontories like Ataraipu, "Devil's Rock" (plate 6), is not exhausted by asserting that they are colonial belvederes, which install Guiana as a visual object "at the feet" (in Schomburgk's words) of the viewer. For such striking natural objects were very much meant to *be seen* as well, and Schomburgk's views—for instance, that of Roraima—place the landmark in a visual vignette meant to invoke the independent existence (and symbolic quality) of the mountain itself. This iconic quality of the landmark translates into a long and rich symbolic life for those images as they were recycled in colonial discourse. The reproduction of Schomburgk's images made these landmarks nothing less than icons of the interior. They stood for the interior, and the interior was possessed through metonymy.[50] Just as such landscape icons have been shown to play a significant role in the iconography of nationhood and the construction of the "imagined community" of the nation, it may be argued they played a similarly constitutive role in the construction of the colonial identity of British Guiana.[51] I will return to this point at the close of this chapter, after looking in more detail at the conventions of landmark representation.

47. Noyes on saturation: ibid., 288.
48. For a discussion of the power of places from which seeing is possible, see Ophir and Shapin, "Place of Knowledge," 12.
49. Duncan and Duncan, "Ideology and Bliss," 23.
50. I find that Pratt uses the notion of metonymic representation as well (*Imperial Eyes,* 125), but she is interested not in landmarks as metonymic objects, but rather in "typical views" as metonymic representations. Where I am discussing specific objects, like Roraima or Kaieteur, she is discussing Humboldt's general landscape categories: forests, pampas, cordilleras.
51. For a similar study relating to Canada, see Osborne, "Iconography of Nationhood."

PICTORIAL CONVENTIONS AND LANDMARK DEPICTIONS

I have already discussed in some detail the fixing of landmarks on the graticule of longitude and latitude, and the technical (as well as rhetorical) means traverse surveyors used to do so. Examining the views of such landmarks demands some consideration of how they were "fixed" within the framework of European pictorial conventions. Ian MacLaren's numerous publications on the place of the sublime and the picturesque in the "imaginative mappings" of Canada in the late eighteenth and early nineteenth centuries constitute the most detailed effort to understand the place of pictorial conventions in travel through, and appropriation of, a colonial space.[52] He goes so far as to suggest that the sublime and the picturesque constituted a kind of "aesthetic graticule" allowing new places to be "fixed" within the field of European notions of landscape.[53] MacLaren has shown that the picturesque's veiled visual promise (of the harmonious relations between man and nature) served to reassure British explorers in new regions and at times caused them to misjudge their surroundings when they saw alien lands in terms of familiar landscapes.

Following Ann Bermingham, who examined the relation of the British picturesque to the history of British agrarian reform and the Enclosure Acts, MacLaren sees the picturesque in the colonial context as the kind of ideological "enclosure act" of colonial representation discussed by Paul Carter.[54] MacLaren presents the British tradition of picturesque landscape painting as emphatically *about* property and territorial control, and therefore he sees the use of the convention by explorers in the Canadian North as a form of "aesthetic appropriation."[55] The imperial picturesque of India was deployed by British topographical surveyors, who saw it both as a natural representation of the Indian landscape and as an invitation to British proprietorship over the pleasing place they saw. Schomburgk's representation of the "naturally" picturesque quality of Guiana worked in similar ways.

This kind of analysis, which implicates the conventions of the picturesque in the history of imperialism, has caused considerable scholarly discord. MacLaren has come under criticism for using definitions of sublime and picturesque that lack sufficient historical nuance, as well as for

---

52. See MacLaren, "Retaining Captaincy of the Soul"; MacLaren, "Aesthetic Map of the North"; MacLaren, "Aesthetic Mapping of Nature"; MacLaren, "Samuel Hearne and Landscapes of Discovery"; and MacLaren, "David Thompson's Imaginative Mapping."
53. MacLaren, "David Thompson's Imaginative Mapping," 93.
54. Bermingham, *Landscape and Ideology;* Carter, *Lie of the Land,* 217–18, 242, passim.
55. For a similar analysis of picturesque landscape depiction in Australia, see Ryan, *Cartographic Eye,* chap. 3.

failing to see the subversions of the conventions at work in the images he examines.[56] The most elaborate critique has come in Kim Michasiw's article "Nine Revisionist Theses on the Picturesque," which undermines the assertion that the picturesque served as a tool for dispossessing native peoples. Michasiw suggests that champions of this argument have neglected to distinguish between two distinct, and possibly conflicting, phases in the development of the picturesque. Stated briefly, these might be called the "traveler's phase"—characterized by William Gilpin and his faithful disciples—and the "propertied phase," in which what had been a set of playful aesthetic conventions that (initially) denied any attempt to "appropriate the *real* land" came to be used as a set of rules for shaping the land itself. Michasiw writes: "Gilpin *assumes* that his readers are powerless tourists obliged by circumstances to leave the landscape as they found it."[57] The "control" that the technique exerted in its countless representational rules was self-conscious, ideational, and assimilated to post-Kantian aesthetics. Only when the conventions were handed over to landscape architects (by class-conscious landowners marking their domains) did the picturesque lose, for Michasiw, its aesthetic innocence.

I review this debate in some detail because I will reflect on the picturesque conventions in Schomburgk's *Twelve Views,* and Michasiw's critique of MacLaren must be answered before I embark on such a project. What is significant about the use of picturesque elements in the work of Schomburgk and other geographical travelers is that they do not fit easily into either of Michasiw's categories. Schomburgk was a traveler, but he was not a harmless tourist. He and other commissioned explorers were aware of the appropriative power of their views and passages, power that resulted from their links to metropolitan institutions. Theirs were appropriative and proprietary passages, involving renaming, landmarking, resource assessment, and the establishment of boundaries and colonial order. They were not merely *representators,* to use Dennis Porter's term for those who made representations of others; these geographical explorers were *representatives* as well—legislators as well as reporters.[58]

Understood in this light, it becomes clear that their use of the picturesque modes could not drift freely as ironic commentary on the contingencies of representation. Even if the images in Schomburgk's *Twelve Views* share Gilpin's preoccupation with "rocks, mountains, and lakes," and even if, like Gilpin, Schomburgk thrills to the Burkean notion of a seductive emotional force of the sublime that lurks in the landscape and

---

56. Belyea, "Captain Franklin in Search of the Picturesque"; Michasiw, "Nine Revisionist Theses."
57. Michasiw, "Nine Revisionist Theses," 84.
58. This distinction and the terms come from Porter, *Haunted Journeys,* 15.

disempowers the spectator, it will still be necessary to recognize that the images deploy European artistic conventions in the assimilation of foreign territory and that the images themselves cannot be separated from the colonial enterprise of codifying alien spaces.[59]

SCHOMBURGK, BENTLEY, AND THE AESTHETIC SATURATION OF SITES

To see how the *Twelve Views* deployed pictorial conventions and aesthetic language to create landmarks, let us examine the images and narrative of the text. My aim is to show that constructing landmarks in the interior of Guiana demanded a process comparable to the *consummatio* of classical rhetoric: a heaping up of multiple arguments to make a single point. Faced with the obligation to construct landmarks in the homogeneous expanses of jungle and savanna, the traverse surveyor deployed an omnivorous aesthetic. The aim was superabundance of aesthetic signification rather than reliance on a particular style. If all conventions could be written onto a point, then perhaps it would not recede into the jungle as the surveyor turned away.[60]

In the text of the book Schomburgk immediately established the coincidence of aesthetic appreciation of landscape, geographical science, and colonial appropriation. He wrote, "It is needless to repeat how I have become acquainted with much that is sublime and beautiful in the interior of Guiana," on expeditions that "had for their object a more extended knowledge of the geography and natural productions of hitherto unvisited regions."[61] While the travels had as their object geographical knowledge, and hence science, Schomburgk concluded the preface by informing his readers of the imperial significance of his geographical labors: "I trust that this publication, and these faithful delineations, may tend to create further attention toward a colony comparatively so little known in Great Britain."[62]

To reinforce the directness and immediacy of his representation, Schomburgk minimized the artistic labor required to produce the book

---

59. Michasiw, although intent on distinguishing Gilpin from the "obscene" and appropriative phase of the picturesque, does not deny that the phases may be entangled, particularly in the colonial context. Michasiw suggests in closing that the two phases may be the "passive and active voices of the same dominative verb." Michasiw, "Nine Revisionist Theses," 96.

60. For a striking account of the significance of a historically weighted object in the interior of Guyana, see Wilson Harris's account of the "anchor" he lost and found on the Potaro. Petersen and Rutherford, "Fossil and Psyche," 186–87.

61. Schomburgk, *Twelve Views,* "Preface," *b* recto. In the following pages I will be citing this text often; if a quotation has no footnote it can be found on the same page as the first note that follows.

62. Ibid.

placed before the reader. "All I have now done," he wrote, "has been to select some of the most striking scenes from the numerous sketches made during my travels in the interior of Guiana."[63] While this is not false, it does somewhat distort the truth: *he* did not draw the sketches that were "made during his travels." In fact, the sketches that formed the first stage in the production of the plates in the *Twelve Views* were done by John Morrison, a draftsman living in the colony about whom little else is known other than that Hilhouse thought him an "execrable dauber."[64] Morrison was briskly shuffled out of the way by Schomburgk, who acknowledged him briefly and explained that the sketches "have been taken under my direction."[65] Morrison did not make it onto the title page, which featured the artist Charles Bentley instead. Bentley had made his debut at the Water Colour Society in 1832 and became an associate exhibitor two years later. Not until 1844 was he made a member, by which time he had established himself as a painter of English coastal scenes.[66]

Bentley put the "artist's finish" on the sketches Schomburgk claimed to have visually "dictated" to Morrison. None of the original sketches has survived, nor have many of Bentley's works in general.[67] Without these for comparison it is difficult to say with certainty in what ways Bentley embellished the images to emphasize their aesthetic character, but even without this evidence a cursory glimpse at the images quickly establishes how closely they conform to the landscape conventions of the period. Although the notions of the "picturesque" and the "sublime" had, by the late 1830s, lost almost all trace of the elaborate and precise meanings assigned them in the eighteenth-century landscape tradition, Schomburgk used the terms with abandon, evoking by them, respectively, a sense of an

63. Ibid.
64. RGS, Correspondence, Hilhouse to RGS, 12 April 1836.
65. Schomburgk, *Twelve Views,* "Preface," *b* recto.
66. Whether Schomburgk settled on Bentley through the agency of one of Charles's kinsmen, "R. Bentley," a member of the RGS, is not entirely clear, but it seems probable. His obituaries in 1854 (when he died within hours of taking ill with cholera) refer to him as a watercolorist of marine subjects, one with bad fortune at the hands of the picture dealers. It appears he died poor. Biography from Redgrave, *Dictionary of Artists of the English School,* s.v. "Bentley," and Ottley, *Biographical and Critical Dictionary of Painters,* s.v. "Bentley."
67. Boase (*Modern English Biography,* vol. 6) mentions that four of his paintings were in the holdings of the South Kensington Museum in 1921. Catalogs of the exhibitions of the Water Colour Society reveal that he sold several works there in his active years (Yale Center for British Art, catalogs). His early work for Ackermann included engraving several plates in 1826 (at the age of twenty) for Grindlay's *Scenery, Costumes and Architecture Chiefly on the Western Side of India.* In volume 1 Bentley executed only "The Fortress of Bowrie," but in volume 2 (published by Smith, Elder, and Cornhill) he did six of the eighteen plates. After the *Twelve Views* Bentley did relatively little work on illustrated travel books, though he prepared a plate ("Crossing by a Sangha") in Roberts's *Hindostan.* Several of his images ("Hull" and "Falmouth") also appear in Tillotson's *Beauties of English Scenery.*

enchanting, pleasing landscape and of an oppressive, frightening one. Bentley, trained in the late 1810s and 1820s, would have been conversant with the more formal aspects of the visual conventions, though in producing a topographical work on commission he would not have seen the manipulation of such effects as his primary responsibility.[68] At the same time, it is clear from the visual works and their accompanying text that the elements of the picturesque and the sublime, even in their somewhat dilapidated later forms, have been manipulated to create the narrative and visual effects of the text.

The artist's task, like that of the author, was to present Guiana as an enchanted land. Schomburgk, waxing poetic on the tropical fertility embodied in the *Victoria regia* (and evoked by the frontispiece), wrote of the Guiana evening: "Thousands of phosphorescent insects flutter among the leaves, emitting a light which, if it does not illuminate, tends to increase the characteristic features of a tropical night, and to realize that idea which the imagination sketches when impressed with the most splendid description of the Arabian tales. Such is the picture which has stamped itself with indelible characters upon my mind."[69]

This Orientalist fantasy of the land was still situated within the comforting landscape triad of the English countryside, for Guiana presented not "jungle" or "pampas" but "glorious scenes of plain, dale, and forest."[70] The first view, that of the Comuti (or Taquiari) Rock (plate 5), confronted readers with the language and visual cues of an enchanted landscape. It was described as a "remarkable pile of granitic boulders" placed above the "highly picturesque" "amphitheatre" of the receding hills. After giving an etymology of the name and situating the landmark by mentioning Charles Waterton's account of (what he called) the "giant of the hill," Schomburgk further underlined the site by recounting the Amerindian explanation of its striking form—it was said to be a petrified water jug. Schomburgk situated that form in the European context too, exclaiming, "What are the famed piles of the Hartz mountains, what the celebrated pedestal of the statue of Peter I, if compared with what I now saw!"[71]

Onto the pedestal of the promontory landmark the reader was invited to step, assuming the commanding stage of a martial statue. Schomburgk's narrative lifted readers, accompanying them on a slippery climb up a ladder of creeping vines to the first "seat" affording a prospect over the whole of the coastal plain of the colony. "So enchanting was this

68. Patrick Moon, curator, Yale Center for British Art, personal communication.
69. Schomburgk, *Twelve Views,* "Frontispiece," B recto.
70. Ibid.
71. This quotation and those above are from Schomburgk, *Twelve Views,* C[3] recto.

view," Schomburgk wrote, "that I was at a loss where to commence in order not to overlook any object in the lovely picture."[72] Readers were invited to search Bentley's illustration with equal care, and in doing so they found an equally enchanted view. A small and seemingly smiling leopard (Bentley clearly had no great ability with quadrupeds) glances out of the Edenic foliage, letting its charmed gaze fall on a parrot perched near the middle of the image. Their locked gaze creates the visual action of the foreground and animates the vegetation-framed vista between them. There is no middle ground at all in the picture, a gesture to the sublimity of the scene: the view over the river and across to the neighboring hills plunges away with dizzying rapidity. The friendly animal images stabilize the view, so that while it may momentarily suggest the vertigo of the climber peeking up over the ledge, the image retains a pastoral feel.

The promontory of the Comuti Rock paled in comparison with that of Ataraipu, or Devil's Rock (plate 6), "another of the geological phenomena which add to the picturesque and magnificent scenery of Guiana."[73] This "natural pyramid" was "one of the greatest natural wonders" in the region, and the site took on a double importance in view of its having been mentioned as a landmark on the Essequibo by Hancock, who placed the camp of Mahanarva near "Toriporo Mountain or the Pyrimidial Rock."[74] Bentley's image here is one of the more successful in the book, featuring a moonlit ledge on which a party of Amerindians have made camp, perched over a lonely, misty valley in the shadow of the jagged tooth of the mountain. A plunging cataract threads its way down the opposite side of the gorge. While the lithography imparts to the image too still and precise a quality to suggest the painterly hand of Salvator Rosa, the scene itself could have been borrowed from one of his canvases.[75] The smoking fire, hunched postures, and feathered nakedness suggest Rosa's *Witches' Sabbath*.[76]

Schomburgk clearly intended this peak—rising 900 feet above the savanna and capped with the bare tooth of 350 feet—to evoke some awe. He describes its "bare head" bursting into sight as the last bit of vegetation fell before the cutlasses of the party. But the fearsome promontory was domesticated, if not by the dreamy gaze of the upright Amerindian gazing out over the valley at the moon in Bentley's image, then by Schomburgk's choice of an epigram: "At a distance of 2 miles the remarkable

72. Ibid.
73. Ibid., D[5] recto.
74. See upper left corner of figure 16.
75. Gilpin considered him the master of "views entirely of a horrid kind." Schama, *Landscape and Memory*, 454.
76. In the British National Gallery.

mass of granite appeared to be one of those eminencies which the poet says, 'like giants stand / to sentinel enchanted land.'"[77]

He got the quotation right in the *Twelve Views*. Someone who was a true devotee of Scott must have corrected the earlier mis-citation in the *Journal*, where Schomburgk's enthusiasm to attach the verses from *The Lady of the Lake* to Ataraipu led him to make "giant" singular (like the mountain) and not plural like the pinnacles around Loch Katrine.[78] The allusion invoked a very familiar enchanted realm: in Scott's poem the lines immediately precede the appearance of the "little skiff" of the poem's heroine, the "guardian Naiad of the strand." The choice of excerpt is suitable in many respects.[79] If Schomburgk's aim was to create a landmark in the Guianas, he could do no better than to attach to the site an image from the romantic work that had so captivated the British.[80]

As human monuments that defy corruption draw those who look upon them into reflections on human destiny and fortunes, contemplation of Ataraipu becomes the occasion for a reverie about the nature of time itself: [81]

> Then the eye again reverted to that monument of unnumbered ages. What changes have occurred since the word which called it into existence was pronounced? Had the earthy substance which probably once surrounded it been washed away by tempests and torrents and left nothing but the column impervious to the tooth of time? Had it risen from the bowels of the earth; the child of that earth's convulsions? Whatever the intermediate cause that may have brought it forth it is a wonderful monument to Him who is Almighty![82]

Remaining aloof from the contention of vulcanist and uniformitarian geological explanations, Schomburgk invokes the conservative premises of natural theology to transform the encounter with the sublime force of nature into a communion with the power of God.[83] It is a posture he adopts with respect to another promontory, the "remarkable basaltic column" called Puré-Piapa (plate 7). The fifty-foot column—"for as such it

---

77. Schomburgk, *Twelve Views*, D[5] recto.

78. Scott, *Lady of the Lake*, 1.14.268–69. Thanks to John Logan at Princeton for assistance on this citation. The line appears in the *OED*, s.v. "sentinel."

79. Scott had a reputation as an evocative poet: "Never, we think, has the analogy between poetry and painting been more strikingly exemplified." Review of *The Lady of the Lake* in *Quarterly Review* (May 1810).

80. See Lockhart, cited by Wilkie Collins in the introduction to Scott, *Lady of the Lake* (Blackwood's English Classics edition). Posthouses in Scotland were crammed with tourists in the wake of publication.

81. See Dettelbach's discussion of Schiller's "significant stones": Dettelbach, "Global Physics and Aesthetic Empire," 75.

82. Schomburgk, *Twelve Views*, 6.

83. For a discussion of the early nineteenth-century use (in geological description) of the eighteenth-century convention of linking a sublime feature to divine power, see Klonk, *Science and the Perception of Nature*, 80–81.

would be taken to be if seen at the distance of a mile, and if the architectural skill of the Indian permitted the idea that it were a work of art"[84]—became the subject of Bentley's fourth view, in which a pair of Amerindians stalk savanna deer from the grove at the base of this ruinlike form. The grassy downs depicted are difficult to reconcile with Schomburgk's description of the arduous trail he took to reach the landmark, a trail of "sharp-pointed rocks, many thirty feet long and scarcely six to eight inches thick."[85] The strange form elicits an excursus on the Amerindian tales associated with such formations, including the account of Puré-Piapa itself, said to be the petrified remains of trees cut in the passage of Makunaima, the mischievous and godlike figure of the Amerindian heroic age. For Schomburgk the site "is no doubt of equal interest to the lover of the picturesque and the geologist." Leaving it to Bentley's balanced composition to satisfy the former with its invocation of vegetation-draped classical ruins, Schomburgk briefly addressed the latter, asking, "Shall we adopt for its origin the theory of elevation, or has the ground which once surrounded it been washed away by tropical rains?" In the end he puts the scientific question aside and closes by drawing over this landmark the presence of his God, as opposed to the Amerindians': "Both theories leave their doubts, while its picturesque appearance and the tropical vegetation which surrounds it, tend only to increase our admiration for the wonderful works of God."[86] The geological history may be shrouded from view, but the presence of Providence is manifest in each landmark, converting bare stones into monuments.

No monument in Guiana could rival Pirara in significance. As discussed in chapter 2, in the *Twelve Views* Schomburgk tells the El Dorado story in detail and in doing so he positions himself as the first explorer to succeed in reaching such sacred ground. Bentley's formalized image (plate 1) bears a structural resemblance to Thomas Girtin's admired *Valley of the Conway* (1800) in that it presents a "low-angle, panoramic response to a deep valley," using a humanized foreground and strong arboreal accent.[87] In Bentley's view a domestic scene occupies the left foreground, in the shade of two vaulting palms. A picturesque hamlet makes up the middle ground, with the tail of smoke and scattered figures suggesting another family gathering. The pink line of the mountains on the horizon provides a glorious backdrop. A path, on which two figures can be seen, links the viewer to the middle ground of the scene, as does the gesture of the Amerindian in the corner, who

84. Schomburgk, *Twelve Views*, F[11] recto.
85. Ibid.
86. Ibid.
87. Hawes, *Presences of Nature*, 58.

seems to point to the village below. The strong horizontal elements of the composition reinforce Schomburgk's assertions concerning the geology of the region, which he claims "leaves little doubt that it was once the bed of an inland lake." The absent inland sea of Parima is read by Schomburgk into the "sea of verdure" that lay between him and the Pakaraima Mountains. Lake Amucu wanders across the view like a chalk district stream.

Schomburgk used Pirara as the prime meridian for all his boundary surveys. However, situating Pirara on the "graticule" of the picturesque and the sublime was a considerably more complicated matter. The true site of El Dorado was suffused with the romantic descriptions of Ralegh as well as the romance of his life. Schomburgk's privileged view over this "penetralia" of South American desire led him to deploy with abandon the language of fantasy. "We stood," he wrote, "on the borders of an enchanted land."[88] In this spirit of romantic flush Schomburgk positioned himself within Bentley's image, more perfectly to evoke both his presence at the site and the picturesque beauty of this inner sanctum of Guiana: "How frequently have I been sitting near those palm trees which we see in the picture occupied by a Macushi family, and allowed my eye to range across the village of motley architecture and the enchanted lake with its verdant isles until it has been arrested by the chain of mountains clothed in bluish tints and the play of extraordinary refractions over a soil exposed to the full influence of a tropical sun."[89]

But Pirara could not so easily be subsumed under the guise of a mountain valley hamlet. To begin with, the very romance of El Dorado was premised on an appalling series of disasters. These could not be excluded from Schomburgk's account, and indeed he catalogs a number of ill-fated expeditions in his text. In doing so he metaphorically sprinkles Pirara with the bones of the "thousands" who were drawn by greed toward the shimmering apparition, an apparition that could only be a "device invented by Satan to lure mankind to destruction." Pirara represents a "waste of human life unparalleled in the history of imaginary schemes."[90] If Satan was the presiding spirit at Pirara, the picturesque village would seem to veil something darker. It is perhaps this that Schomburgk witnessed from that same spot under the blowsy palms:

---

88. Schomburgk, *Twelve Views,* [7] recto.
89. Ibid., 8. This invocation of the character of the refracted light calls to mind Uvedale Price's 1794 *Essay on the Picturesque,* which established the central importance of the "affect" of the picturesque image, an emphasis that, for Price and later his opponent Richard Payne Knight, placed visual *quality* over the depicted *objects* as the defining element of the style. Price called for images that "irritate" the eye by the plays of light. See Belyea, "Captain Franklin in Search of the Picturesque," 5.
90. Schomburgk, *Twelve Views,* [7] recto.

> I shall never forget the splendid spectacle I witnessed one evening after darkness had set in, when toward the north the whole horizon was illuminated: for the grass on the savannas which had been burning for the last four days had communicated the fire to the mountain chain, which now blazed at a distance of many miles. A thunderstorm approaching from the northwest much enhanced the sublimity of the scene, and mingled its forked lightening with the fiery columns which, as if arranged in battle array, seemed to storm the heights of the Sierra, and the vivid lightening and the rolling of the thunder were the batteries employed for the on-set.[91]

Schomburgk's own reference to the sublime leaves no doubt about the character of the scene he invokes, which suggests nothing less than a titanic clash of caliginous forces—the sooty fire and the black clouds dueling with tongues of flame and bolts of lightning. Carter shows how in colonial discourse, particularly that of explorers, "thunder and lightning in particular could be interpreted . . . as a malevolent *genius loci* bent on resisting civilization."[92] Schomburgk's view across Pirara must be read as suffused with those colonial anxieties—all the more so in view of the very real struggle he saw emerge at Pirara, a conflict of colonial interest to which he gave a manifestly Manichaean interpretation.

Schomburgk's depiction in the *Twelve Views* of the Olympian clash of forces over the Pirara savanna must be read in light of conflict between Brazil and Britain over the village. While Bentley's picturesque image captures the idyll of Schomburgk's outbound passage—when he found a thriving Amerindian settlement and a newly arrived missionary—the text describes the tragic later return through the same site, when it had been abandoned as a result of Brazilian depredations. Staying in Pirara the first time, Schomburgk had a foretaste of the disaster to come when he witnessed the arrival of a Brazilian slaving party that had raided the nearby Ursato Mountains. He recounted this scene in the *Twelve Views:* "I saw with the deepest sorrow that the number of those led away into slavery consisted of forty individuals, namely eighteen children under twelve years of age, thirteen women and nine men, of whom only four were less than thirty . . . , the sensation which these cruel proceedings caused among the Indians at the new mission cannot be described."[93] Schomburgk's sorrow was inflected by his implication in the event: he had earlier seen the press-gang on its way upriver and had persuaded his acquaintance, the ranking Brazilian officer at the fort, to deflect their imminent raid from Pirara. When Schomburgk wrote that he had used

91. Ibid., 8.
92. Carter, *Lie of the Land,* 262.
93. Schomburgk, *Twelve Views,* 9.

"some influence in saving the new mission at Pirara from the evil effects and subsequent miseries of a *descimento*," this meant he had, in effect, directed the blow elsewhere.[94]

By May 1839, when Schomburgk arrived back in Pirara after his sojourn into the territories of Venezuela and Brazil, his fears had been realized. He wrote of the village, "We found it occupied by a detachment of Brazilian national guards." The seeds of civilization had been swept away in a dark flood of violence, symbolized best by the inversion of the providential focal point of the settlement: "The church, in which formerly hymns to the praise of our Lord had been sung, and where the first seeds of Christianity had been sown among the benighted Indians, was now converted into a barracks, and the theater of obscene language and nightly revels."[95] Schomburgk used his narrative to subvert Bentley's pastoral image. Reinforcing the desolation of Pirara, stripped of its population and charm, Schomburgk deployed the rhetorical trope of the *sermocinatio*, answering the question put to him by an imagined future traveler:

> The traveller who may pass from the present village of Pirara to the place of embarkation on the rivulet Pirara, will observe a spot which evidently shows that it was once the place of human habitations; but posts, on which the vestiges of fire are observable . . . are all that remain of this once-happy Macushi settlement. His guides will tell him that on one dark night a lawless band of slave hunters from the Rio Branco surprised the poor inmates, and after having set their huts on fire, carried old and young away to die far from their native land in slavery and bondage.[96]

By rhetorically reducing Bentley's scene to rubble, Schomburgk placed the domestic bliss of the picturesque image at risk. The narrative transformed the pastoral image of rural productivity and contentment into a scene of imminent disaster. For Burke, the contemplation of the peril of the innocent and vulnerable was a hallmark of the sublime.[97] The foregrounded child nested in maternal arms becomes a figure symbolizing the urgency of a boundary settlement. Schomburgk ended his text on Pirara with a peroration on the need for territorial delineation: "May the moment soon arrive when the boundaries of the rich and productive colony of British Guiana shall be clearly defined."[98] Citing William Cowper's popular paean to British life, Schomburgk invokes "The Task" Britain

---

94. Ibid.
95. Ibid., 10.
96. Ibid.
97. For a discussion of vulnerability and the sublime in the colonial context, see MacLaren, "Samuel Hearne and Landscapes of Discovery."
98. Schomburgk, *Twelve Views*, 10.

faces at Pirara.[99] When the boundary is set: "Where Britain's power is felt / Mankind will feel her blessings too."[100]

Nowhere else in the *Twelve Views* does the imperial program become as explicit, but the figure of the storm as the malevolent genius loci returns with particular force in the section on the mountain called Roraima (plate 8). Bentley's illustration follows the structure of the depiction of Pirara almost exactly: a slight elevation, visually weighted on the left by a prominent stand of palms and a domestic scene of Amerindians. Again, one of the Amerindians lies recumbent, emphasizing a certain languorous ease; again, the three-part horizontal composition is strongly marked. The crowning layer in this image, however, is not the gentle hills of the Pakaraima but rather the rugged and vertical walls of Roraima's summit. The scene appears to be dusk, since the western horizon glows with particular intensity, but the light has been carefully manipulated to allow the striking form to be *both* backlit *and* bathed in a pinkish glow, as if the mountain were facing a bright sunrise to the east. Schomburgk concedes that such a view, if it was had at all, was very brief. Within moments of their first view of the much-rumored range, "dark opaque clouds hovered around their summits, which, chased by the morning breeze, produced sudden changes of light and shade on the mural precipices, that they appeared perpetually under new shapes and colours." This scene of fantasy was short-lived, however, for the clouds prevailed, and Schomburgk reported eight days of bad weather.[101]

No picturesque view across the valley at these strange mountains could rival the exhilaration of an approach to their base, which Schomburgk described as a descent into a cold and troglodytic realm: "Large trees, rooted in clefts and overhanging the glens, added to the somber character of the scene" as the party advanced to the base of one of the soaring cataracts (Kamaiba) that descended from the top of the 1,500 foot wall. The scene afforded "on a large scale, that which the spring of the Brocken in the Hartz mountains offers in miniature."[102] The approach to the base of the mountains, represented by Bentley as a fairy-pink citadel, took Schomburgk and his narrative into a dank cul-de-sac: "An oppressive solitude prevailed; there was no sign of animal life, only the noise of falling waters was heard, which served as a guide to direct our steps hither." As proximity to the mountain grew, so too did the apparent rage of the forces of nature. On the threshold of the fall, "the thunder clouds which had been threatening, passed the mountains and

---

99. On the poem, see Priestman, *Cowper's Task*, and Feingold, *Nature and Society*, 121–54.
100. Schomburgk, *Twelve Views*, 10.
101. Ibid., 14.
102. Ibid., 15.

enveloped us almost in darkness: the rain fell in torrents and thunder and wind appeared to vie with the cataract in producing the greatest uproar."[103]

In the midst of this scene, "the forest opened, and as if it had been called forth by magic, a perpendicular wall stood before us, from which the Kamaiba, swelled by the torrents of rain, precipitated itself with thundering noise into a spacious basin below." What had been the telos of the journey cast out the travelers just moments after they arrived:

> The summit of the wall was perfectly hidden from us; even the cliff opposite the one on which we stood was seen occasionally as through a veil, illuminated by vivid flashes of lightening. Numerous blocks, apparently torn from these gigantic walls, were lying in great confusion around, conveying the possibility that a similar accident might now occur, an idea which was strengthened by the uproar of the elements, and the idea of being near to these cliffs was so fully impressed upon me that, instead of enjoying the romantic scene, I felt oppressed and a wish to escape from it.[104]

The symbolic significance of the storm at the falls is reinforced by Schomburgk's description of the retreat: no sooner had the party cleared the thick jungle at the foot of the cliff than they "observed a sunny landscape" and thus were obliged to acknowledge that "the thunder storm was perfectly local." The storm was exactly a *genius loci* whose powers were firmly rooted at the base of the wall. As if to underline the dark heart of the landmark, Schomburgk quoted the burden of the Amerindians who chanted into the night outside his tent: "Of Roraima, the red-rocked, I sing, where with daybreak, night still prevails."[105]

The sublimity of the foot of Roraima, and the encounter with the malevolent *genius loci* of the storm, is entirely absent from Bentley's view. In place of any suggestion of the "cauldron, with water foaming and bubbling within it with uproarious noise" that Schomburgk discovered in the maw of the cataract, Bentley has chosen to emphasize the classical forms that Schomburgk invoked in his attempt to describe Roraima. Just as each promontory thus far in the *Twelve Views* has been described with the metaphorical language appropriate to a classical ruin—the "column" of Puré-Piapa, the "pyramid" of Ataraipu, and the "amphitheatre" of the Comuti Rock—so too did Roraima and its neighbors receive such treatment. Schomburgk wrote: "They convey the idea of vast buildings, and might be called nature's *forum*." Extending the analogy, they become the ruins of a colossal colosseum: "Associating them with those splendid remains of man's gigantic conception and ex-

103. Ibid.
104. Ibid.
105. Ibid.

ecution, we can imagine what the forum would have been, if its columns had been raised to a height of one thousand five hundred feet, and if it had covered an extent of ten miles."[106] Closer examination of Bentley's illustration, as well as an accompanying woodcut made from a different perspective, shows that he, the colorists, and the engravers have emphasized this aspect. Bentley would have known the work of his predecessor in the Water Colour Society, John "Warwick" Smith, who died the year before Bentley first exhibited there. Smith's five-year residence in Rome resulted in a large number of Colosseum images, several of which were exhibited at the society in Bentley's lifetime.[107] Not only did the artists' brick tones and engulfing vegetation suggest contemporary images of the Colosseum, but Bentley emphasized the mountain's form by means of a parallel with the fortified Amerindian encampment in the left middle ground of the lithograph. This camp is colored and shaped to echo the forumlike range.

Ruins did more than link such images to the picturesque conventions of Gilpin's "judicious hammer" (his catch phrase for the artist's aesthetic use of ruination). In the context of Schomburgk's landmark construction, visual and textual allusions to ruins were part of the effort to root the interior with some form of history (see chapter 2).[108] The apparent historical "shallowness" of the Americas in general (and of the Guianas in particular) constituted an obstacle both to portraying the place as picturesque and to anchoring landmarks in multiple frames of reference. A geological curiosity was a geological curiosity, but unless it could be aligned with the history of previous European passages or Amerindian traditions, or failing that, at least some geological history, the site itself possessed no depth. It lacked the overdetermined quality of a "natural" landmark. Without multiple significances, a rock remained a rock. Without architectural ruins of greater magnitude than collapsed banabs (huts), Guiana called on the explorer and artist alike to *see* in terms of ruins.

Nowhere is this ruin-mindedness more apparent than in the view of Esmeralda in the *Twelve Views* (plate 9). Bentley structured this image, which depicts the site of the sacred consummation of Schomburgk's Humboldtian identity, according to the same conventions as the views of Roraima and Pirara. Again the Amerindian family occupies the foreground in the shadow of the requisite palms. Very much like the image of Pirara, the hamlet is situated in a broad and flat middle ground, linked by

---

106. Ibid., 16. Emphasis mine.
107. Hawes, *Presences of Nature,* 146.
108. For a discussion of the place of ruins in efforts to restore lost golden ages, see Jackson, *Necessity for Ruins,* 101–2.

a path to the domestic scene on the left. What stands out in the image, however, are the unmistakable forms of Aztec ziggurats that dominate the mission huts and are dominated in turn by the brooding form of Mount Duida.[109]

It takes close examination for the viewer to realize that these Aztec ruins are not ruins at all but merely very curious *natural* formations. Lest they be missed, Schomburgk drew these "grotesque forms" to readers' attention, pointing out that they appeared "like vast edifices in ruins." Esmeralda, despite its status as a mini-El Dorado (its name derived from a chimerical wealth of emeralds), received a half-hearted romantic treatment by Schomburgk, who confessed that the torments of mosquitoes militated against the enjoyment of the "romantic scenery and situation," leaving the explorer with a sense of the mission as a place "of proscription and chastisement."[110] Nor does Bentley's view dispel this image. The view of Esmeralda is unique in the *Twelve Views* in depicting a gathering storm. Gray thunderheads shroud the left side of Mount Duida and give Esmeralda a dreary, if not ominous, appearance. The path from the picturesque promontory of Comuti Rock—the first of the *Twelve Views*—to the looming storm over Duida—the sixth—suggests a long aesthetic journey.

AESTHETIC CIRCUITS

If Esmeralda presents the bleakest view of the twelve, it is significant that as the sixth it represents the midway point in Schomburgk's circuit. Esmeralda then, figuratively, if not in terms of kilometers, embodies the point at which Schomburgk "turns around" and begins heading back "home," back to *British* Guiana. In reality, of course, his path out of Esmeralda took him farther to the southwest and hence farther from Georgetown, but from Esmeralda forward he would be within the colonial context of Venezuelan and Brazilian missions and forts. Whereas to reach Esmeralda from Pirara called for Schomburgk to cross "seven hundred miles of country not yet trodden by any European, as far as we know,

---

109. The stones look very much like the pyramid of Cholula, which lies in the shadow of Chimborazo in the frontispiece of Humboldt's *Atlas géographique et physique du nouveau continent*. It is also likely that Schomburgk was familiar with the work of Lloyd Stephens and Frederick Catherwood, who were at the time revealing to American and European readers the extensive Mayan ruins of Chiapas and the Yucatán. Though their major publication was several years away, it is certain that Schomburgk knew Catherwood from an early date. In 1845 Schomburgk, then on the board of directors of the heavily capitalized Demerara East Coast Railway, recommended Catherwood to be the principal engineer of what was to be the first railway in South America. See Von Hagen, *Frederick Catherwood*, chap. 10; Von Hagen, *Search for the Maya*, chap. 17.

110. This quotation and those above are from Schomburgk, *Twelve Views*, 22.

and to suffer privations and fatigues of every description," the passage from Esmeralda down the Casiquiare to the Rio Negro and back up the Rio Branco posed a very different problem.[111] In consulting with a local at Esmeralda about his route, Schomburgk fretted that there appeared to be no way to avoid the Casiquiare and the Rio Negro. This was bad not because of hostile Amerindians or difficult terrain, but rather because "we had no decent clothes left in which to appear among civilized people, neither had we any money, which on the Rio Negro would be requisite in order to buy what we wanted for our sustenance."[112]

Schomburgk, to his chagrin, had to take what amounted to the "civilized" route home, traveling through mission villages in full explorer's kit, bivouacking somewhat ridiculously at populous settlements on the way. The *Twelve Views* largely omits this portion of the journey, offering only a glum view outside the deforested settlement of Fort São Gabriel, which Schomburgk could describe only as a "melancholy scene" and a "picture of utter desolation" (plate 10).[113] Within the narrative of the scene from São Gabriel, Schomburgk brings himself home, summing up the arrival back at São Joaquim and shortly thereafter at Georgetown. From São Joaquim to Esmeralda had taken five and a half months; Esmeralda to São Joaquim via the Casiquiare and Rio Negro, fifty days.

That Esmeralda and Fort São Joaquim constituted the desolate and melancholy extreme of the aesthetic circuit taken in the *Twelve Views* recapitulates at the scale of Schomburgk's entire circuit a pattern noticeable in the text for each view. MacLaren has asserted that "during a single season of surveying Vancouver and his men had run the gamut of eighteenth-century landscape aesthetics" and that later nineteenth-century explorers of Canada repeated this pattern—wandering from picturesque scenes to sublime terrors before making their way back.[114] Schomburgk presented his journey in a similar light. Esmeralda, the "apex" of his expedition from the geographical standpoint, presents a stormy face in Bentley's view. The lone cross on the hill has a forlorn quality, overshadowed by the silhouette of Duida. None of the other *Twelve Views* is as bleak. The images of the deepest interior—where Schomburgk has "crossed over" into the dark realms of former Iberian possessions—are the *least* picturesque. The final images of the *Twelve Views* depict domestic scenes of Amerindian village life and hence constitute a "return" to the picturesque.

111. Ibid., G[21] recto.
112. *JRGS* 10 (1841): 244.
113. Schomburgk, *Twelve Views*, 22.
114. MacLaren, "Aesthetic Mapping of Nature," 41.

What is particularly interesting about the *Twelve Views* is that this same trajectory—picturesque to sublime and back—maps not only the circuit through Guiana, from Guiana *British* to Guiana *Iberian,* but also the structure of Schomburgk's narrative *at each site.* The Comuti Rock at first presents a picturesque appearance, nestled in its "amphitheatre." Not until the climb up the slippery lianas does its treacherous character become apparent. But ultimately the glorious sunset from its "pedestal," like Bentley's view, leaves no doubt about the serenity and enchantment of the scene. A similar reading could be offered for each landmark. The column of Puré-Piapa, from a distance so like the column of Trajan, retains its delightful charm until the explorer is upon it. Then, no sooner does he observe the nest of an elegant stork at its summit than a shot rings out, and this delicate maternal figure teeters on the lip of the nest and falls, victim of the overeager marksmanship of an Amerindian escort. But this tragic scene is soon forgotten in a reverie inspired by Schomburgk's conjectural history of the curious monument. Nowhere is the pattern more strongly marked than in his description of "noble" Roraima: the early morning's first glimpse is bathed in a palette of pastel hues, but the nearer approach becomes a descent into a black cauldron buffeted by boulders seemingly cast down from the heavens. No sooner does the party depart than the skies clear and Roraima becomes the picturesque ruins of an Olympian colosseum.

To the degree that Schomburgk succeeds in hanging on every landmark the language of both the picturesque *and* the sublime, he succeeds in saturating each site with aesthetic references. The landmarks are places of manifold significance: each site is layered with an exuberance of aesthetic associations; each evokes the full gamut of sentimental experience. What at one moment has been couched in the enchanted language of the picturesque—suggesting leisurely habitation, familiar landscape forms, inviting plays of light, and fertile vegetation—is at the next moment hastily overwritten with the weight of the sublime, replete with its intimation of peril and invocation of forces beyond human control and without regard for human welfare.[115] This aesthetic overdetermination served as an essential aspect of landmark construction in the interior. Obliged to construct landmarks in the homogeneous expanses of jungle and savanna, the traverse surveyor marshaled multiple meanings onto single sites: the more the better. By invoking both the sublime and the picturesque and concatenating aesthetic, historical, and geological signifi-

---

115. Hugh Ridley points out that "facile sublimity" was avoided by colonial writers, who "wanted to show themselves to have been shaped by the experience of the colonial landscape." Ridley, *Images of Imperial Rule,* 63.

cance, the explorer performed a geographical *consummatio,* heaping meaning onto each landmark in the hopes that it would endure.

FROM LANDMARK TO ICON

The success of Schomburgk's efforts, aided by Bentley, can be gauged by the enduring legacy of their "views." They succeeded in constructing landmarks that, through cycles of borrowing and reproduction, became icons of the interior of the colony. To begin with, we might look at the *Victoria regia* itself, the lily depicted on the frontispiece and cover of the *Twelve Views*.[116] While it might be argued that a flower is hardly a landmark, in that it lacks the fixity of a geological formation, the *Victoria regia* was, in fact, fixed on maps of the colony.

Looking closely at the RGS key map that serves as the cartographic field for the *Twelve Views,* we see that the frontispiece image of the *Victoria regia,* though without coordinates, has indeed been placed on the map. The upper Berbice is marked with one of the colored indicators that signal a point depicted in one of the views. The point is titled "*Victoria regia* discovered" (plate 11). Through this treatment, the frontispiece image becomes a stylized depiction of the moment of discovery of the flower itself, a point reinforced by the accompanying text that describes that moment. Dogged by the disappointments of the journey up the Berbice, the explorer recalls arriving "at a point where the river expanded and formed a smooth basin" (not unlike the frontispiece). Spying a white speck on the eastern bank, Schomburgk—adding drama to an ineluctably static scene—urges the crew on and seizes the immobile flower, this "vegetable wonder" of the New World.[117] I have already discussed the "staging" of this scene in connection with Schomburgk's rocky relationship with his sponsors in 1837 (chapter 3). To establish how the flower became an imperial icon of British presence in Guiana—and how its blossom bound the explorer to the queen just as its taproot bound the queen to her territory—I turn to how *Victoria regia* took that name and became a Victorian sensation.

When, in 1851, Walter Fitch and Sir William Hooker joined to publish an elephant folio pamphlet titled *Notes on Victoria Regia, or Illustrations of the Royal Water-Lily in a Series of Figures Chiefly Made from Specimens Flowering at Syon and Kew,* they labored their way through the roll call of disputants with claims to priority. The naturalist Tadeás Haenke, explorer in the service of Spain (none of whose private papers survived his

---

116. The modern name of the plant is *Victoria amazonica.*
117. Schomburgk, *Twelve Views,* "Frontispiece," B recto.

demise), narrated an encounter with the flower to a Peruvian mission priest at the end of the eighteenth century. Aimé Bonpland had written from Montevideo of his dismay on seeing what he considered *his* flower titled as *Victoria regia* on the stylish European fans of the *blanquitas* of Rio Pardo. Dr. Eduard Pöppig's 1832 specimen of *Euryale amazonica* (as he called it) might still be floating in alcohol at one of Europe's botanical gardens, but it had not turned up. Alcide d'Orbigny, who had actually deposited specimens at the Jardin des Plantes about 1834, not only believed the flower was a *Euryale* (though he confessed that nothing remained of his sample but a piece of leaf, "un peu détériorée") but thought Schomburgk's maneuver a botanical heist of the first order.

In the autumn of 1837 popular publications in Britain had announced Schomburgk's discovery and dedication.[118] Unfortunately for botanists, they did so under a variety of names. If the genus *Victoria* was easy, there seemed to be considerable confusion about whether the species name was *regia, regina,* or *regalis.* In one place it was announced differently in the text and the plate caption;[119] in another it was listed differently in the text and the index.[120] In the end, putting aside the other claimants, Hooker ruled for Schomburgk's propriety as primary discoverer, on the grounds that "without his drawings, descriptions and observations, and strenuous endeavors commenced by him to encourage the introduction of the living plant to English Gardens, we should still have been very much in the dark about it."[121] As for the name, Hooker decided that *Victoria regia* had priority, having been used in the twenty-five imperial folio pamphlets prepared privately by John Lindley, fellow of both the RGS and the Royal Society, to whom the RGS gave Schomburgk's botanical correspondence and sketches.[122] An exemplar of this volume, featuring a colored quarter-scale illustration (plate 2), was presented to the queen in November 1837. In it Lindley recalled first suspecting that the flower was merely a species of *Nymphea,* but closer investigation converted him: here was a new

118. *Athenaeum,* 9 September 1837; *Gardener's Magazine,* October 1837; *Magazine of Zoology and Botany,* October 1837.
119. *Magazine of Zoology and Botany,* November 1838. Hooker, copying Gray (*Annals and Magazine of Natural History,* 2d ser., 6, no. 32 [1850]: 146–47), wrongly gives the year of this publication as 1837.
120. *Athenaeum,* 9 September 1837, 661.
121. Hooker, *Notes on Victoria Regia,* 6.
122. This was a change of heart for Hooker, who had advertised his forthcoming *Notes* in *Hooker's Journal of Botany and Kew Garden Miscellany* 2, no. 21 (1850) as "The Victoria Regina." This was the species name used by John Edward Gray, of the Botanical Society, in the *Athenaeum,* 9 September 1837. The reason this earlier publication was deemed to have been preempted by Lindley's *later* private pamphlet lay in the analysis of the minute books of the RGS, where it was recognized for the first time that the RGS gave Schomburgk's materials to Lindley before sending them to Gray at the Botanical Society. See *Annals and Magazine of Natural History,* 2d ser., 6, no. 36 (1850): 491–94. Gray was also a member of the RGS. See Secord, "Extraordinary Experiment," 353.

genus (which accorded well with offering the flower to Her Majesty, since it was poor form to lower royal sobriquets to the status of mere species).[123] He composed a notice to accompany the illustration: "It appears to me that the object of the discoverer will be best attained by suppressing the name of *Nymphea victoria* under which he [Schomburgk] had proposed to distinguish the plant, and by embodying Her Majesty's name in the usual way, in that of a genus. I propose to call it *Victoria regia*."[124]

Word of Lindley's private publication quickly got around to newspapers; when the substance of the memoir was reprinted in the "Miscellaneous Notices" of the *Botanical Register* in February 1838, a note was appended referring to "the great interest having been excited by stories told in newspapers of this extraordinary flower."[125] Whole issues of several botanical magazines were given over to the flower.[126] Not until 1849 would anyone in Britain get the opportunity to see it, however, since none of Schomburgk's attempts to bring back shoots or seeds succeeded.[127] For ten years various travelers and colonists tried to "repatriate" a living sprig of the plant without success; Wardian cases opened at Kew stank of rot. Not until 1849, when Hugh Rodie (better known for his work on the greenheart tree, *Ocotea rodie*) abandoned the practice of packaging seeds in mud or soil (and placed them in vials of pure river water) did seeds survive the mail and germinate in Britain. These were quickly distributed to the "tropical propagation houses" of Chatsworth (the estate of the duke of Devonshire), Syon (the estate of the duke of Northumberland), and Kew Gardens.[128] In these houses, where coal stoves gently heated the underside of the tanks, the plants flowered with startling rapidity, stated to be the fastest of any plant recorded, going from a seed to a spread of nearly twenty by twenty feet in seventy-nine days.[129] The duke of Devonshire had the first flower and Kew's came last, though one of Kew's first was cut, boxed in a specially prepared case "large enough to allow its full expansion,"[130] and swept down the Thames to the queen by commissioned steamer. She was said to be delighted with her encounter with a fresh specimen.

123. Note that Schomburgk had once before gotten himself entangled in a maladroit effort to seek patronage with a dedication: he had tried to dedicate his first public map, of Anegada, to the English king. See chapter 3, note 31.

124. Cited in Hooker, *Notes on Victoria Regia*, 10.

125. Ibid.

126. See in particular *Curtis's Botanical Magazine*, 3d ser., 3 (January 1847 [misprinted on first page as 1846]).

127. Though Schomburgk put a batch of (as it turned out, dead) seeds up for sale in 1841. He had purchased the seeds from a local boatman in Berbice. See *Floricultural Cabinet and Florists' Magazine* 9, no. 99 (January–February 1841): 119–20.

128. Figure 19 has been reproduced from *Gardener's Chronicle and Architectural Gazette* 35 (31 August 1850): 548–49.

129. Hooker, *Notes on Victoria Regia*, 11.

130. Ibid., 14.

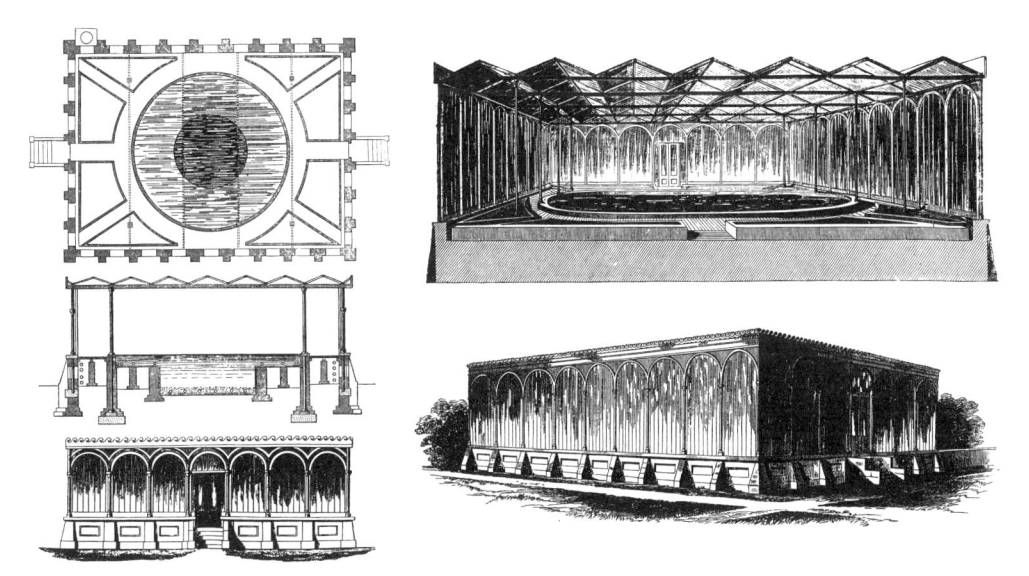

Figure 19. The *Victoria regia* house at Chatsworth: Paxton's design, a forerunner of the Crystal Palace.

As were tens of thousands of Britons. It is not easy to characterize the full extent of Victorian infatuation with Schomburgk's "vegetable wonder."[131] The flower and the tale of its romantic provenance blossomed in Britain amid a garden-obsessed people whose fascination with flowering plants was outstripped only, and that briefly, by their fascination with aquatic life and aquariums.[132] The *Victoria regia* sent its rhizomatous roots into fertile terrain in its new environment: it needed a vast aquarium and a greenhouse as well, the Victorians' two most fetishized spaces for the propagation and domestication of the natural world. The private gardens of Chatsworth and Syon opened their doors to the public at appointed times, and Kew saw a steady stream of visitors. By 1851, a mere eighteen months after the first germinations, Hooker estimated that tens of thousands of Britons had seen the flower. The American press kept pace and called for funds to be deployed for erecting *Victoria regia* propagation houses, in the hopes of "seeing its superb blossoms expand in this country."[133] At Chatsworth a "Victoria house" (fig. 19) was erected to meet the special needs of the flower, which thrived only in lightly circulating

---

131. For a taste of the enthusiasm, see *Floricultural Cabinet and Florists' Magazine* 16, no. 15 (March 1848): E[1] verso (with colored plate). An announcement of the first flower in Britain can be found in the same journal the following year: 17, no. 36 (December 1849): 308–9.

132. Rehbock, "Victorian Aquarium in Ecological and Social Perspective"; Woods and Warren, *Glass Houses*.

133. "The New Water Lily," *Horticulturist, Journal of Rural Art and Rural Taste* 5, no. 6 (December 1850): 275. See also "Mr. Downing's Letter from England," *Horticulturist, Journal of Rural Art and Rural Taste* 5, no. 5 (November 1850): 217–24.

soft water at high temperatures.[134] Kew planned a new house with two tanks to meet the same purpose.[135]

Joseph Paxton—gardener, botanist, imaginative structural artist, and the duke of Devonshire's much envied director of works—had his hands in much of this, engineering glasshouses that became *Victoria*'s English homes. He coaxed that first British blossom out of the pond he oversaw at Chatsworth, and later he delighted visitors by placing his daughter ("42 pounds") on a large leaf, which "bore the weight extremely well" (fig. 20).[136] The incident inspired Douglas Jerrold to versify:

> On unbent leaf in fairy guise,
> Reflected in the water,
> Beloved, admired by hearts and eyes,
> Stands Annie, Paxton's daughter.[137]

Two additional verses played out the metaphorical significance of the moment: she is unweighted by cares, and flowers crowd to her feet. This fairy scene became a powerful icon when Paxton rose from being a respected jack-of-all-trades in the world of society gardens to become a national hero: in May 1851 the Crystal Palace opened, the centerpiece of the Great Exhibition and a staggering seventeen-acre structure of glass and iron, and just about everyone in Britain knew that its design was Paxton's—dashingly sketched on a sheet of blotting paper in the boardroom of the Midland Railway in June 1850. This romantic tale, of a "gardener's boy" triumphing over the greatest architects and engineers of Europe, became still more irresistible when in November 1850 Paxton offered an account of his inspiration for the design to the Fine Arts Society: what he showed them was the structural ribbing on the underside of the massive leaf of *Victoria regia*. "Nature was the engineer," he declared, adding, "Nature has provided the leaf with longitudinal and transverse girders and supports that I, borrowing from it, have adopted in this building."[138] The flower itself, not surprisingly, was a centerpiece of the exhibition, the flower within the flower, the seed of genius.[139]

In the wake of the Great Exhibition, a clique of Victorian poets settled on the flower as the very *form* of the royal dedication, titling their col-

---

134. For a description of the stove and specially designed waterwheel used at Chatsworth, see "Mr. Downing's Letter from England," note 133 above.

135. Hooker, *Notes on Victoria Regia*, 18.

136. Ibid., 17. Figure 20 has been reproduced from "Mr. Downing's Letter from England" (see note 133 above). A similar image appeared in the *Illustrated London News*. Later, adults were photographed on the leaf too, but these were trick shots, the leaf supported from underneath. Postcards of such images are currently available at Kew.

137. Violet R. Markham, *Paxton and the Bachelor Duke* (London: Hodder and Stoughton, 1935), 181.

138. Quoted in ibid., 182. For a more recent treatment of this story, see John McKean, *Crystal Palace: Joseph Paxton and Charles Fox* (London: Phaidon, 1994).

139. Hooker, *Notes on Victoria Regia*, 13.

Figure 20. The sensation of *Victoria regia* on show. Reproduced from volume 5 of *The Horticulturist*, 1850; courtesy of the New York Public Library.

lected works after the flower and composing a dedicatory prologue that underlined the romance of the lily's origins. A certain elision of Schomburgk's nationality proved necessary:

> When on the shining waters of the west
> An English traveller saw the Queen of Flowers
> He sought a name whereby might be expressed
> The chiefest glory of this world of ours
> *Victoria regia*—Never happier name
> A flower, a woman or a Queen could claim.[140]

Nor was the dedication complete without a picture of the flower, a picture that matches Bentley's from the frontispiece of the *Twelve Views* in considerable detail.[141] It is impossible to say for certain, of course, that Bentley's image was copied. In fact, given the ubiquitous depictions of the lily in the 1850s and the small edition of the *Twelve Views*,[142] it is more

---

140. From the dedication of Procter, *Victoria Regia*.
141. Bentley did not depict the "scales" of the leaves as radiating from the center as those in the 1861 image do, but the composition of leaves, buds, and flowers is nearly identical.
142. Likely five hundred or fewer, given the number of subscribers and the standard run for Ackermann's editions at the time. Ford, *Ackermann*, 123.

likely that it was not. But the picture of the *Victoria regia* in the *Twelve Views* was one of the earliest printed images of the flower available to the garden-owning elites who would come to dominate its cultivation and hence initiate the popular fascination. The volume, after all, was dedicated to the duke of Devonshire, who became a key propagator of the *Victoria regia* craze.

In this respect the view of the flower in the *Twelve Views* must be seen to lie at the origin of a long chain of visual citations that served to establish the *Victoria regia* as an emblem not only of imperial splendor, but of tropical fertility as well. Its record-setting growth rate and phenomenal girth made the plant a symbol of the agricultural productivity possible in the South American colony. Schomburgk could not help but adduce his prided flower as evidence of the "vigour and luxuriance" of the colony's "vegetable productions," writing: "What could better give an idea of the luxuriance and richness of vegetation in Guiana, than the splendid *Victoria regia*, the most beautiful specimen of flora in the western hemisphere?"[143]

As much as the flower could be construed as, in Lindley's felicitous phrase, "embodying" Her Majesty—in its ivory purity on opening as well as in its regal place in the flora of the Western world—the same flower embodied the colony itself, both in beauty and in productivity. Thus the flower served as a delightful conjunction of state and head of state: it was the crown of Guiana's Crown lands and the crown of Guiana's crowned queen as well. As an icon of the colony the flower served as a floral metonymy for the territory: it was a figure of the colony that could be given to the queen and live in a special house in her garden.

As a colonial landmark, the flower existed in a special way at the point of Schomburgk's discovery, even as it was increasingly recognized to be nearly ubiquitous in the colony and across much of the rest of lowland South America. The RGS map in the *Twelve Views*, which placed Schomburgk's encounter at its precise geographical coordinates, was doubtless the source of inspiration for William Downing, who in preparing his cadastral map of Berbice in 1844 placed an engraving of "the Victoria Regia of Schomburgk" at the top. Conveniently, that coincided with the uppermost portion of the river, the very point of Schomburgk's discovery. Downing's image (fig. 21), complete with the background hill, will be quickly recognized as a considerable enlargement of Bentley's depiction of the flower in the *Twelve Views*.[144] Nor did the realization that the flower was not endemic (unlike many species in Guiana) diminish its symbolic

---

143. Schomburgk, *Description of British Guiana*, 30.
144. Figure 21 has been reproduced from RGS, Map MS, "British Guiana."

Figure 21. Downing's *Victoria regia:* the flower fixed on the map. Courtesy of the Picture Library of the Royal Geographical Society.

importance. In fact, an apocryphal story emerged in the late nineteenth century that the true significance of the flower's discovery lay in its proving the unity of the river systems of Guayana and Amazonia, "which was at one time so strenuously denied."[145] Under this interpretation, *Victoria regia* was reinvented, becoming for Schomburgk what the Casiquiare canal had been for Humboldt: proof of the incongruous hydrography of the Americas. Moreover, it was easier to transport than the canal, and considerably better looking.

The *Victoria regia* thus had a curious status as a landmark. It was written on the colony as a point, even as it was recognized to be choking up canals over much of tropical South America. This profusion did not diminish its significance as an icon of the colony and of the colony's link to the empire, nor did it change the fact that the flower, in addition to sugar, came to be many Britons' primary association with their only colony in South America. Schomburgk (and Bentley) set this process in motion.

As they did with other icons of the colony's interior. Several of the sites depicted in the *Twelve Views* passed through similar chains of reiteration on their way to becoming the predominant symbols for British Guiana. Ataraipu and Roraima serve as good examples. Not only did small wood

---

145. Webber, *Centenary History,* 189.

engravings very similar to Bentley's views of these two sites appear in the *Journal* in 1840 (as well as in the German edition of Schomburgk's *Travels*),[146] they also found their way into a number of later colonial publications. When George Bennett prepared his *Illustrated History of British Guiana* (published in Georgetown in 1866), he confessed that he had "initially wanted to reproduce Schomburgk's 'Interior of Guiana'" (meaning the *Twelve Views*).[147] Instead he produced a pastiche taken from a variety of authors, but images from the *Twelve Views* still dominated the visual material of the book: ten of the twelve were presented, each with its associated longitude and latitude. The photographic reproductions were clumsy, owing to their much-reduced format and the printing in black and white of the original hand-colored images.

The Reverend William Brett, likely the best-known nineteenth-century missionary in the colony, avoided the same pitfall when he set to work preparing his popular *Indian Tribes of Guiana,* published in London in 1868 with a number of colored prints, all lithographs. These were put on stone by the M. and N. Hanhart press after sketches done by Brett from "Schomburgk's" originals (fig. 22). Bentley was not acknowledged. Ataraipu was reproduced without appreciable change (the hammock in the right corner was emptied), but Brett redrew Roraima considerably (fig. 23), in keeping with his focus on the indigenous tribes. Bentley's tight and marginal humanized foreground has been centered and amplified. Four figures in classical poses have been replaced by a crowded domestic scene where nine Amerindians and their two dogs settle down after bringing home a pile of game. The strange colosseum-like village that had echoed the background in Bentley's view has been left out, and in its place rise a pair of more conventional banabs.

The landmark images of the *Twelve Views* proved durable. As late as 1931, the colony's *Centennial Handbook* reproduced one of the images from the *Twelve Views* (that of Pirara). For the most part, however, travelers who made trips through the interior of British Guiana later in the nineteenth century—men like Sawkins, Brown, Im Thurn, and Boddham-Whetham—chose to illustrate their own publications. This does not mean, however, that they very often invented or discovered their own landmarks. In the metaleptic cycles discussed in chapter 2, explorers were drawn to—and fixed, drew, versified, photographed, and narrated in turn—the sites Schomburgk had established as the landmarks of the interior.

146. Though it is not possible to say for certain that one was copied from the other, since they could both have been prepared from a third and earlier sketch, possibly that by Morrison himself.

147. Bennett, *Illustrated History of British Guiana,* "Preface."

Figure 22.   Brett's visual citations from the *Twelve Views*. Reproduced from the 1868 *Indian Tribes of Guiana*.

Figure 23. Brett's Roraima, redrawn from the *Twelve Views*. Reproduced from the 1868 *Indian Tribes of Guiana*.

The mysterious summit of Roraima received broad public attention in the 1870s in the wake of an article in the April 1877 *Spectator* that pleaded on behalf of Schomburgk's mountain, "Will no one explore Roraima, and bring us back the tidings which it has been waiting these thousands of years to give us?"[148] At least one leisured man of adventure, John Boddam-Whetham, thrilled to the call and embarked the following year, returning to publish a tome of 363 pages, illustrated with engravings and a route map, detailing the author's approach to the foot of this "sermon in stone" that turned its "weird and solemn" back on the climbing party.[149] In the wake of the publication of *Roraima and British Guiana*, Roraima became the object of a set of ambitious climbing efforts, but the much-discussed scheme of making an assault by balloon proved unnecessary. In December 1884 the Oxford-educated anthropologist Everard Im Thurn found his way to the top via the (relatively) easy approach from the southwest, an event heralded at the British Association and the Scottish Geographical society, among other places.[150]

148. Cited in Boddam-Whetham, *Roraima and British Guiana*, v.
149. Ibid., 243.
150. Im Thurn had been made a curator of the Guiana Museum through the patronage of Joseph Hooker. See *Proceedings of the Royal Geographical Society* 7, no. 8 (1885): 534.

Im Thurn, whose fame was inseparable from his conquering this icon of the "forgotten world," was loath to undermine any of the mystique surrounding the mountain. In fact, he reiterated a supposed Amerindian tale that looks suspiciously like a colonial fantasy of visibility and conquest. "The Indians vow," he declared, "that when white men are about it—the mountain is enveloped from base to summit in the densest mist."[151] However, while anxious to insist on Roraima's privileged place in the landscape, Im Thurn could not help but "dispel a notion which seems to be not uncommonly held about Roraima," to wit, that it was at all *unique* or even very *outstanding* in the landscape of western Guiana. Roraima was "the most famous" of the flat-topped mountains of the region, to be sure, but it was only "perhaps" the tallest or the most striking. Its status as a landmark was not exactly, as Im Thurn struggled to admit (and yet conceal), *natural*.[152] To salvage its particularity, Im Thurn fell back on Schomburgk and Bentley's characterization in the *Twelve Views*. Citing Schomburgk, Im Thurn reiterated that the mountain resembles "when seen from a sufficiently distant height, an indescribably vast natural forum—a forum in ruins."[153]

Once its summit was reached, Roraima seemed to gain rather than lose its cryptic appeal. Its antediluvian plateau of stunted trees and bizarre rock formations led to further expeditions and culminated, at the beginning of the twentieth century, in Roraima's being enshrined as the inspiration for the mysterious "Maple White Land" of Sir Arthur Conan Doyle's hugely popular "The Lost World." This tale, about the ultimate (Johannes) Fabianesque journey out from the metropolis and back into time, described a death-defying trek to a flat-topped mountain so remote that it harbored a prehistoric fauna, which, brought to London, wreaked havoc in the metropolis. Roraima's fame had crystallized, as had its privileged place in the colonial imagination as a symbol of the interior of British Guiana. Wrote a colonial commentator in the 1950s: "How many Englishmen owe their first knowledge of the existence of British Guiana to its familiar and picturesque Roraima and

---

151. Im Thurn, "Roraima," 258.

152. Ibid.

153. Ibid., 259. Additional nineteenth-century material on Roraima includes Flint, "The Ascent of Roraima"; Flint, "My Return from Roraima"; Quelch, "Journey to the Summit of Roraima"; Im Thurn, *Among the Indians of Guiana*, 82–86; Whitely, "Explorations in the Neighborhood of Mount Roraima"; Perkins, "Notes on a Journey to Mount Roraima"; Smith and Bentham, "Contributions towards a Flora of South America"; and Salvin and Godman, "Notes on Birds from British Guiana." At the beginning of the twentieth century a European woman reached the summit for the first time, and she wrote a popular account of her climb: Clementi, *Through British Guiana to the Summit of Roraima*. Other twentieth-century literature includes Warren, "Report of the 1971 British Expedition"; Zahl, *To the Lost World;* and MacInnes, *Climb to the Lost World*.

Kaiteur [sic] Jubilee stamps?"[154] The *Victoria regia,* of course, also appeared on those issues (plate 12).[155]

Kaieteur, perhaps the metonymic icon of Guiana par excellence, cannot be traced to Schomburgk or Bentley. Charles Barrington Brown, a surveyor and assistant under James Sawkins on the Geological Survey of Jamaica, spent the years 1867 to 1872 in British Guiana, performing geological surveys over many regions of the colony visited by Schomburgk.[156] Brown paid attention to Schomburgk's landmarks, repositioning them slightly, redrawing them, and even, as at Ataraipu—where Brown recorded pondering, as his predecessor had, the vast passages of time—having similarly programmatic reveries about them. While Brown thus negotiated his route with respect to Schomburgk's (and Bentley's) landmarks, and further layered those sites with significance, he also had the occasion to erect a landmark out of his own passage. In 1870, descending the upper Potaro, Brown was led to the top of Kaieteur Falls, a 700 foot vertical drop in a river normally more than 200 feet across (plate 13).[157] It was a "discovery" that roused even the exploration-surfeited British audience of the 1870s.[158] In the colony it led to excitement and considerable confusion: How had no one heard about it before?

Kaieteur, like Roraima, Ataraipu, and Pirara, became the object of a number of pilgrimages in the latter part of the century, and, since it was closer to Georgetown than any of the others, it was visited even more. A satirical paper in the colony published a versified version of Brown's encomium on the falls' grandeur and ended the poem with a "note from the editor" mocking the Kaieteur craze:

> In reply to numerous enquiries in regard to charges for families, single persons and pet poodles at the Great Kaieteur Hotel . . . there will not be the slightest difficulty in the world [getting there]. . . . Immediately on the arrival of Sir C. B. Brown's telegram announcing his great discovery, His Excellency sent to England for two monster Balloons . . . for the conveyance of illustrious strangers and unillustrious colonists to the seat of the Great Fall.[159]

Kaieteur was also the subject of the first photographic trek into the interior, and at least one entire book, in 1872, was dedicated to recounting an

---

154. Roth, *Roth's Pepper Pot,* "Books in the Bush," 130.
155. Plate 12 has been reproduced from the author's collection.
156. Stafford, *Scientist of Empire,* 82–83.
157. Plate 13 is the frontispiece in Brown, *Canoe and Camp Life.*
158. The discovery of Kaieteur is narrated in Brown, *Canoe and Camp Life,* 203–23; Brown and Sawkins, *Reports;* and *Proceedings of the Royal Geographical Society* 15 (12 July 1871). On the occasion of the Guiana centenary the *Geographical Journal* printed a brief essay called "The Discoverer of Kaieteur Fall," accompanied by some of Brown's sketches of the site. *Geographical Journal* 80, no. 5 (1932): 436–38.
159. CRL, Long Collection, scrapbook K; Brown was never knighted.

expedition to its top.[160] It has remained the ubiquitous symbol of the interior of Guiana from the end of the nineteenth century to the present day.[161] The curious status of waterfalls as geographical landmarks, and as experiences in the path of the geographical explorer, will conclude the next chapter.

EPILOGUE: WIDESPREAD VIEWS

Geographical explorers structured accounts of their expeditions—and their maps—around landmarks, sites that gained significance as they were layered with meaning through representation and narrative. Books like the *Twelve Views* served this critical function in the construction of colonial territory. An elegant illustrated text that drew on technical advances in printing, rhetorical devices in travel writing, visual conventions in the depiction of landscape, and an omnivorous aesthetic, the *Twelve Views* had the power to evoke the geographical character of British Guiana. Much of this power hinged on the use of a micro/macro strategy that situated detailed topographic views in a cartographic field. By "fixing" these perspectival images by means of geographical coordinates, while inscribing the "remarkable" objects in the views with historical, mythological, religious, and aesthetic meanings, the *Twelve Views* established a set of key points in the interior—landmarks of the colony. These sites can be seen as nodes of colonial meaning in the homogeneity of jungle and savanna as well as cartographic landmarks in the blank field of a terra incognita. As promontories, these nodes were nothing less than the "sentinels" Schomburgk intimated them to be: they afforded an overview of the space of the colony. The *Twelve Views* anchored the points Schomburgk had fixed on his traverse survey. By establishing landmarks at those points, the text set in motion the saturation of an alien space with European meanings, a process accelerated by the appropriation, transmission, and reproduction of Bentley's images and elements of Schomburgk's

---

160. Webber, *British Guiana*. Im Thurn included a section called "How to Visit the Kaieteur" in his *Among the Indians of Guiana*.
161. It is the subject of A. J. Seymour's 1940 poem "The Legend of Kaiteur," recently reprinted, with accompanying commentaries, in MacDonald, *AJS at Seventy*. One of the most interesting texts on the fall is Walter Roth's apocryphal "indian tale" about the toponym, presented as "The Story of Atakaleet." Roth reported discovering a weathered Dutch manuscript while on a surveying trip. The manuscript supposedly recorded the uncorrupted Kaieteur legend transcribed from a Patamona "soothsayer" in the seventeenth century. Roth later made it clear that the story of the textual "discovery" was a fiction. See Roth, *Roth's Pepper Pot*, 42. Aloysius De Weever's 1903 geography textbook ("for the use of schools") collected the stories around British Guiana's iconic landmarks for the purpose of informing the youth of the colony and the West Indies about the region. See De Weever, *Text Book of the Geography of British Guiana*, 26, 34–35.

Figure 24.   Details from the Tallis atlas map: the views reviewed. Collection of the author.

Figure 24   continued

narrative. The elite readership of the text meant the landmarks of the traveler became the views of the colonial official. Through the visual citations that followed, those same landmarks became icons, and even metonymic figures for the colony itself.

Perhaps the best evidence for the path and outcome of this process lies in a look at the map of British Guiana engraved by John Rapkin for inclusion in John Tallis's 1851 *Illustrated Atlas and Modern History of the World* (plate 14). This handsome volume, sold on subscription in sixty-six parts at one shilling each, was immensely popular in Britain and became one of the first popular atlases to sell successfully on both sides of the Atlantic.[162] It remained in print until 1865.[163] Tallis secured this appeal by having the volume edited by Montgomery Martin, the authoritative chronicler of Britain's rise to imperial prominence and a name familiar to middle- and upper-middle-class readers with an interest in the colonies. The other key to the volume's success was Tallis's skillful use of the new technique of steel engraving to produce an ornate volume at a reasonable price. Tallis's atlas featured more up-to-date information, more detailed text, and more atmospheric illustrations than any competing offerings. James Wyld's *New General Atlas* of a decade earlier was overshadowed by the new technology, as was John Arrowsmith's *London Atlas of Universal Geography* of 1842 (which recycled plates from the 1830s as late as 1858). Tens of thousands of Tallis subscriptions sold, and complete books remained popular well into the next decade.[164]

Montgomery Martin, writing his popular *History of the British Colonies* back in 1834, had to strain to come up with something good to say about British Guiana. A quotation from William Guthrie's *Geographical Grammar* of 1827 offers a glimpse of the dim view taken of the colony taken before Schomburgk's explorations. In just two paragraphs, British Guiana was summed up as consisting almost entirely of "blacks and people of colour," as well as "savage tribes dispersed over the country." These cavort almost entirely naked, practice vile polygamy, and "treat their wives in an arbitrary and imperious manner." The climate was deemed insalubrious.[165] This account was reprinted unchanged in 1843 and was echoed by other chroniclers of Britain's overseas possessions. The text of the Tallis atlas, however, reflected a metropolitan change of heart. Under the editorship of the same Martin who had given the colony short shrift a decade before, British Guiana received considerably better treatment in 1851. Readers were encouraged to envision the "picturesque prospect" of the colony, adorned with Dutch houses, "well-tilled plantations, excellent brick-made and avenued roads, with numerous white bridges, affording a most pleasing indication of an industrious and intelligent community."[166]

Is it possible to argue that this new image of the colony was largely a result of exposure to Schomburgk's *Twelve Views*? A closer look at the Tallis map reveals that its geographical content has been copied from Schomburgk's map, the same one that served as the key map in the *Twelve Views*. Confirmation that this was indeed the source lies in the vignettes

162. Smith, *Victorian Maps*.
163. It was reprinted again in 1989. See Potter, *Antique Maps of the Nineteenth-Century World*.
164. Jackson, "John Tallis."
165. Guthrie, *Geographical, Historical, and Commercial Grammar*, s.v. "British Guiana."
166. John Tallis and Company, *Illustrated Atlas*, s.v. "British Guiana."

set in the margins of the territory. Bentley's illustrations of Ataraipu (plate 6) and the Christmas Cataracts (plate 15) have been copied to make these marginal illustrations (fig. 24). The Tallis map borrowed more than the map and the images from Schomburgk's *Twelve Views*. It borrowed, to a degree, the book's micro/macro representational strategy too, placing local views in the framework of a cartographic depiction. Schomburgk's cartographic landmarks found their way back onto the map, as icons of British Guiana for the British public.

CHAPTER FIVE

Marks of Passage: Landmarks and the Practice of Geographical Exploration

RE-VIEWED

The geographical explorer, charged with the task of generating a map of colonial territory out of a solitary traverse survey, constantly faced the tension between the local view and the broader overview. In his article "Inventing America: A Model of Cartographic Semiosis," William Boelhower argues that the act of mapping demands "the concealment of place and image."[1] The landmark might anchor the surveyor's fixed points, but only by sublating local space within the global context of the cartographic field could the map achieve its authority, becoming an *overview* instead of merely a *view*. Far from being some product of contemporary theory's lust for paradox, the reality of this erasure was not overlooked by those who thought about surveying in the period. Major Basil Jackson addressed the invisibility of locales and practices in his *Course of Military Surveying* of 1838: "When examining a plan, how rarely do we think of the labour with which it has been produced—the triangulation to establish certain points as landmarks—the arduous business of surveying every yard of road and stream—the ability and care necessary when sketching the forms of the ground—and the minute attention required for innumerable minor details."[2]

It was an inability to suppress such details that contributed to Schomburgk's nearly failing in his career as a geographical explorer in 1837. When his reports were edited, detailed memoirs on local conditions were excised.[3] Called on to evaluate the first year of Schomburgk's work, the members of the Society's "Committee on the Guayana Expeditions" caught the flavor of these concerns when they wrote "that in the pursuit of *natural history* Mr. Schomburgk appears to have lost in great measure that the Society's principal object is *Geography*."[4] "Great measures" were

---

1. Boelhower, "Inventing America," 484.
2. Jackson, *Course of Military Surveying,* "Introduction," 6.
3. A closer study of these deletions is the subject of Burnett, "Science and Colonial Exploration."
4. RGS, Correspondence, Schomburgk file, "Report of Committee on the Guayana Expeditions." Emphasis mine.

exactly what Schomburgk had failed to make. His local focus and sentimental enthusiasm meant that he sent back reams of detail concerning his experience of place, and hence that he continually failed, in the eyes of the Society, to rise sufficiently out of his personal views to *see* the place itself. If he did not do so, how would the geographers of the Society "grasp" the overview of the place? Schomburgk had become enmeshed in *local* politics, glimpsed *local* features, and collected *local* specimens: things of as little (geographical) interest as a particularly nice vein of pipe clay or a possible source of raw material for millstones.[5] He had entangled himself in the damning "insect-eye attention to minutiae" that was the opposite of the imperial overview.[6] He was seeing and doing at the wrong scale. Continued the committee, "Mr. Schomburgk seems only to be directed to ascertaining the *topography* of the colony instead of endeavoring to seize its striking *geographical* features."[7]

It was this critique—the assertion that he had afforded a preponderance of local topography in his narratives (and his sketches and maps) without providing any of the geographical "overviews" he had been employed to "seize"—that nearly ruined Schomburgk's career (chapter 3). And he knew as much. After reading of the Society's decision to terminate his support, he penned a desperate letter to its secretary before he set out on the ill-fated Corentyne expedition in 1836. In it he bemoaned his financial straits and complained of the "shock to my constitution" that had kept him in his hammock for weeks. From this perspective, ill, alone, and facing ruin, the prospect of severance from his metropolitan affiliation, and the umbilical cord of respectability and security it provided, would have been frightening indeed. The distinction between topographical views and geographical oversight became, in a manner of speaking, a matter of life and death.

Against this background, the synthetic structure of the *Twelve Views,* completed immediately on his return to London, takes on a meaning beyond the efficiency of its strategy for making the colony visible. Considering Schomburgk's harsh lesson about the limited scope for the personal "view" in geographical exploration, a text like the *Twelve Views* must be seen as a gesture of reconciliation—as a successful insinuation of the eye of the explorer into the cartographic field of the RGS's own map. By employing local views to anchor his fixed points, Schomburgk's landmarks marry the local image to the geographical line, the view to the overview.

---

5. *JRGS* 7 (1837): 289, 293.

6. Carter, *Lie of the Land,* 229.

7. RGS, Correspondence, Schomburgk file, "Report of Committee on the Guayana Expeditions." Emphasis mine.

By placing his places back onto the map, he bridged the geographical/topographical rift that nearly stranded him.

TRACKS OF PASSAGE

If the micro/macro strategy made such a reconciliation possible, it was less successful in resolving the other tension the surveyor faced: that between the active participation with the place (which he experienced) and the static representation of the place (which he did not experience but was charged to provide). The quotation from Major Jackson above draws attention to more than the "innumerable minor details" of river and road that must disappear to make a map. He mentions too "the labour of the surveyor," which similarly becomes invisible in the passage from place to cartographic representation. Comparing John Rapkin's map of British Guiana in the Tallis atlas (plate 14) to the *Twelve Views* map (plate 4) (from which it was copied) illustrates this process of erasure. Much of what appears as dotted lines in Schomburgk's map has been redrawn by Rapkin as solid. This applies particularly to the rivers. The Corentyne, the Mazaruni, the Siparuni, the Buro-buro, and others appear on Schomburgk's map as stippled fluvial hypotheses (plate 11). This graphic convention distinguishes the rivers on which Schomburgk has himself traveled (presented as solid) from those he has heard about, whose course he has inferred from other maps or conversations with local informants (presented as dotted lines). The comparison of the two maps is telling: the map Schomburgk drew—the one included in the *Twelve Views*—displays a symbolic distinction reflecting the contingencies of the traverse surveyor's personal perspective;[8] in the reproduction and popularization of that map the distinction was erased. By this convention of cartographic consistency—all lines have the same weight—the explorer's map was represented as an increasingly homogeneous geographical surface. It is difficult to imagine a clearer example of the cartographic erasures that distance the map from the passages that made it possible. The differences between points of passage and points of inference are elided, presenting a field of even neutrality. The hatching of these stitchwork strokes reflected a map in the making; sutured into solid lines they strengthened the authority of the cartographic artifact. What was a representation of an engagement with place became a representation of place itself.

This stabilized image, proliferated by steel-plate reproduction, super-

---

8. For identification of this map as originally drawn by Schomburgk himself, see Coote and Bolton, "Chronological List of the Principal Maps of Guiana," in *Venezuela Arbitration, British Case Appendix*, 5:375, no. 147.

seded the contingencies of the earlier speckled route map. It is a striking instance of the central theme of Paul Carter's *The Lie of the Land:* imperial power relies on installing the "off the ground" perspective afforded by representations (what he calls *mimesis*) while denying the significance of participation with local places, *methexis*. The appropriation of place through representation has been the subject of numerous studies, and different metaphors have been employed to describe the process.[9] Bruno Latour has used cartography as an example of the way the right representations "make domination at a distance feasible."[10] A miniature of a spatial reality affords, he argues, a mastery over place and process impossible to match in the absence of such representational structures.[11] The map of the colonial possession afforded a form of mastery by mimesis.[12] This mastery had dimensions both practical and symbolic. On the practical side, Jackson himself captured the spirit of a map-minded military, always facing expanding territories and recalcitrant natives: "What would be the science of war, unassisted by plans?"[13] On the symbolic side, an essayist commenting on the improved cartographic science of the early nineteenth century reflected the possessive potency of maps that allowed the winged gaze of "an eye situated above the sphere":

9. This literature is too vast to cite concisely, but it includes Said, *Orientalism;* Greenblatt, *Marvelous Possessions;* Pratt, *Imperial Eyes;* Edney, *Mapping an Empire;* Ryan, *Cartographic Eye;* Miller, *Empire of the Eye;* and Richards, *Imperial Archive*. See also above, chapter 4, note 35.

10. The map constitutes his illustrative case of the *immutable mobile,* a form allowing the local, the recalcitrant, the ephemeral, the stubbornly fixed to be collected and consolidated at what emerges as a *center of calculation*. See Latour, *Science in Action,* 223. The RGS and its associated geographical institutions could be described in these terms. In 1843 the Society adopted a proposal to have a large blank book labeled "Desiderata" placed in the meeting room, "wherein every Member, or friend of a Member, may insert such queries or suggestions concerning particular objects of research, as may occur to them from their own sources." *JRGS* 14 (1844): cxxv.

11. In dramatizing how a networked structure operates to provide a form of domination, Latour uses the example of the scale model. Latour offers three forms of mastery: by physical domination, by network of allegiance, or by precedence in time. Although he points out that a scale model aids the last of these, it can be argued that the map of foreign territory facilitates all three. A choice instance of the map as center of calculation would be Giovanni Domenico Cassini's archival *planisphère terrestre,* inscribed on the third floor of the west tower of the Paris Observatory, 1690–96.

12. Of this process Greg Dening writes: "The map-readers in bureaus and salons needed to make the globe a real world and the real world a map for the strategies of Empire." Dening, *Islands and Beaches,* 269. For a discussion of the mimetic conception of the map, see Harley, "Historical Geography and the Cartographic Illusion," 82–84. For mimesis, miniatures, and possession, see Stewart, *On Longing*. Following Greenblatt and Carter, I use "mimesis" rather than simply "representation" in order to suggest the movement and performative uses of such representations, which share some of the characteristics of a "simulation." See Greenblatt, *Marvelous Possessions,* 6–7 and 120. For Carter, the term links the representation to the "stage conventions" that are his concern. Carter, *Lie of the Land,* 6–7, passim.

13. Jackson, *Course of Military Surveying,* "Introduction," 9. For modern discussions of maps and the military in the nineteenth century, see Harley, "Maps, Knowledge, and Power," 282, and Edney, "British Military Education."

> The use of astronomical instruments ascertains the bearings of hills, of temples, of lakes; the picturesque and splendid style, in which maps of vast continents are delineated, by De la Rochette and our Arrowsmith, delivers to the reader a bird's eye view . . . by which the eye could travel with accuracy from empire to empire over distinct divisions of the globe. . . . The reader travels with grateful joy, and with a speed unallayed by fears and by dangers, from mountain to mountain, from a lake to a fountain, from a fountain to a distant embochure, from a settlement in the wide forest of the interior to a populous haven.[14]

The previous chapter argued that the *Twelve Views* made possible an experience very much like this one. It afforded its readers a fearless "flight" over the surface of the map and allowed them to swoop down in certain places for panoramic views of key sites. For Carter this passage would exemplify the European proclivity for "leaving the ground" that he considers the original sin of imperialism. By negating the "lay of the land" the Icarian ideal of territorial representation abetted dispossession, exclusive enclosure, and absentee proprietorship.[15] Those whose relationship with the land was rooted in intimate participation with its form, productions, and seasonal cycles—those who "walked on the land"—came to be displaced by those whose relationship was rooted in representations of the land—those who had "left the ground." Carter points out that the flattening of the land into a map turned walking the land from a way of living into little more than a symbol.[16] The *peripateia* of one foot following another across the contours of the terrain became merely a line on the map. Instead of being an exploration—an active engagement with place—the passage became a fixed inscription: the place became a stage.[17] The dialogues of foot and ground, light and eye, breath and breeze all vanished, and with them the territorial claims of wanderers and nomads, those whose relationship with place was rooted in participation (methexis), not representation (mimesis).

Carter is looking to escape what he sees as the postcolonial legacy of this imperial refusal to walk the ground. It is no surprise that he looks to colonial explorers for traces of another approach, traces he hopes will serve as a point of departure for a different contemporary attitude toward place. Explorers' experience of place shared little with the ideal of the overview. Walking the land was their way of life. So alien was the overview to the interior explorer that when Charles Waterton set out to give his readers some orientation to Guiana, he mockingly adopted the

---

14. E. H. B., *Geographical, Commercial, and Political Essays*, 38.
15. Carter, *Lie of the Land*, 358.
16. Ibid., 359.
17. Carter's ideas about the "stage conventions" of European landscape representation were anticipated by J. B. Jackson's 1980 essay "Landscape as Theater." See Jackson, *Necessity for Ruins*, 67–75.

aliferous gaze, writing, "We will now ascend in fancy on Icarian wing and take a view of Guiana in general."[18] The wings sat awkwardly on the shoulders of a man self-styled a "wanderer": after a brief bird's-eye sketch of the region, Waterton announces he must "return the pinions we borrowed" and get back to the ground. Forging into the interior, Waterton seizes on a classical allusion more suited to his peripatetic view of Guiana. Falling from the Icarian to the Stygian mode, he describes the jungle path with an allusion to Ovid, Orpheus, and Hades: "Arduus, obliquus, caligine densus opaca"[19] (laborious, steep, impenetrable, shrouded in darkness).

Remembering that the explorer constantly experienced blinkered views, we might reconsider the dotted lines on Schomburgk's map. The dotted geographical line can be seen as an instance of Carter's notion of the *macchie*—the "brush-handprints"—that are his metaphor for a system of representation that refuses to cover its tracks. The Venetian art of *macchiare* relied on the physical gesture of the blot and on the layering of colored patches that left the "footprints" of the artist's brush across the canvas. As he puts it, the aim of the *macchiare* was to "produce forms, rather than reproduce them."[20] This kinetic quality, the active process of moving over the surface, and the textural ambiguities of a surface tracked by the representational passage, combine to make the art of the *macchiare* a figure, for Carter, of the possibility of "an art of peripateia." The *macchie* dramatize the methexis of image making rather than hiding praxis under the polished surface of a mimetic representation.

Two kinds of dotted lines appear on Schomburgk's map: first, the dotted line of Schomburgk's route; the other, the stippled hypotheses that mark untraveled rivers. Both serve as *macchie* on the Cartesian surface of the map: the first as metaphoric footsteps, the second as the blurry intimation of the unresolved and uncertain, a depiction of the limited eye of the explorer within the oversight of the map. The long tradition behind the metaphoric use of the dotted line as a path—its place as a signifier of the *peripateia* of place—is reflected in the following excerpt from a handbook on map drawing published in 1846: "The round dotted line is very generally useful. It may be employed for walks through the demesnes, pleasure grounds, and gardens, especially where the scale is small and the work is close. For this purpose the round dotted line looks well, and sparkles among the woods and garden grounds."[21]

The place of these dots in blurring geographical certainty, in bracket-

---

18. Waterton, *Wanderings,* 205.
19. Ibid., 268. The line is a variant of Ovid's *Metamorphoses,* book 10, line 54, which reads *obscurus* for *obliquus* in the Loeb Classical Library. The translation is mine.
20. Carter, *Lie of the Land,* 165.
21. Wilme, *Handbook for Mapping,* "Analysis of Map Drawing," s.v. "dots."

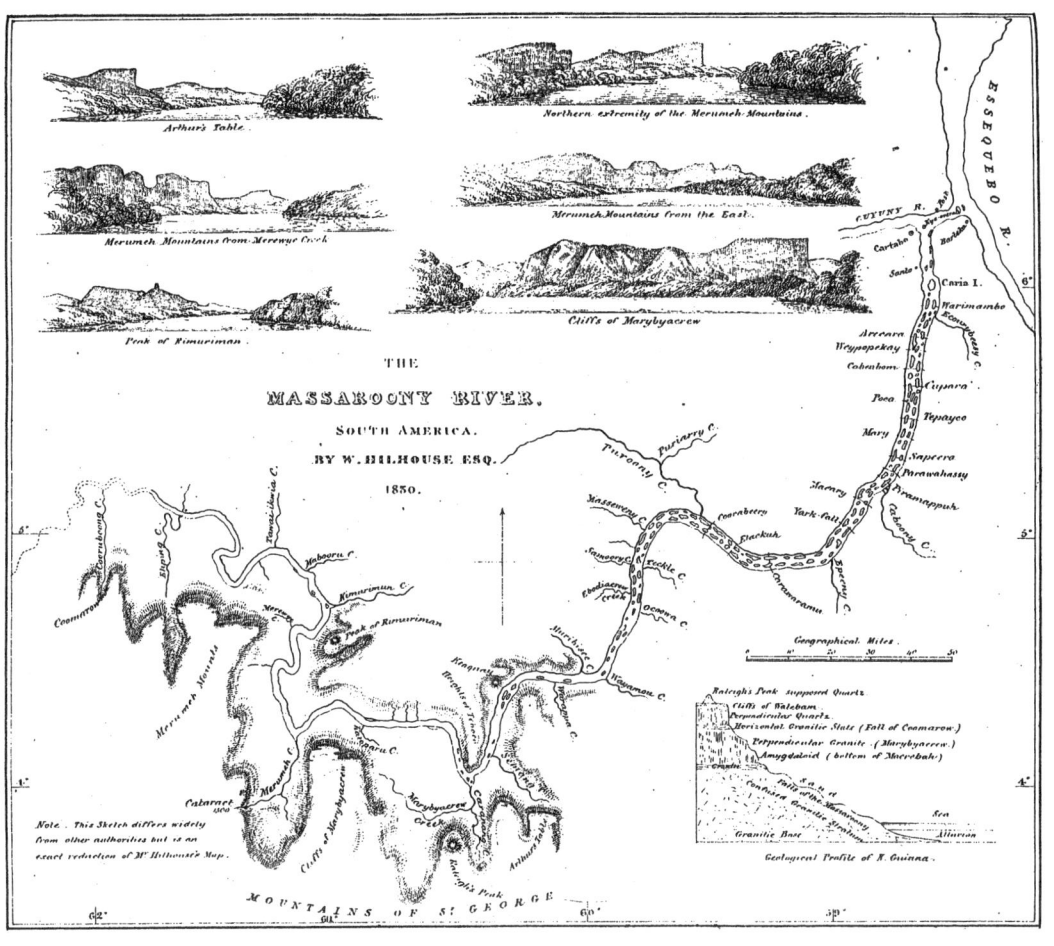

Figure 25. Hilhouse's map of the Mazaruni. Reproduced from the Royal Geographical Society's *Journal*, 1834.

ing the precision of vision and fixity on which the map's objectivity relied, can best be shown by relating Schomburgk's map to those of his predecessors. The Mazaruni, on Schomburgk's map, appears under the aspect of the *macchie* (plate 11). But this is not merely the statement that Schomburgk has not been there, for there are numerous rivers (those, say, to the west, fixed by Humboldt) that have been traced solidly, though they are not accompanied by the small round dots that signal Schomburgk's passage. In fact the Mazaruni had been surveyed by William Hilhouse, and his map appeared in the fourth volume of the *Journal* (fig. 25). That Schomburgk left the Mazaruni shrouded by a scrim of *macchie* indicated not only that he had not been there, but that he did not trust Hilhouse's survey. Dots could be instruments of metalepsis.

The *macchie* then, as overlapping symbols of the footstep and the limited glimpses of the explorer's eye, indicate Schomburgk's negotiation of the place he traveled. They constitute traces, embedded in the map itself, of his idiosyncratic spatial encounter. As Schomburgk's maps were redrawn and became, via Rapkin, a map of the colony for the consumption of the British public, these traces vanished. British Guiana—indifferently accessible and stabilized into a stage for the future drama of empire—replaced a depiction that left clues to the role of Schomburgk's feet, eyes, and imagination in constructing the place. The *Twelve Views*, with its succession of views, each formally divided into a foreground, middle ground, and backdrop, certainly participated in the installation of the colonial stage conventions that Carter describes. But just as it has proved possible to detect traces of the explorer's methexis within the surface of the map, such traces of Schomburgk's active engagement with the place can be found in the images and text of the *Twelve Views* in general. To do so we must turn to the place of the landmark in the *peripateia* of the traverse surveyor.

LANDMARKS AND METHEXIS

In 1853 Edward Belcher gave the technical meaning of a "landmark" in the context of surveying and navigation as "crossed lines of direction."[22] This definition reinforces the place of the landmark in the *movement* of the explorer. For Edwin Hutchins, discussing the structure of Western navigation, the landmark represents the site of a superposition of an internal structure (the compass rose) and an external structure (a promontory, a tree, a tower). The superposition places the landmark and the compass onto "a common image space" and thus transforms the landmark into something more than itself by assigning it a directional meaning.[23] For explorers, then, landmarks constitute a way forward; they take on meaning in relation to directions.

Perhaps the best image of the significance of landmarking in the actual, immediate *practice* of geographical exploration is provided by the bush technique, used by guides in Guyana, of marking trees and rocks to avoid getting lost. No traveler in the bush, picking his way along an overgrown path, fails to notch trees with a swing of the cutlass, ensuring that, should he need to retrace his route, a set of distinctive blazes will lead back to the starting point. One colonial commentator in British Guiana wrote that a man of leisure might doodle with his pen on a blotter, but the bush guide

---

22. Belcher, *Treatise on Nautical Surveying*, 99.
23. Hutchins, *Cognition in the Wild*, 123.

"doodles with his cutlass by making several light chops on the nearest available surface" as a way to negotiate his passage.[24]

The importance of landmarks as a way to see one's path through a place is reflected in asides by geographical explorers: Schomburgk's comment on orienting himself with respect to Ataraipu; Charles Brown's comment that the citadel-like Ayangike Mountain "is a regular landmark in this district, being visible from our camp near the head of the Aruwari, from this spot, and from our camp in the Siparimer."[25] As long as Brown could see Ayangike, as long as Schomburgk could see Ataraipu, they knew where they were. Perhaps the epitome of the geographical explorer's landmark was the peak near the Watuwau River, which exemplified the place of the landmark in the surveyor's task of seeing and transcribing the place through which he passed. "The rock appears to be white quartz," Schomburgk wrote, "and being colder than the atmosphere, is constantly moist, in consequence of the condensation of vapours":

> As soon therefore, as the sun reaches a certain height, and throws his rays under a certain angle upon the moist rock, it shines with a dazzling white, and may be seen at a distance of fifty or sixty miles. Such rocks shine periodically according as the sun has a N. or a S. declination. A similar rock lies on the side of one of the mountains of the Pacaraima, bearing 29 degrees W. from Pirara; it shines only from May to August . . . such rocks may be called natural heliotropes, and have served me in my geodetical operations in lieu of that instrument to determine their exact situation.[26]

For the explorer landmarks afforded—subject to the contingencies of atmosphere, season, and time—rays of direction, orienting glimpses in the disorienting jungle and savanna.

This reliance on landmarks as orienting points explains the oppression and anxiety interior explorers described feeling when they were deprived of such points of reference.[27] Brown wrote of the "imprisonment" of the canopy forest.[28] Schomburgk described a sort of madness produced by the sensory deprivation of long stretches with few openings between the trees. Between Roraima and Esmeralda he wrote: "The monotony of the scene was only broken by occasional glimpses of the rugged and broken ridges of sandstone mountains in the distance, which, to our imagination, assumed a thousand fanciful forms."[29] On reaching a slight elevation

---

24. Roth, *Roth's Pepper Pot*, 19.
25. Brown, *Canoe and Camp Life*, 382.
26. *JRGS* 13 (1843): 48.
27. For an anthropologist's account of disorientation and fear in the bush of Guyana, see Roopnaraine, "Behind God's Back," 56ff.
28. Brown, *Canoe and Camp Life*, 374.
29. *JRGS* 10 (1841): 226.

Schomburgk watched as his hopes for some orienting observations fell under the shadow of the obscuring storm: "In the evening we were visited by a severe thunderstorm, with such cloudy weather that my hopes of an extensive view from this elevated site were entirely frustrated; it was equally impossible to procure astronomical observations."[30]

When they are understood as the obligatory orienting points of a successful passage through unfamiliar terrain, the concrete importance of promontories like Roraima to Schomburgk's exploration becomes clearer. Bentley's enchanted depiction of the mountain must be read against a passage of Schomburgk's narrative omitted from the *Twelve Views* but published in his report to the RGS. In the neighborhood of Roraima Schomburgk and his party came upon a stark reminder of the necessity of clear sight in navigating the savannas: near the Zuruma River the party was "shocked by the sight of a skeleton of a human being."[31] A passing group of Macushis told the man's story: blind, he had become disoriented on his return from the provision grounds. Schomburgk ended the paragraph on this disconcerting discovery with a precise observation of the latitude by altitude of Archenar. The skeleton represented what happened to a traveler who could not see landmarks; Schomburgk fixed the site's position.

The skeleton was a powerful reminder of the potential Icarian consequences of the labor to construct the Icarian perspective: the explorer was constantly in danger of collapsing into the object of his geographical interest. He might, like Horstman, keep going into the interior and fail to come back, becoming an interloper or "going native." He also might die. This ultimate anxiety was not the only danger. Without landmarks, without external points of reference, the explorer ran the risk of collapsing within as well. As Richard Schomburgk wrote of the oppressive monotony of a leg of the Takutu voyage, it was only the view of the Pakaraima, glimpsed irregularly at a distance, that "lightened" the burden of blinkered movement.[32] It was the visibility of external objects that kept the explorer from despair and afforded the focusing devices he could use to compose his view of the terrain.[33] If these external points of reference failed, he found himself in that "sublime void which threatened either to take him prisoner or drive him insane."[34] The invisibility of landmarks annihilated movement, made all directions feel the same, and gave the explorer no respite from himself. Richard Schomburgk wrote of the dizzying experience of the jungle in terms that invoked the difficulty of super-

---

30. Ibid., 227.
31. Ibid., 212.
32. M. R. Schomburgk, *Travels*, 2:53.
33. MacLaren, "Retaining Captaincy of the Soul," 59.
34. MacLaren, "Aesthetic Mapping of Nature," 49.

imposing any navigational framework on the merciless homogeneity: the dense stands of trees stretched "their trunks to heaven and their far reaching branches to every point of the compass."[35] By reaching to every point, they pointed to none. How disorienting that space could be he discovered in a nearly fatal lesson. Pursuing a howler monkey into the bush, he lost his way and spent two delirious days and nights awaiting discovery. After felling the mother monkey and collecting her youngster, Richard realized his mistake: "But where was home? On my right, on my left, in front, or behind me? Everything I asked myself about it remained without reply, because in my violent haste, I had taken no notice of the way and had now lost its direction. *Without fixing my sight on anything at all, without breaking a twig,* I had just followed on . . . and now looking perplexedly around, could find no outlet from the labyrinth."[36]

After circling fruitlessly in the bush he was forced to bed down, in the rain, beside the lifeless body of the primate he had killed. He wrote, "It was an awful time, the horror of which was still further increased by the wild flights of imagination caused by the fit of fever that I already felt within my limbs."[37] This macabre image, of the delirious man huddled against the corpse of a monkey, was a tableau of the interior of Guiana without landmarks. By failing to negotiate the space properly—fixing objects as lines of direction, marking twigs to create a path—Schomburgk not only lost his way, but nearly lost his life and sanity too. On being found, he remained in a delirium for several days.

With stakes so high, it is little surprise to discover that the accounts of geographical explorers demonstrate a meticulous attention to the inscription and legibility of landmarks. These landmarks not only designated routes into and out of unfamiliar regions, they also marked the paths of previous European passage and hence indicated the boundaries of the terra incognita the explorer sought (chapter 3). Given that the traverse surveyor's identity hung on reaching some terra he could legitimately call incognita (and ensuring that it could never be called incognita again), it is no surprise to see the care with which explorers monitored and made landmark inscriptions. Schomburgk not only transcribed carved initials and Amerindian rock engravings wherever he found them, he also engraved his own initials at sites.[38] More elaborately, he often deposited bottles containing accounts of his passages and the names he gave to sites.[39] Such caches turned the moments of passage into permanent marks and

---

35. M. R. Schomburgk, *Travels*, 1:92.
36. Ibid., 218–19. Emphasis mine.
37. Ibid.
38. RGS, Journal MS, Schomburgk, file 1.
39. *JRGS* 15 (1845): 26; M. R. Schomburgk, *Travels*, 1:305 and 2:14.

left fixed records of a mobile exercise. They were the subject of a chapter in Francis Galton's *Art of Travel,* which advocated enclosing the bottle in a hole worked into a rock and sealed with molten lead.[40] At times a toponymic occasion demanded overlooking inscriptions, as when Schomburgk, anxious to give the impression that he was far out of range of ordinary colonial traffic, ignored the numerous dated carvings of colonists' initials at the falls he would dub the "Christmas Cataracts." There were seven sets of initials carved in the rocks, with dates as early as 1803, but referring to them would have significantly undermined Schomburgk's Christmas christening.[41]

This depositing of marks led to an archaeology of previous passages. The aim was to establish the thresholds of terra incognita. On the upper Huena River, Schomburgk recorded the ruins of a settlement the Amerindians called "the last place of the white man"; he deemed it the remnant of an intrepid Dutch trader.[42] "Several names and initial letters" on trees near Yucuribi Falls in the region of Waraputa indicated an unknown European passage.[43] A lemon tree near Pirara implied a very early passage, perhaps even Spanish.[44] Negotiating the upper Corentyne in 1872, Brown discovered the remains of the village and provision grounds that marked the starting point of Schomburgk's descent of the river in 1843.[45] Near that point, at the upper end of the Berbice, Brown "had a group of trees cut down, and a mark placed on a tree, in order that when I ascend this river I can recognize the spot, and so determine my true position."[46]

In the practice of interior exploration, landmarks served as ways to configure passage through the terrain and to inscribe the passage on the territory itself. They were points of reference, objectives, and inscriptions of territorial history. Returning to the images in the *Twelve Views,* we can see some of these traveler's meanings reflected in the text and even in the images themselves. Comuti Rock, for instance, was not merely a romantic "amphitheatre," it was an occasion for Schomburgk's vindication. On his first expedition in the interior of Guiana, Schomburgk had passed the same rocks and had written in his report to the RGS that they "rise perpendicularly to a height of 100 feet, forming a very remarkable feature."[47] The editors of the *Journal,* at this point already disappointed with the first exploits of their Guiana explorer, inserted a footnote that eroded the cred-

---

40. Galton, *Art of Travel,* 120.
41. MQ, Brown MS Journal, 2 February 1872.
42. *JRGS* 12 (1842): 181.
43. M. R. Schomburgk, *Travels,* 1:255.
44. Ibid., 2:340.
45. Brown and Sawkins, *Reports,* 213.
46. Ibid., 170.
47. *JRGS* 6 (1836): 231.

ibility of Schomburgk's account: "Dr. Hancock ascended these hills . . . [in 1810]. 'I measured their height' says Dr. Hancock, 'and sketched this curious natural monument with the red sandstone of the place, having lost my pencil. The height of the column is not so great as it appears from the river; it measured only fifty feet.'"[48]

The footnote, which both called into question Schomburgk's enthusiastic measurement and emphasized the resourceful graphical abilities of his rival and predecessor, wounded Schomburgk's pride. He read his own report undercut by the note. It is thus no surprise to find Schomburgk offering Bentley's illustration of the Comuti Rock as the first of the *Twelve Views,* accompanied by a precise measurement of its height: 160 feet. Reporting to the Society in 1839 in the wake of his long circuit in the interior, Schomburgk detailed climbing Comuti Rock *twice* and ascertaining its elevation on the way both into and out of the bush, "fully confirming my estimate of them on my first ascent of the Essequibo, which had been doubted."[49] This passage shows that Schomburgk gave pride of place to the Comuti Rock both because it symbolized for him the "gate" of the interior—the portico of the "amphitheatre" that opened onto the terra incognita—and also because of its place in an important incident of metalepsis: by 1840, when the account of the full circuit was published in the *Journal,* the editors let Schomburgk's measurements stand; Hancock's citation was footnoted as an error.[50] Schomburgk's treatment of Ataraipu in the *Twelve Views* worked in a similar way. He wrote, "Dr. Hancock saw it from a distance, but did not approach it within twenty miles; and its situation has been so erroneously given that there is a difference of eighteen miles of latitude between its assumed and its real position."[51] Schomburgk "re-placed" the mountain, even as he replaced Hancock as the authority on the geography of Guiana.

These passages show that the landmarks that served as icons of the interior for a distant readership played a role in the explorer's attempts to negotiate a relationship with the place and its history. In fact, no landmark represented in the *Twelve Views* is devoid of metaleptic significance for the explorer: Waterton and Hancock preside at the Comuti Rock; Hancock mis-placed Ataraipu; Schomburgk draws Ralegh's presence over Pirara and even, in a stretch, over Roraima;[52] Humboldt, of course, is the genius loci of Esmeralda. He presided too (in spirit) over the sources of

48. Ibid.
49. *JGRS* 10 (1841): 159.
50. Ibid.
51. Schomburgk, *Twelve Views,* C[5] recto.
52. Im Thurn, "Roraima," 256.

the Orinoco, which he had not reached but had sought. Schomburgk failed to attain them as well, but he placed his failure in relation to Humboldt's and so drew Humboldt's presence over the event, making a landmark out of a thwarted view. Stopped by rumors of hostile tribes, Schomburgk recorded, "Baron Humboldt himself was prevented from fixing [the source] by a similar misfortune, frustrated, he says, by the hostile Indians above Esmeralda . . . who had thus so unexpectedly thwarted my own views."[53] Schomburgk shared even the aporias of his larger than life predecessor.

The sources of rivers in general, although they served as the ideal telos of an expedition embarking upriver from the shore (they constituted points whose existence could be deduced, even if unknown, providing goals within a terra incognita), proved troublesome to "fix." How could one say for certain what was *the* source in the absence of a convenient lake or inland sea? The rivers of Guiana anastomosed into meandering swamps, reconnected along navigational channels grooved by the Amerindians, or vanished altogether, depending on the season. Schomburgk circumlocuted the problem on the Essequibo by claiming to have reached "one of the sources," a claim he substantiated by explaining that the boats could go no farther. On walking ahead, he reached a spot "surrounded by high trees interwoven with lianas, so much so that we could not get sight of sun or stars."[54] In this conveniently grotto-like (and unobservable) spot, "we hoisted the British ensign, which we fixed to one of the trees, there to remain until time destroys it."[55] After a drink from the "unadulterated waters," they headed back. In Guiana, river sources did not make boats turn around; boats' turning around made river sources. As for the "unadulterated waters" of the spring, the rhetoric accorded ill with Schomburgk's later claim to be able to identify the sources of Guiana rivers by the characteristic "blackwater" (swamps) he began to notice on approaching the end of the fluvial line.

Later, Brown followed Schomburgk's path to the source but could find no approach to the point Schomburgk claimed to have reached. As for the flag, its nonexistence could easily be ascribed to the ravages of the tropics and time, which together frustrated so many attempts to mark points. Inscriptions, and with them historical meanings, relapsed into the trackless bush. This was a fact that could cause the explorer a kind of spiritual pain, according to Humboldt, even as it made it difficult to keep track of where one was. "This absence of memorials," he wrote, placed continually be-

---

53. *JRGS* 10 (1841): 232.
54. Ibid., 171.
55. Ibid.

fore the traveler that greatest fear: "The void ... has ... something painful to the traveller, who finds himself deprived of the most delightful enjoyments of the imagination."[56]

So significant were historical presences to the interior explorer that *potential* landmarks could be rejected if it seemed impossible to assign them a provenance and a figurative weight in the explorer's historical and symbolic imagination. Finding a "remarkable rock rising solitary to a great height" near the sources of the Takutu, Schomburgk observed that it had "much the form of the natural pyramid of Ataraipu." On inquiring of his guides, however, Schomburgk was "unable to learn whether any traditions were connected with it."[57] Without "traditions" the rock, for all its prominence, was no landmark; the party did not make a trip to it, nor did Schomburgk depict it or fix its position. Without a tradition the point had no gravity: neither the power to attract the traveler nor the weight to anchor the map. Perhaps the best example of such a "nonlandmark" is afforded by Schomburgk's veritable denigration of one site encountered near the Watuwau. Seeing a mile-long wall running to the east, Schomburgk wrote:

> It reminded me of the granitic wall of the Caquire, near Esmeralda; but where was the majestic Duida—that landmark which guides the voyager on the Orinocco for hundreds of miles—with its cloud-topped summit and colossal walls of sandstone? Wurucokua could vie with it neither in height, nor in historical interest; no Humboldt had botanized or executed geodetical measurements it its vicinity; no Spanish legend told of treasures of diamonds or emeralds buried in its bowels![58]

As such, Wurucokua was really no place at all. Schomburgk and his party passed on, and while they doubtless used the mountain to orient themselves in their search for the sources of the Takutu, it was a traveling landmark, not fit to be an icon of the interior. For future travelers, of course, the meaning of Wurucokua would lie in Schomburgk's passage. He had installed history at this particular landmark even as he denied its history.

AMERINDIAN NEGOTIATIONS

Monuments weighted with Humboldt's presence were few in the interior of Guiana; near the village of Annai, Schomburgk sought out the landmark of a less celebrated European passage. In 1828 two men—Smith, a

---

56. Humboldt, *Personal Narrative*, 2:287. Cited in Bunkse, "Humboldt and an Aesthetic Tradition in Geography," 142.
57. *JRGS* 13 (1843): 52.
58. Ibid., 47.

trader from Caracas, and Gullifer, a lieutenant in the Royal Marines—set out from Georgetown on an expedition to the interior of British Guiana. Despite their relative anonymity, the site they made known—a grotto lake—held particular significance for explorers in Guiana. The two men traveled up the Essequibo and crossed over to the Rio Negro, where Smith took sick with fever and died. Lieutenant Gullifer completed the journey to the Amazon and took passage to Trinidad, where, shortly after arrival, he hanged himself from a beam in the steeple of a church.[59] According to Gullifer's diary of the journey, he and Smith had encountered a Cariban tribe whose chief had given them accommodations near Annai. Gullifer claimed to have witnessed during their stay an intertribal confrontation that culminated in ritual *itoto* cannibalism.[60] He even suggested that he had been tricked into participating. Seeking to bathe, the travelers made their way to a nearby cave, where they found a spring-fed pool. Gullifer wrote that they had been warned the water harbored a vengeful spirit that would kill them, so they bathed publicly, to dramatize the power of their Christian beliefs. Wrote one explorer who read the journal during a visit to Gullifer's brother in Georgetown: "They laughed . . . but sure enough, they were both 'clods of the valley' before twelvemonth expired."[61]

That explorer tried to secure the manuscript journal for the library of the RGS but did not succeed. It later came into the hands of Robert Schomburgk,[62] who made several references to it in his own writings, including an allusion in the *Twelve Views,* where he recalled "those two unfortunate travelers, Lieut. Gullifer, and Mr. Smith, both of whom lost their lives in the attempt to visit the interior of British Guiana. The circumstances connected with their death are of so melancholy a nature that they deeply excite our sympathy."[63] Those circumstances excited Schomburgk's interest enough that, while staying near the village of Annai (whose natives, he claimed, recalled the passage of Smith and Gullifer), the Schomburgks inquired about "the enchanted pond." The locals denied any knowledge of the place. Richard wrote that he and his brother sought for the site in vain in the neighborhood.[64]

The story of Smith and Gullifer offers an insight into a particular kind of landmark in the interior. When they laughed off the warnings of the Amerindians about the "water-mama" spirit that lived in that pond, or

---

59. *JRGS* 2 (1832): 71.
60. For discussion of ethnohistorical sources that substantiate the possible truth of this account, see Whitehead, *Lords of the Tiger Spirit,* chap. 8, "Cannibalism and Slavery."
61. *JRGS* 2 (1832): 72.
62. *JRGS* 10 (1841): 184.
63. Schomburgk, *Twelve Views,* 10.
64. M. R. Schomburgk, *Travels,* 1:289.

perhaps more correctly when Gullifer wrote about laughing and others rewrote their laughter, a bid was being made to refute not just a single superstition but an Amerindian cosmology and the relationship with the territory that it implied. Their laughter was an assertion that the space of the interior was familiar, an extension of their expectations; their laughter was, in effect, a denial of the lay of the land and an effort to install the stage conventions of colonial space. When they bathed in water that was supposed to kill them, they were staging a small theater of possession, unseating a genius loci.[65] Little wonder, then, that the fulfillment of the prophecy of the place-myth—that they would die—fascinated the interior explorers who followed them. Explorers knew the risks of denying the lay of the land.

The encounter with the Amerindian spirit landscape was an inescapable part of any penetration into the interior, and the dynamics of that encounter were essential to the process of exploration. Nineteenth-century interior explorers quickly discovered that nearly every rock, cataract, and valley they encountered was associated with Amerindian legends.[66] Some places were possessed by particular spirits; many rock formations were thought to be the petrified remnants of animals, plants, men, and food from a heroic age of giant ancestors.[67] Geographical explorers moving through Guiana confronted a landscape already written over with meanings. After his first expedition Schomburgk wrote a letter to the colonial governor, reporting: "The Indians . . . stand in perpetual fear of evil spirits, etc. who are imagined to be constantly ready to injure or destroy them. They suppose high mountains, large rocks, cataracts, etc. to be inhabited by them. . . . Indeed, the Indian is a professor of Demonology."[68]

The terra incognita of the interior, though sparsely populated with Amerindians, was seen by Europeans as thickly colonized by Amerindian spirits. The explorer could not escape entanglements with the "demon-

---

65. See Dening, "Possessing Others with a Laugh," in Dening, *Mr. Bligh's Bad Language,* 262 and 393. See also above, chapter 2, note 138.

66. Im Thurn, *Among the Indians of Guiana,* 364.

67. For contemporary anthropology touching on place-myths of Amazonia and Guayana, see Guss, *To Weave and to Sing,* chap. 3; Roopnaraine, "Behind God's Back"; Williams, "Forms of the Shamanic Sign"; Shapiro, "From Tupa to the Land without Evil"; Whitehead, *Lords of the Tiger Spirit;* Roe, *Cosmic Zygote;* Roe, "Of Rainbow Dragons"; and Mentore, "Relevance of Myth." Little contemporary work exists on this particular tradition of rock formations and Makunaima (Mentore, Roopnaraine, and Menezes, personal communication). My understanding of place-myths in Guyana has been informed by fieldwork in the Pakaraima Mountains in the autumn of 1995, particularly by the following informants: Amai Williams, interview, Georgetown, 16 and 17 December 1995; Desrey Fox, interview, Georgetown, 14 November 1995; and Nurse Sago, interview, Paramakatoi, 2 and 3 December 1995. Thanks to Terry Roopnaraine for assisting me with these contacts.

68. Cited in Menezes, *Amerindians in Guyana,* 16.

ology" of the land. Writing of his work on the western frontier of British Guiana in the late nineteenth and early twentieth centuries, Theodore Koch-Grünberg sought to give European readers a sense of the spirit landscape through which he had traveled:

> Belief in spirits and demons is deeply implanted in both tribes and that is no wonder when one takes into account the grandeur of the mountainous area in which these people live. The lofty rocks in whose grotesque shapes man's fancy sees all possible animal and human shapes, the roaring cataracts that at all times tumble down hundreds of meters. . . . all this leads to a belief in the supernatural that expresses itself in the numerous myths and legends, and which also affects the European when he lives a longer while in this wonderful scenery.[69]

No nineteenth-century explorer of British Guiana failed to mention that the landscape of the Amerindian was suffused by spirit presences;[70] several concurred with Koch-Grünberg that a passage through this landscape worked something of a spirit conversion upon the European explorer. Salient among these was Richard Schomburgk, who wrote of the region on the approach to the sources of the Takutu: "I . . . now fancied myself transported to a veritable fairyland, where the world turned into stone was passionately awaiting the wizard's wand for deliverance, so as to resume undisturbed once more the active life that a mysterious spell had brought to a sudden stop."[71] This recognition of the hallucinatory power of the landscape placed the explorer at risk; it was a short step from empathy to the kind of collapse—the "going native" (or mad)—that disqualified the explorer. Koch-Grünberg's compatriot Karl Ferdinand Appun failed. A solitary naturalist, he died in 1872, confined in the penal settlement of the colony, raving of being pursued by avenging native spirits.[72] I have argued elsewhere that Charles Waterton's disturbing taxonomic monster—the "non-descript" he claimed to have shot in the interior—represented an allusion to the threat of collapse into a spirit landscape.[73] Perhaps to guard against such ever real possibilities, several explorers maintained a distance from the spirit realm as dismissive as that of Hilhouse, who ridiculed Amerindian "superstitions" as "nearly as absurd and obscene as the mythology of the Hindus."[74]

The "demon landscape" introduced a complex dialectic into the work

---

69. CRL, RBR, Roth Collection, Roth's manuscript translation of Koch-Grünberg's *From Roraima to the Orinocco*, 12.

70. Selected examples would include M. R. Schomburgk, *Travels*, 2:148; Brown, *Canoe and Camp Life*, 138; and Im Thurn, *Among the Indians of Guiana*, passim.

71. M. R. Schomburgk, *Travels*, 2:52.

72. His work on the region includes Appun, *Unter den Tropen*. Webber (*British Guiana*, entry for 26 April) reports staying with him in the Kaieteur valley shortly before his death.

73. Burnett, "Terra Incognita."

74. *JRGS* 2 (1832): 244.

of all explorers, but particularly traverse surveyors, who, as we have seen, drew on Amerindian traditions in constructing landmarks. On the one hand, the explorer was obliged to believe (and demonstrate) that Amerindian claims about geographical spirits were devoid of substance.[75] On the other hand, the explorer's identity as intrepid and courageous, his heroic character, hinged on the imminent potency of the "hostile forces" he confronted.[76] This ambivalence required subjecting native place-myths and "superstition" to ridicule while at the same time offering some subtle intimation of their power. In this light, the preservation (by later explorers like Hilhouse and the Schomburgks) of a story like that of Smith and Gullifer can be better understood. Telling the story, in an aside or a footnote, enabled explorers to hint at the strange potencies and reason-defying causalities and casualties of the interior without obliging them to let slip their unshakably derisive attitude toward native superstition. At the same time, Amerindian accounts of powerful points in the landscape served as convenient starting points for the spatial differentiation necessary for constructing landmarks. Smith and Gullifer's story of the lake at Annai was hardly unique. A parallel story helped distinguish Ataraipu, which was not by chance known as "Devil's Rock" in the *Twelve Views*. The tale of the foretold death of the eighteenth-century traveler Tollenaer, who approached the cursed promontory, was dutifully preserved and handed down in local colonial lore.[77]

Robert Schomburgk himself simultaneously scorned and substantiated the threatening potency of interior points in his retelling of the "incident at Karinambo," a narrative of his geological collecting. Finding an unusual rock formation, Schomburgk set about breaking off samples. He described the anxiety of the crew as they watched him pound the rocks, noting that they "ascribe them to the Great Spirit." He claimed they "momentarily expected to see fire descend to punish our temerity."[78] Despite being weakened by fever, Schomburgk broke off what he understood to be petrified limbs of Carib archancestors. His overzealous disdain for "native tradition" earned him the censure of the editors of the RGS *Journal* (and later of his idol, Humboldt), who requested that travelers refrain

---

75. The effective practice of both imperialism and science demanded an unshakable faith that all situations to be encountered had a best explanation in the agent's own terms. This certainty constituted the first and indispensable element of the process of bringing new people, new knowledge, and particularly new territory into the purview of European culture. Interior explorers, at the periphery of that culture and responsible for effecting the appropriative acts necessary to maintain its continuity and expansion, were positioned at a critical spatial and intellectual frontier.

76. The more the place posed such an alien threat, the more it was a "wilderness" in the strictest terms: a place of unsettled and unsettling experiences. Walter, *Placeways*, 215.

77. *Brazil Arbitration, British Counter-case Notes*, 66.

78. *JRGS* 6 (1836): 276.

from defacing such local monuments. But even overplayed, the act dramatized an explorer scorning the indwelling spirit of the place.[79]

Or did it? Despite his cavalier depiction of himself hammering, Schomburgk subsequently intimated that the monuments might indeed be capable of protecting themselves. He wrote that those specimens, along with all his other geological samples, were lost when the corial that was carrying them struck a rock and sank on the route back to Georgetown.[80] To hint at the meaning of this event, Schomburgk wrote: "The Indians, not having forgotten the theft of part of their forefathers' limbs, reproached me with having caused the loss of the corial and put them in imminent danger of drowning in the cataract."[81] His remarks, tinged with irony but leaving open all possibilities, underline that exploration demanded ongoing negotiations with a spirit landscape, and with Amerindians, for whom events were ominous and unseen presences monitored passages.

These negotiations of the power and place of spirits would not have been so important had Amerindian spirits not been such a ubiquitous part of an expedition into the interior of Guiana. Traverse surveyors could not ignore them because the place-myths of Amerindian landmarks controlled places, shaped routes, determined stops and detours. Where Amerindians would not go, European explorers could rarely penetrate.[82] While Amerindian place-myths could thus determine the course of an interior expedition, they also provided occasions for the explorer's narration of an intrepid passage. Schomburgk, writing in the *Twelve Views*, mentioned a rumor associated with a supposed lake atop Caruma Mountain, south of Pirara. It was said to harbor a curse identical to that of Smith and Gullifer's pond, bringing death to visitors within the year. Such tradition, Schomburgk noted wryly, "no doubt prevented its nonexistence from being ascertained."[83] Schomburgk's toils to reach the top

---

79. And the censure itself merely reinforced this scorn in a different key. Sacred Amerindian sites had to be "preserved" *as monuments,* hence as memento mori of Amerindian tribes that were seen to be "hastening as by a divine decree, to complete extinction." *JRGS* 15 (1845): 104. For more on the function of custodial and antiquarian rhetorics in the colonization of South America, see Pratt, *Imperial Eyes,* and Taussig, *Nervous System,* "Violence and Resistance in the Americas."

80. I narrate this in the epilogue of Burnett, "Science and Colonial Exploration."

81. BOD, MS, Schomburgk, "Fragments of Indo American Traditions." Schomburgk recorded a similarly ominous incident on a cataract on the Wanumu, where the boatmen insisted a spirit would exact a toll of every passerby. The anxiety of the crew was not appeased by the death of Schomburgk's dog in the rapids below the fall: "The loss of the dog did not appear to satisfy the spirit of the waters," he wrote, "for one of the bark canoes was swamped" in the ascent. *JRGS* 15 (1845): 67.

82. Schomburgk himself was thwarted at least twice in his travels by crews unwilling to go forward because of their beliefs about what lay ahead: in 1837, during his attempt to reach the Sierra Accarai by the Corentyne, and in 1838, approaching the sources of the Orinoco.

83. Schomburgk, *Twelve Views,* E[13] recto, n.

Figure 26. "Palace of the Maranika-mama": jousting with the genius loci. Courtesy of the Picture Library of the Royal Geographical Society.

of the mountain (and use it as a point of triangulation) not only fixed its position but also enabled him to unseat the myth: "No lake was to be seen, nor is the second tradition to be much trusted, that of death within the year as the penalty for treading upon the summit of Caruma; for while I write this, two years have elapsed since I enjoyed a splendid prospect."[84] The explorer's body itself became the bond of imperial territory, and Amerindian myths were a foil for the intrepid explorer.

Foil could yield to lance. Schomburgk made a detour to visit a cavern on the Corentyne said to be the "habitation of one of their spirits (the mother of the salamander tribe) called by the Arawaks Maranika-mama." In the company of several Amerindians, Schomburgk and a companion traced a small brook until they "stood on the head of the famed cavern, the palace of the Maranika-mama." Schomburgk made a sketch in his notebook of the curious sandstone mouth of the spring (fig. 26).[85] Then, while the accompanying Amerindians "stood at a respectable distance," Schomburgk called for a pole, and as they expressed "the greatest horror," he "shoved the pole into the cavities." He noted their chagrined silence as they watched "nothing make its appearance" and saw that "these celebrated holes which they said extended for miles underground, as the Maranika-mama herself was upwards of four miles

---

84. Ibid.
85. Figure 26 has been reproduced from RGS, Journal MS, Schomburgk, file 2, "Rough notes on the Courantyn and Berbice expeditions," 57.

long, namely from Look-out Island to Oreala, were scarcely so many yards deep."[86]

Such a literal jousting with the genius loci was rarely necessary. The Amerindian tribes of the Guianas had (and in fact, still have in certain areas)[87] a tradition of "ritual blinding" that accompanied the approach to powerful sites or the dwellings of powerful presences. Numerous explorers remarked on the ritual, in which paddlers and members of the crew would pause as they came upon petroglyphs or salient rock formations and squeeze into their eyes the caustic juice of tobacco, lime, or capsicum, often mingled with water running down their paddles.[88] The ritual figured in the first pages of the first report that Schomburgk sent back to the RGS, which featured a rich misprint: "On first gaining sight of these hills, the Caribee Indians, who had never ascended the river so far, had to undergo an initiatory sight [*sic*, read rite]—which consisted in squeezing tobacco juice into their eyes."[89] The erratum is charged; the *sight* belonged to the explorer. For explorers, this curious habit of their guides afforded an opportunity to unseat the local spirits merely by looking on their dwelling places. Such instances—where Amerindians blinded themselves to the sights that European explorers narrated themselves surveying—are numerous. Richard Schomburgk captured the moment best: "As we never adopted these precautionary measures, but continued to direct our vision on these wonders of nature, they naturally expected nothing else than our immediate annihilation."[90]

This was, of course, an imperial fantasy, a landscape in which the old adage "master of all I survey" could be meaningful in its most literal sense. As for the notion that Amerindians saw their ancestral spirits exorcised by the blue-eyed gaze of "paranghiris," it is difficult to square with the longevity and persistence of those spirits, whose associated locales remain sources of local myth in the Guyanese interior. To try to reconstruct what Amerindians understood when explorers like Schomburgk looked unblinking on the dwelling places of hostile spirits is beyond the scope of

---

86. This quotation and those above are from RGS, Journal MS, Schomburgk, file 2, "Rough notes on the Courantyn and Berbice expeditions," 57.

87. See note 67 above.

88. *JRGS* 6 (1836): 231. Im Thurn wrote: "Before attempting to shoot any cataract for the first time, or the sight of any new place, and every time a sculptured rock or striking mountain stone is seen, Indians avert the ill-will of the spirits of such places by rubbing red peppers each in his or her own eyes." Im Thurn, *Among the Indians of Guiana,* 368. Guss discusses the history of this practice among the Yekuana (Schomburgk's "Maiongkongs"), who call the herbs rubbed in the eyes *maada.* See Guss, *To Weave and to Sing,* 62. For a fuller discussion of pepper and "forest ogres" among the Shipibos and Waiwais, see Roe, *Cosmic Zygote,* 220–30, particularly 224. Ogres and rock landmarks (and links to Macushi Kanaima) are discussed on 227ff.

89. *JRGS* 6 (1836): 229.

90. M. R. Schomburgk, *Travels,* 1:257.

this book.[91] What is significant here is the role such scenes played in the process of exploration. These stories allowed the explorer to take possession of Amerindian landmarks. By unseating—in their own minds, in the minds of their readers, and in the minds they imagined for the local inhabitants—the indwelling spirits of the land, explorers made the first step toward transforming terra incognita into territory. The marginal spaces that had bristled with myth and rumor were reproduced as an extension of familiar ground. Spirit names became toponyms attached to hills, lakes, and cataracts. Named and fixed, they became reference points for future travelers and for the maps they used. The interior expedition turned native myths into landmarks of the colonial territory.

ICARIAN ANXIETIES: THE FALL AS LANDMARK

Landmarks that demanded negotiating Amerindian place-myths introduced particular tensions into the work of the exploring surveyor. Because these sites posed threats, they were charged points in the passage. So were waterfalls. Waterfalls and cataracts resonated with the Icarian anxiety of the geographical encounter, the potential for collapse into the place itself. What colonial administrators saw as merely impediments to potential commerce became, in the context of the riverine exploration, nothing less than potential vortices for the circulating explorer. The concrete threat of accidents—overturned canoes, lost supplies—was accompanied by a more general and nameless menace. No story illustrates this better than that of Charles Reiss, a young colonist who accompanied Schomburgk on his expedition up the Berbice. Reiss drowned when his corial pitched over the edge of the Christmas Cataracts. The crew had emptied all the boats in preparation for the dangerous and highly skilled task of running them down the rapids. Understandably, Schomburgk reported that he had been "much surprised when Mr. Reiss expressed his intention to go down in the corial, in order to see better how she would go down."[92] He was not a good swimmer, and Schomburgk wrote that he thought he had persuaded him to think better of the idea. This only set the scene for the moment of highest drama in the *Twelve Views*:

91. The act of ritual blinding aimed to help travelers in unfamiliar territory avoid seeing those things that should not be seen: the mischievous tricks and seductions that the spirits of the most hidden areas used to guard their realms. Given the mishaps and deaths that were the constant companions of the European explorers, their Amerindian companions would have had little reason to see their spirit anxieties anything but confirmed. Catholic and (more recently) Wesleyan Methodist missionary work has considerably reduced these practices in the Pakaraima, though eye irritation has not entirely disappeared as a part of bush practice in the region. Nurse Sago, interview, Paramakatoi, 2 and 3 December 1995.

92. *JRGS* 7 (1837): 337.

> When the corial hove into sight the first object that struck me was Mr. Reiss, standing on one of the thwarts in the corial, when prudence would have dictated that he should sit down. From that moment to the catastrophe not two seconds elapsed.... The shock, when her bow struck the surge, caused Mr. Reiss to lose his balance, in falling, he grasped one of the iron staunchions of the awning. The corial was upset and her inmates, thirteen in number, were seen struggling with the current, and, unable to stem it, were carried with rapidity towards the next cataract.

While the twelve hale boatmen managed to cling to the rocks and pull their way to shore, the thirteenth man thrashed:

> My eyes were fixed on poor Reiss: he kept himself above the water but a short time, sank, and reappeared; and, when I had hopes that he might reach one of the rocks the current of the next rapid seized him, and I fear he came in contact with a sunken rock: he was turned completely around and sunk in the whirlpool at the foot of the rapid.[93]

His body surfaced later.

On 13 February Schomburgk conducted the burial ceremony, in which the deceased, his hammock a makeshift shroud, was interred in the land he had come to explore. In narrating the scene in the *Twelve Views* Schomburgk did not neglect to mention that the "small tablet, which he himself had brought, in order to engrave his name, and leave it as a remembrance in case we reached the Sierra Accaray mountains," became the marker of his grave. The site of the entombment, Schomburgk sketched (in words) with topographic precision: like a gate, "two aged trees stand on the western bank of the river, just opposite to the place where our companion was drowned." There Schomburgk ordered "a path to be cleared to rising ground" beyond the reach of highest water. A pile of stones was erected "on a level spot, where mora trees and palms—the latter a symbol of the Christian faith—form an almost perfect circle ... under which rests our lamented companion to await his Maker's call." His tablet was fixed to a tree:

> Drowned,
> February 12, 1837
> Charles F. Reiss,
> Aged 22 years.

Instead of leaving a cache, Charles Reiss became one. Instead of his claiming a landmark in the Sierra Accarai, a landmark claimed him. A cairn marked a new tumulus, a new point of meaning in the colonial landscape of Guiana. Dutifully, Schomburgk placed the point of Reiss's death on the key map of the *Twelve Views* (plate 11).

---

93. This quotation and the one above are from Schomburgk, *Twelve Views*, 28.

The point became landmark.[94] In his private journal Brown recorded his search for the cairn and the inscription when he came to the Christmas Cataracts thirty-five years later, almost to the day: "I searched the sand beach where we camped amongst the scattered trees there for signs of his grave, but found none."[95] Whatever elaborate landscaping Schomburgk and his men had undertaken had relapsed into the thick brush of the Berbice. Reiss's tumulus had become a landmark in memory only. Instead of displacing the local spirits around the falls, Reiss had joined them. He had become, and remained, the genius loci of the Christmas Cataracts. Brown added a note in his journal about the cataract: "None but mad men would attempt to run it." Schomburgk suggested as much in his report to the RGS describing the incident, where he recorded the content of his conversation with Reiss the night before the tragedy: "After our scanty meal, we were rather surprised when Mr. Reiss indulged in a melancholy strain, and observed, 'he knew he should die young.'"[96] Wrote Schomburgk, "We ridiculed the idea." The odd foreshadowing suggests that the menace of the fall went beyond rapids and slippery rocks.

Perhaps what Reiss experienced was not madness proper but some romantic misprision of the landscape, the sort Ian MacLaren blames for the tragedies that befell some North American explorers in the nineteenth century. Perhaps Reiss let his romantic sensibilities obscure his vision. Perhaps his urge to "*see better how she would go down*" betrays the scopic drive of the tourist displaced to the wilderness. Perhaps he thought the fall was picturesque, or sublime. What is certain is that waterfalls had a special place in the imagination of the interior geographical explorer. In the first place they represented, particularly in Guiana, a disjunction between the civilization of the coast and the "wilds" of the interior: Schomburgk called the Wonotobo Falls the "gates" to the realm of the Amazons;[97] Brown wrote of being "cut off . . . from the world by the two great barriers" of the Essequibo falls.[98] Still more significantly, the fall could not but symbolize the collapse into the place that was the supreme anxiety of the geographical explorer.

That this was so is revealed by the numerous instances of explorers' en-

---

94. Nor was this at all unique in the history of exploration. Another choice instance appears in the work of David Livingstone, whose geographical calculations in the Mwambezi region could be subjected to a twentieth-century planimetric analysis precisely because the landmark point called the "Anchorage" has remained an identifiable site in the landscape. It was the grave of Livingstone's traveling companion Richard Thornton. See Bosazza and Martin, "Geographical Methods of Exploration Surveys in the 19C."
95. MQ, Brown MS Journal, 1 February 1872.
96. *JRGS* 7 (1837): 336.
97. Schomburgk, *Twelve Views*, 24.
98. Brown, *Canoe and Camp Life*, 238.

acting metonymic disasters at the cataracts. Im Thurn, on reaching the top of Kaieteur, told of "the unbearable pain" he experienced on watching one of the men hurl a bromeliad over the lip of the sheer drop.[99] He described watching "small white butterflies" that seemed "irresistibly attracted by the fall" and, flying lower and lower, "flitted in the rainbow-tinted spray; and then they were sucked in by the water."[100] Schomburgk, in the *Twelve Views*—which contains views of three falls—described a still more spectacular scene of reenacted disaster. The Purumama Falls on the Parima River featured a churning whirlpool of particular intensity at its base:

> In order to witness its power we sent some Indians up above the falls and ordered them to cut down one of the largest trees that line the bank. How great was our admiration when, after having been directed to mid channel, it approached the chasm and with its numerous branches was hurled down the precipice! Scarcely had it approached the whirling gulf when it was sucked under, and when it reappeared, it was already in the middle of the outflowing stream, a naked trunk.[101]

The anthropomorphic language leaves no doubt what was on the mind of the onlookers.

The structure of the event follows quite closely the symbolic structure of the waterfall outlined by Patrick McGreevy in his "Reading the Texts of Niagara Falls: The Metaphor of Death."[102] His reading of the structure of Niagara breaks the waterfall into a set of physical and symbolic elements: the "brink" (the boundary between life and death that inspires soul searching and perhaps affords a glimpse of the hereafter); the "plunge" (the liminal state between life and death); the "abyss" (the yawning vortex of another realm); and finally, the "rising mist and rainbow" (suggesting peace and hope behind the veil). As McGreevy shows, the fall afforded nineteenth-century visitors a rich landscape for contemplating final things. It seemed to open a window onto the opacity of death, a window that framed a certain seductive silhouette.

In visiting Kaieteur in 1873 Lt. Col. Edward John Webber recorded his first sentiments on reaching the brink: "This is far grander than Niagara!"[103] In riverine exploration, in which the river *was* the path, the gravitational force of the fall (the pull of the water) impressed itself on the mind. This was particularly the case at Kaieteur, whose visitors knew that the first European to come to the fall had done so *from the top*. Brown had reached the upper Potaro on foot and was descending it in light ca-

---

99. Im Thurn, *Among the Indians of Guiana*, 67.
100. Ibid., 79.
101. Schomburgk, *Twelve Views*, F [17] recto.
102. McGreevy, "Reading the Texts of Niagara Falls."
103. Webber, *British Guiana,* entry for 30 April.

noes when he learned of the fall ahead. His narration of bringing the woodskins (bark canoes) to shore, as the roar of the fall grew louder ahead, impressed itself on his readers. Brown retold a tale, supposedly recounted by an Amerindian at the brink, of "old Kai," the mythic figure who, precipitated over the fall with all his goods, bestowed on the place his name: death was embedded in the toponym itself.[104] "Lying in a position facing the fall," he wrote, "I had only to open my eyes, and there was its grim, white, ghostly outline before me."[105] Facing this specter he "spent a most uncomfortable night," tossing to the sound of its "ceaseless roar."[106]

Schomburgk, already having recounted the oppression of standing at the foot of the Kanaiba Falls near Roraima,[107] described the oppressive force of the brink very clearly. In narrating an encounter with the Wonotobo Falls on the Corentyne, he revealed his intimacy with the metaphoric power of the fall and its peculiar gravitational force. Just over three months before Reiss's death, he and Schomburgk had shared the silencing pull of the fall. Schomburgk wrote:

> I stood surprised—the sight of the foaming waters below, the unceasing noise of the cataract, which made every attempt fruitless to communicate my feelings to my companions, rendered the impression of this scene powerful almost to oppression. I became giddy, and retired quickly to prevent myself from joining the dance of the whirling white crested billows.

Of the pull, he continued:

> I have stood in much more perilous situations without ever feeling the slightest sensation of vertigo, and I ascribe it in the present instance to those masses of water unceasingly rolling in the abyss below, which seemed to urge me to follow them.[108]

At one level this *Twelve Views* scene—like Reiss's death itself and the attendant funerary rituals—can be interpreted within the framework laid out in the previous chapter. What could be a better example of an invocation of the sublime at a landmark than this passage? What could be better for establishing the iconic qualities of the site than to attach to it an account of its mythic seductive force? The narration of Reiss's death too could be construed as part of the effort to weight the site of the Christmas Cataracts with multiple aesthetic associations: the picturesque occa-

---

104. For a (verse) rendition of this tale, see "The Old Man's Fall," in Brett's *Legends and Myths*, 200. See also the citations above, chapter 4, note 161.
105. Brown, *Canoe and Camp Life*, 215.
106. Ibid.
107. Hence, in McGreevy's scheme, standing in the "abyss" of death itself. McGreevy, "Reading the Text of Niagara Falls," 60.
108. Schomburgk, *Twelve Views*, 26.

sion of the camp Christmas that gave them their name must be juxtaposed with a macabre invocation of the sublime.[109] At Wonotobo, just as the tumbling waters cause a picturesque scene of billowing mist, the same site is distinguished by the fearsome sublimity of its dark gravitational field. The *consummatio* of imperial landmarking might be said to be heaping high the reader's aesthetic plate at Wonotobo and the Christmas Cataracts. Can anything be said about this same passage within the framework of the tactical considerations of an explorer negotiating the terrain? Of the active engagement with the land that has been the subject of this chapter?

One reading of Schomburgk's dizzy recoil from the fall understands the moment as romantic hyperbole. The gesture evokes in European readers the familiar notion of the Burkean sublime, which derived its emotional power, like the sizzling light and heat of Humphry Davy's arc lamp, from maintaining precisely the proper distance between the elements. Too far, and the spell was broken; too close, and the viewer collapsed into unmitigated terror and could no longer feel the radiant forces before him.[110] Schomburgk's gesture, then—his recoil—is a nod to European aesthetic sensibilities. But another reading might see in the same jerk of the head evidence of the idiosyncratic perspective of an explorer struggling in an alien place. Schomburgk was an epileptic, subject to fits so violent it could take four men to restrain him.[111] Leaning over the brink, dizzy, he may well have recoiled in fear of himself.

PICTORIAL CONVENTIONS AND EXPLORATORY SUBVERSIONS

In examining Bentley's illustrations for the *Twelve Views* in the previous chapter, I argued that the aesthetic conventions of European landscape art played an important role in structuring the images and in making them readable as depictions of familiar space. But I also acknowledged that this treatment of the visual products of European exploration has come under criticism. In particular, Barbara Belyea has suggested that this sort of reading (exemplified by MacLaren) can overlook how exploration artists subverted the very picturesque conventions they deployed. For Belyea, these kinds of images show that the artists, attuned to their

---

109. Of course the associations here are not merely aesthetic: it would be interesting to develop an account of how pain, suffering, and (particularly) death worked in the consolidation of colonial territory. (For a start on this, see Noyes, *Colonial Space,* 134.) Some notion of "mingling" one's labor (and hence sweat, blood) with the land lies at the heart of the earliest accounts of creation of "property" out of *terra nullius*. Could colonial cemeteries be examined in this light?

110. On the Burkean sublime and the "collapse" of the explorer, see Riffenburgh, *Myth of the Explorer,* chap. 3, "Decade of Change."

111. Goodall, "Diary July to December," 47.

dual mandate to be "professionally accurate while being artistically conventional," embedded within their images ironic commentaries on the unsuitability of European notions of the picturesque to the experience of the expedition: a bucolic river scene, on second glance, reveals a sweaty struggle to paddle upstream; the requisite classical repose of the humanized foreground becomes risible when juxtaposed with a canoe about to be swamped within arm's reach. If the task of this section is to examine the *Twelve Views* for traces of the distinctive and dynamic perspective of the explorer, then a version of Belyea's analysis—a ferreting out of subversions of the artistic conventions within Bentley's images—would seem to advance the cause. Are such readings possible? I will suggest a few ways they might be, but I will ultimately settle on a distinct interpretation.

How could one make the case that the picturesque has been subverted in Bentley's views? One might point to the view of Comuti Rock. The river that, according to the conventions of the picturesque, should be a fluvial link between backdrop and foreground here appears as a labyrinthine delta of meandering channels (plate 5). In Bentley's image of Schomburgk's "melancholy" Brazilian Fort São Gabriel, the foreground boats, on closer examination, are struggling *up* the rapids (plate 10). Perhaps Esmeralda's misty shroud (plate 9) should suggest the *macchie,* a scrim of uncertainty hanging over all of Schomburgk's journey after the snapping of his watch chain. If the nebulous (smoke, mist, storm) intimates clouded vision—and in doing so gives the lie to the fantasy of imperial oversight—then perhaps each of Bentley's images not only contains a veiled allusion to the undoing of the imperial eye but contains that allusion in the form of the veil.[112] For not only is Mount Duida about to vanish behind the scrim of the storm, but almost all the *Twelve Views* contain some visual quotation of an obscuring nebula: in the foreground of Roraima the white smoke of the campfire creates a ghostly aporia in the middle of the classical scene of repose (plate 8); the same can be said of the fire burning in the corner of the depiction of Ataraipu (plate 6), where the smoke mingles with the mist of the valley to create an image of enchantment, or perhaps of obscurity.

Sometimes, however, a campfire is only a campfire. And that is the problem with the attempt to craft such elaborate readings of the images

---

112. Carter, *Lie of the Land,* 276. Belyea calls on the picturesque-subverting function of smoke and clouds when she rereads George Back's "Manner of Making a Resting Place on a Winter's Night." She points out how the centered campfire creates a column of smoke in the midst of the image, a column that obscures the vanishing point. For Belyea, the resulting image portrays "a paradox foreign to both the picturesque and the sublime, that of claustrophobic immensity." Belyea, "Captain Franklin in Search of the Picturesque," 14.

in a text like the *Twelve Views*. Not only does the suggestion of such lurking symbolist significance stretch the bounds of the topographical tradition in which the images were produced, the interpretation also requires that we ignore what we know about the distance—geographical and genealogical—between Schomburgk's actual *view* and Bentley's image. Schomburgk "dictated" to Morrison, Morrison was copied by Bentley, and Bentley in turn was copied by Coke Smith, P. Gauci, or George Bernard, the lithographers. In the colored exemplars, these images received treatment at the hands of anonymous piecework colorists. With this much distance between impression and image, with this set of reworkings of the image, how could a close reader be sure of spotting a subversive signal amid the aesthetic noise of multiple image variations?

Instead of following MacLaren's curious claim that the picturesque conventions "caused" tragic misprision of place, or Belyea's tenuous assertion that sly gestures subvert those conventions, I suggest that the role of pictorial conventions was to bring the explorer home as they repatriated his views of a foreign place. Through the cycles of preparation, Schomburgk watched over the reparation of the intractable elements of the alien environment from which he had returned. Negotiating the placement of bird and bush with Bentley, Schomburgk saw Comuti Rock become the picturesque and enchanted site that it failed to be when he stood atop it and fired his musket into the void, hearing no echo: the shot was "lost, there being no object to return the sound."[113]

Bentley's view of Purumama Falls showed no trace of the violent game of metonymic disaster that had been played on its brink with the felled tree, and Wonotobo Falls (where Schomburgk flinched) was graced with an arching rainbow (fig. 27). So was Bentley's view of the Kundanama Junction (plate 16). Pirara, peaceful under the pink glow of the Pakaraima, became the El Dorado it had emphatically failed to be when he and the party were forced to weather the rainy season there, when soggy vermin of all descriptions sought refuge in the huts and gnawed the botanical and ethnographic collections to ruin. The encounter with recalcitrant places, the obscurities, and the disappointments of a passage all could be repaired by Bentley's brush: Esmeralda received the ruins it deserved, Roraima revealed itself, the stork of Puré-Piapa was resuscitated. For the traveler, as Carter points out, "a well-judged homecoming de-

113. Schomburgk, *Twelve Views*, [3] recto. Schomburgk's contemporary, John Ruskin, could not veil his scorn for those whose attempts to fill the landscape with some echo of themselves reduced them to inarticulate fusillades of rifle fire: "When you are past shrieking, having no articulate human voice to say you are glad with, you fill the quietude of . . . valleys with gunpowder blasts, and rush home, red with cutaneous eruption of conceit, and voluable with convulsive hiccoughs of self-satisfaction." Ruskin, *Sesame and Lilies*, 53. Cited in Schama, *Landscape and Memory*, 506.

Figure 27. Wonotobo Falls: teetering on the brink. Courtesy of the Yale Center for British Art, Paul Mellon Collection.

pended on a sensitivity to local conditions."[114] A sensitivity to the conditions on the ground (snake, quicksand, broken twig) could get an explorer back to the coast, but only sensitivity to the local conditions of metropolitan society could bring him all the way home. The *Twelve Views* made Schomburgk's views right, allowed them to be shared with the elites of British and colonial society. By doing so it not only made British Guiana visible, it brought Robert Schomburgk back to Europe.

EPILOGUE: PLACE AND ERASURE

In this chapter I have examined the notion of a landmark as one of the tactics of an explorer negotiating unfamiliar ground. While it is possible to tease traces of idiosyncratic and methetic meanings from the images, the map, and the text, I have suggested that perhaps it is most fruitful to understand the place of books like the *Twelve Views*, in the traveler's spatial practice, as that of a homecoming, a "repatriation of perspective,"

---

114. Carter, *Lie of the Land*, 86.

effected through the progressive editing of images and writings. Through this process not only did the places depicted come to conform to the aesthetic conventions of the European readership, so too did the views of the explorer himself. The idiosyncrasies, the blind spots, the very *perspective* of the explorer were reacculturated. His disturbing sights were erased as the place was drawn.

But if the *Twelve Views* cleared the fog, fixed what had been broken, made visible what had remained hidden, and repatriated the explorer, it could not repatriate his remains. Remains remained. Bentley's view of the Christmas Cataracts (plate 15) expressed as much. A close examination of the image discloses a definite break with the conventions of the topographic depiction: to the right of the fateful cataract, on the exact horizontal axis of the image plane, stands a solitary figure in a white shirt, dwarfed by vaulting trees, set apart from the camp scene at the right margin and the busywork of the foreground. The minute figure, manifestly European, stares at the rushing water of the fall, his arms folded, as if in contemplation (plate 17). There can be little doubt that the figure is transfixed by that peculiarly powerful and metaphorically rich landmark before him. He watches *to see better how she goes down.* Is the figure Reiss, the thirteenth man, on the day before his death? Or is it Schomburgk, pondering his friend's demise and turning over in his mind the Icarian anxieties of the interior explorer?

In no other image in the *Twelve Views* does any depiction of a European explorer appear. This invisibility of the geographer within the geography is the very premise on which geographical epistemology is built: "For the surveyors to be shown working *within* the landscape would subvert the entire ideology of geographical observation."[115] The geographer *as* landmark, rather than maker of landmarks, *as* tumulus, instead of maker of tumuli, this too subverted the essential separation between the geographical explorer and the land. In this image Bentley shows Reiss (or Schomburgk) *in* the landmark. It is exactly where Reiss, for one, would remain. Fittingly, the strange image, depicting too literally the view of the explorer, was corrected before it became part of the Tallis atlas map of British Guiana (fig. 24). Here the image of the Christmas Cataracts has been edited down and the curious figure deleted; the trace of subversive methexis has been erased in the production of a map of the colony of British Guiana for the people of Britain.

---

115. Edney, *Mapping an Empire,* 74. For other accounts of the rarity of such self-referential images, see Belyea, "Captain Franklin in Search of the Picturesque," 21, and Godlewska, "Map, Text and Image," 19–22.

CHAPTER SIX

Boundaries: The Beginnings of the Ends

DISPUTED BOUNDARIES/DISPUTED MAN

There are red lines on the key map of Schomburgk's *Twelve Views* (plate 4) and red lines on John Tallis's 1852 map of British Guiana prepared by Rapkin for the *Illustrated Atlas* (plate 14). They do not stand for the same thing. On Schomburgk's map the red line highlights the *macchie* of his passage: it marks a route. In Tallis's atlas map the red line circumscribes the colony: it marks a boundary. The aim of this chapter is to understand what relationship holds between Schomburgk's routes and the boundaries of the colony. The three narrative sections that make up the core of this chapter correspond to the geographical regions of Schomburgk's boundary surveys. In chronological order these are the northwest boundary with Venezuela, the southwest boundary with Brazil, and the eastern boundary with Dutch Guiana. Each of the three regions has the dubious distinction of having become a case study in the history of boundary disputes and arbitrations.[1] Portions of all the boundaries traced by Schomburgk between 1841 and 1843 remain in dispute to the present day, though for different reasons.[2] The Venezuela boundary question has proved the most acrimonious and durable of these conflicts, in part because the discovery of gold in the disputed region in the 1850s raised the stakes a great deal. At the end of the nineteenth century tensions over the Venezuela boundary reached a crisis point, and when a British gunboat was reported off the Orinoco, the conflicting claims came to the attention of a United States government that was consolidating its influence in Latin America and the Caribbean. Seen as a potential infringement on the Monroe Doctrine, British claims to the upper Cuyuni were treated as truculent by the United States, which sided with the Venezuelans, secured mining concessions in the disputed zones for United States companies, and pressured Britain into submitting to arbitration.[3]

---

1. Guyana is reproduced as one of the thirty-three maps in Prescott, *Political Frontiers and Boundaries*, 200–212.
2. See Ireland, *Boundaries, Possessions, and Conflicts in South America*, 152–58, 234–43, 245.
3. For a detailed study of the crisis point and its resolution, see the recent reevaluation of the incident by Marshall Bertram in his *Birth of Anglo-American Friendship*. For a closer look at the business interests tied to the dispute, see Kreuter, "Empire on the Orinoco."

The textual production of this process gives pause to the most voracious historian. Supported by their respective governments, teams of historians, archivists, geographers, international legal scholars, and diplomats spent half a decade preparing more than twenty encyclopedia-sized volumes. Each side offered a case, argument, countercase, counterargument, atlases, and multiple volumes of evidentiary documentation culled from four countries and many more languages. The arbitration tribunal, consisting of two United States and two British chief justices as well as a Russian scholar-diplomat, spent 216 hours considering the merits of the two sides before it handed down a decision that awarded 94 percent of the disputed territory to Great Britain.[4] Broadly speaking, the "Schomburgk line" was upheld, though Barima Point was a consolation prize awarded to Venezuela, as I discussed briefly in chapter 2. The award never sat easily on the territory. Pursuit of natural resources and later accusations of backroom dealing, as well as politically expedient populist uses of boundary anxiety by beleaguered governments on both sides of the line (and elsewhere), have kept the dispute alive.

At the turn of the century, however, it was possible to see the arbitration process as a success, particularly since it resolved United States–British tensions that for a time looked likely to lead to military conflict.[5] And so, shortly after the 1899 award had been accepted by both governments, a second process of comparable scale was undertaken in the attempt to settle British Guiana's southwestern boundary with Brazil. Again tens of thousands of pages of disputation and documentation were amassed, again disputants ransacked archives in the quest for "cartographic proof," and again a decision came down (in this case the work of a single arbitrator, the king of Italy) without any articulation of the legal or historical premises used in its making. A considerable portion of this line too conformed to that surveyed by Schomburgk in 1842. While the "eastern" dispute with the Dutch (and subsequently with the independent nation of Surinam) has never precipitated the same voluminous textual production, it also has remained unresolved.[6] A look at a map of Guyana in a recent geographical study of political boundaries reveals that less than a quarter of the current boundary of the country can be called entirely undisputed. Venezuela's claim extends over almost 70 percent of modern Guyana.[7]

---

4. Braveboy-Wagner, *Venezuela-Guyana Boundary Dispute,* 106.
5. Bertram, *Birth of Anglo-American Friendship,* 27–29.
6. Its nonresolution hinges on the seemingly intractable question of Schomburgk's understanding of what the Corentyne River was in 1843. Prescott, *Political Frontiers and Boundaries,* 214.
7. There are a number of recent (Venezuelan) studies of the Venezuelan claim: Tamayo et al., *Guayana Esequiba;* Fernández, *Historia y el derecho en la reclamación;* Delgado, *Venezuela y Gran Bretaña;* and Bermúdez, *Estrategia imperial británica en la Guayana Esequiba.*

No synthetic history of the boundaries of Guyana exists, but Peter Rivière's recent microhistory of Pirara (introduced in chapter 2) has cast new light on the subject of the Brazilian boundary.[8] Working from the archival sources in Brazil and Britain, Rivière has reconstructed the events of the Pirara dispute: the founding of the mission by Thomas Youd, its subsequent abandonment and occupation by Brazilian forces, the decision by the British to send an occupying force into the region to accompany Schomburgk's surveying party, and the ultimate official abandonment of the region by both sides. The title of the study—*Absent-Minded Imperialism*—gives away the thrust of his analysis. For Rivière, the archival sources give the lie to a long-standing tradition of partisan conspiracy theorizing about Pirara specifically and British territorial interests in Guayana more generally. Schomburgk as an individual lies at the center of these debates, and it is no surprise that Rivière has turned from this study of Pirara to a biography of Schomburgk himself. Several Venezuelan and Brazilian scholars who have treated the events and the period have demonized Schomburgk.[9] The expanding ambit of his boundary claims for the colony over his years of exploration has led to a thesis that Schomburgk "evolved" during his service to Britain from unassuming man of science to land-hungry imperial booster. In the felicitous phrase of one of the boundary dispute documents, Schomburgk (at Pirara, no less!) caught the virus of an "idée raleighienne d'agrandir l'empire Britanique."[10] Schomburgk's putative humanitarian concern for the Amerindians disguised a British land grab.

Rivière suggests that the reverse is equally likely,[11] and his sources undermine the assumption that Britain operated from a coordinated imperial strategy of domination. Just because he refuses to see Schomburgk as merely a monomaniacal imperial point man does not mean the book minimizes Schomburgk's responsibility for the Pirara incident. After failing to find within the records any trace of a "hidden agenda" of British territorial expansion, Rivière is left to consider Schomburgk's "personal" role in the events. Schomburgk's agency cannot be overlooked, Rivière concludes, in any attempt to make sense of the events that led to the boundary survey and military occupation of a lonely savanna outpost. Af-

---

8. For a full bibliography on the boundary disputes, see Braveboy-Wagner, *Venezuela-Guyana Boundary Dispute*. See also Benjamin, "Review of Braveboy-Wagner"; Menezes, "Background to the Venezuela-Guyana Boundary Dispute"; Burnham, *Guyana/Venezuelan Relations;* and Jeffrey, "Bid for El Dorado."

9. These would include Ojer, *Robert H. Schomburgk,* and Reis, *A Amazônia e a cobiça internacional.* On the latter I am working from Rivière's discussion of the text. Another relevant work in Portuguese was impossible for me to use directly: Farage, *As muralhas dos sertões.*

10. *Brazil Arbitration, British Counter-case,* chap. 6, 121.

11. He argues, most recently in his "From Science to Imperialism," that Schomburgk's primary motivations were humanitarian.

ter all, the survey produced a boundary that Britain for the most part declined to reinforce or fight to maintain.

DELIMITATION, DEMARCATION, "DECONSTRUCTION"?

The absence of a boundary line in the *Twelve Views* map must be understood to be as much a plea as an omission. In fact, *Twelve Views in the Interior of Guiana* poses the question of territoriality at the same time as it constructs that territory. The book was not called "Twelve Views in the interior of *British* Guiana." Was Pirara in British Guiana at all? What about Roraima? Or Puré-Piapa? At the end of his first four years of explorations, Schomburgk penned an explicit call in the RGS *Journal:* "May the moment soon arrive when the boundaries of the rich and productive colony of British Guayana shall be decided by a government survey!"[12] It was a boundary he himself hoped to draw. By emphasizing the humanitarian threat to the Amerindians of Guiana, Schomburgk made the *Twelve Views* into a lengthy request for a boundary for the colony. In the event that this message—the thinly veiled subtext of the *Twelve Views* as a whole—had been missed by an incautious reader, the text finished by invoking the desolation of Pirara and the sacrilege that had occurred there: "Although the first chapel erected to the worship of the true God in the interior of Guiana is now abandoned in consequence of Brazilian aggression and intolerance, the time may be approaching when its walls will again resound with hymns of praise of Him who is Almighty."[13] A wood-engraved view impressed the image on the reader's mind: a forlorn clapboard chapel at the nearby Brazilian fort, dilapidated and abandoned (fig. 28). Here was no picturesque effect of the "judicious hammer," but the tragic effect of the benighted blows of papistry, sexual license, and slavers' greed.

In April 1840, in part on the strength of a recommendation of Governor Henry Light, but more broadly as a result of all of Schomburgk's work to draw the boundary to the attention of colonial officials, Schomburgk was appointed boundary commissioner, and over the summer of that year he began preparations to bring together the equipment and resources he would need to return to British Guiana in this new capacity. Lord Palmerston, who had given the operation his blessing, recommended that Schomburgk's own outline for the boundaries of the colony be adopted as the preliminary boundary line to be demarcated.[14] Palmerston advocated the erection of "permanent" markers (by Schomburgk)

---

12. *JRGS* 10 (1841): 190.
13. Schomburgk, *Twelve Views,* 38.
14. Submitted in July 1839 to Governor Light in the form of a memoir and transmitted by him to the Colonial Office. Rivière, *Absent-Minded Imperialism,* 67.

Figure 28. An abandoned Brazilian chapel: intimating Portuguese impiety. Courtesy of the Yale Center for British Art, Paul Mellon Collection.

along the limits of the colony advocated (by Schomburgk), and recommended that a map, marked with the locations of the markers and the boundary that linked them, be sent to the various neighboring countries, less as an absolute statement of right than as grounds for further discussion. He wrote: "It would then rest with each of the three governments above mentioned [Brazil, the Netherlands, and Venezuela] to make any objections they might have to bring forward against these boundaries."[15]

In January 1841 Schomburgk arrived in the colony, commission in hand, equipped with an array of fine new instruments and accompanied by his brother (funded by the Prussian government, on Humboldt's intercession, to collect botanical specimens) and an expedition artist. After nearly four months negotiating local politics, Schomburgk set out on the first leg of the boundary surveys, the survey of the northwest frontier with Venezuela.[16]

---

15. Ibid., 68.
16. The plantocrats in the Combined Court refused to pay their share for an expedition they saw as benefiting only Britain. Local officials in Georgetown denied Schomburgk access to the land where he wanted to erect a platform for his baseline observations. Rivière, *Absent-Minded Imperialism,* 85.

Two and a half years later, Schomburgk completed the circumambulation of the colony when he arrived back in Georgetown after a perilous descent of the Corentyne, the eastern boundary. It was his final expedition into the interior; he left Georgetown for the last time in May 1844 and completed his map in London.

The unusual extent of Schomburgk's role in the boundary definition defies the traditional terminology used by political geographers and social scientists in discussing boundary formation. According to these sources, boundaries are the product of three separate stages: *allocation,* or the initial political division of territory; *delimitation,* or the selection and definition of boundary sites; and *demarcation,* or the "construction of the boundary in the landscape."[17] These tasks are seen as separate, operating with respect to different communities and agencies. Allocation and delimitation are understood to be political projects, the subject of diplomatic negotiations. Demarcation, by contrast, is understood to be essentially a technical project, in which surveyors install markers that make visible a preexisting line, invented and specified by political forces. While some interchange between these communities is inevitable—delimitation depends on access to accurate topographical maps, and surveyors may recommend alterations in the line of delimitation as a result of further field investigation—the concept of institutional separation between the making of boundaries *in principle* and the making of boundaries *on the ground* lies at the core of the analytic approach taken by social scientists discussing boundaries.[18]

This separation has no meaning in the context of the boundaries of British Guiana. Schomburgk's memoir on the boundaries of 1839—in which he emphasized the three pressing reasons to fix the boundary (protecting Amerindians who considered themselves British, claiming the Pirara portage for the development of trade, and securing the rich fishing grounds to supply the colony with food) and outlined the "natural" boundary he believed best represented the colonial limit—served as Lord Palmerston's "point of reference" in commissioning the boundary survey. Schomburgk was instructed, in effect, to demarcate the boundary that he himself had delimited. As we shall see, Palmerston's idea that this demarcation constituted a "proposition" of British territorial rights (as opposed to a "preemptive assumption" of those rights) proved difficult for British Guiana's neighbors to accept, and it resulted in considerable confusion over exactly what Schomburgk was doing. Was he installing a boundary

---

17. Jones, *Boundary-Making;* Prescott, *Political Frontiers and Boundaries,* 13.
18. The most extensive historical investigation of boundary making in the context of state formation is Sahlins, *Boundaries.*

or merely making a topographical survey of the boundary areas in order to form the "base map" for future negotiations? Backpedaling on this question by British officials while Schomburgk was in the field increased the doubt and suspicion that surrounded the surveys.

While the extent of Schomburgk's role in defining the boundary has never been overlooked, his actual practices in doing so have been. Rivière reconstructs chains of diplomatic correspondence and shows that the missteps at Pirara were based on misinformation. He thus emphasizes Schomburgk's freedom and unravels a story about the logical relation between politics and political activities. But when Schomburgk leaves his base camp at Pirara and heads into the bush on one of the legs of the survey itself, he drops out of the picture, resurfacing later to send off a report or receive a communication. Pirara is Rivière's historical stage. When Schomburgk is not there, he is "out surveying," generally to return within a paragraph.[19] This chapter examines those interstitial episodes, episodes that quite literally occur "between the points" defined by Schomburgk's landmarks. I have several aims in "following" Schomburgk's boundary surveying expeditions.

The first is to offer further evidence for the idiosyncratic character of the boundary surveys. Just as Rivière has insisted that a certain "personal" dimension not be omitted from an attempt to explain what at first glance might appear to be a centrally coordinated imperial venture to extend British territory, my examination of the passage, sight lines, and inscriptions that constituted that survey will dramatize just how personal the boundary could be, even as its making constituted an act of international significance. A second concern in examining Schomburgk's boundary surveying will be to see how the construction of the boundary drew on—and came into conflict with—the spatial practices of the traverse surveyor that have been the subject of previous chapters. Just as constructing landmarks required "weighting" sites in the interior—with history, aesthetic associations, precise coordinates, and significance in botany, geology, or other disciplines—the inscription of the boundary lines drew on these same techniques. If landmarks were, as I have argued, nodes of colonial significance in a foreign field, boundary lines emerged when these nodal points could be connected by passages, natural features, or simply the sight line of the geographical explorer. At the same time, the traverse-surveying explorer like Schomburgk defined himself by his ability to cross boundaries, surpass obstacles, and overstep limits. This made for trouble in defining boundaries.

In the three narrative sections that follow, I aim to pay sufficiently close

---

19. Rivière, *Absent-Minded Imperialism,* 85, 114, 151.

attention to the detailed practices of the boundary survey to show that the unified boundary line traced on the surface of the map is a fib. The continuity and evenness of the line suggest a continuity of character and a uniform etiology for the boundary itself. This, of course, belies the local geographical discontinuities as well as the shifting character of the practices that constituted that boundary. The line was made by wandering through space. In some regions the place of the line could be seen from a promontory, or embodied in a range of hills. In other places it had to be inferred from a supposition about hydrography or geology—if there is a river *here,* then it has sources *over there.* In some places the whole line devolved onto a key point, in other places the line was everything between two points. By insisting on these details, by drawing attention to the passages and sight lines that called the boundary into being, the hearty, blood-red circulation of the boundary as cartographic sign—and hence as political signifier—starts to meander and eddy, inosculating like Guiana creeks. Courses are seasonal. They depend on wind, rain, and clouds.

What is the point of this exercise in cartographic deconstruction?[20] For an author like Denis Wood the purpose of this sort of treatment of any feature on a map—the attempt to undermine its "well-heeled image"; the setting of the cartographic field against its own footnotes and marginalia; the refusal to allow the assessment of the map to remain on the field of positional and geometric accuracy (the field in which the map's "transparent" relation to the territory is most easily maintained)—must ultimately be to disclose the "interests" inscribed in the map. If maps are territorial "weapons," most often deployed by the strong and rich against the weak and poor, then the tools of cartographic deconstruction can be used to reveal who is doing what to whom and how they have covered their tracks. If the map links the territory with all that comes with it, then undermining the relationship between the map and the territory might be used to protect those caught in the web of territorial obligations installed by cartography. This treatment of maps relies on yoking the textual tools of Jacques Derrida to the power-knowledge formulations of Michel Foucault. It is not clear that this unruly team can draw scholar and reader toward a constructive and progressive politics of liberation from cartographic oppression. Scholars like Wood and Brian Harley, advocates of these ideas, have been criticized by those who have questioned their command of the French poststructuralist thought grounding the project.[21]

20. Harley, "Deconstructing the Map."
21. Belyea, "Images of Power"; Belyea, "Review Article of Denis Wood's *The Power of Maps.*" Subversions of the "'mimetic fallacy' of cartographic representation" have in fact been hailed as essential to a postcolonial political program. See Huggan, "Decolonizing the Map," 408.

The project of unsettling the symbolic unity of a boundary line need not run aground on the subject-denying premises of actual deconstructionists, nor need it be a paranoid search for the hegemonic forces that have made the world over to reproduce their interests and authority.[22] I think the work of Thongchai Winichakul suggests how. His study of the cartographic consolidation of Thailand demonstrates how the inscription of a boundary around the geographical "body" of Thai territory was instrumental in the construction of Thai nationality. By reifying a "geo-body"[23] for Thailand, the boundary did more than transform indigenous ideas about space and Thai identity. The boundary also made possible an anachronistic projection of "Thailand" back into Thai history. So real did Thailand's geographical body become that Thai history became a story of territorial loss, even though those "losses" were in fact integral to the process whereby the geo-body was invented. As Winichakul shows, the national fretting over lost provinces depends on the assumption that premodern Thailand had a bounded geo-body of the same kind as the current one, only bigger—"complete." This was never the case. Instead, the premodern structures of overlapping authorities and tributary local leaders came to be replaced by the European notion of a political boundary: a vertical plane perpendicular to the surface of the earth. There were losses, but not to "Thailand" (which did not exist); rather the losses were to the local authorities, who were entirely erased, and to their populations, caught in a seismic political and geographical shift.

By getting "under" the cartographic boundary to reveal its role in distorting history, Winichakul has shown how disturbing the symbolic solidity of a boundary line can be a first step in a constructive analysis. Rather than turning a historical investigation into a reductionist pursuit of self-interested agents, Winichakul's "deconstruction" of the map shows how it has functioned as a conceptual tool, how acceptance of the map as a natural depiction of territory has distorted the understanding of Thai history and influenced the formation of Thai identity. As in life, absolute winners and losers prove difficult to find.

22. For an interesting recent statement on this very question, see Sparke, "Between Demythologizing and Deconstructing the Map." Sparke's claim that deconstruction can be best deployed "as a disruptive supplement to the project of demythologization" (by militating against complacency) is consistent with my approach. Sparke (ibid., 4) is noncommittal on the fidelity of his position to Derrida. For an account of what the trajectory from demythologizing to deconstruction might look like, see Duncan and Duncan, "Ideology and Bliss."

23. Of the neologism, Winichakul writes: "Geographically speaking, the geo-body of a nation occupies a certain portion of the earth's surface which is objectively identifiable. . . . But the term geo-body is used to signify that the object of this study is not merely space or territory. It is a component of the life of a nation. It is a source of pride, loyalty, love, passion, bias, hatred, reason, unreason." Winichakul, *Siam Mapped*, 17. For a discussion of the organic metaphor for the state as applied to South America, see Hepple, "Metaphor, Geopolitical Discourse and the Military."

I have a similar aim in looking at the boundary line that codified the territory of British Guiana. There are parallels with Winichakul's case study: Guyanese nationality has been very much constructed around a shared anxiety over the integrity and future of the geo-body of the nation (anxiety that has been manipulated by political leaders, particularly since the early 1960s); Guyanese history has ended up caught to a large degree in the (circular) justification of just what Guyana is and has been.[24] Just as Winichakul's revisionist history of Thai geography has a tacit goal of undermining the militarized nationalism that justifies itself with reference to the geo-body it purports to defend (and in the name of which it calls for sacrifices of human life), my attempt to reveal the idiosyncrasies out of which Guyana's boundaries arose ultimately aims to question the status of the boundary as a line valued more highly than lives and limbs. And not, of course, just the boundary of Guyana.

At the same time, opening up the boundary—seeing it as an uneven line, cobbled together out of lines of sight and transitory passages—places the spatial encounter of an explorer right at the root of the map's most forceful inscription. Nothing is a better instance of the "off-the-ground perspective that makes possible the identification of ground-grooving with the process of exclusive enclosure and dispossession" than the map inscribed with national boundaries.[25] By showing how this line—which claims to make the land what it is—is in fact a product of the explorer's unrolling encounter with the land itself, I hope to suggest not that the line can be ignored, but that its fragility can call into question other boundaries, not least that well-marked terminus in the European imagination, the pale between nature and culture.[26]

CONSUMMATIO, LANDMARKS, AND THE "NATURAL" BOUNDARY

In chapter 4 I suggested that the construction of geographical landmarks depended on the capacity of the geographical explorer to evoke the significance of a given site in multiple frames of reference, to use a geographical version of the *consummatio* to "heap up" meanings on a single

---

24. The importance of Guyana's outward focus, its look to the West Indies for its most significant political and cultural associations, should not be overlooked. Work on Guyanese nationalism includes Despres, *Cultural Pluralism and Nationalist Politics,* and Williams, "Nationalism, Traditionalism, and Cultural Inauthenticity."

25. Carter, *Lie of the Land,* 358.

26. Three recent edited volumes have examined this issue in different ways: Cronon, *Uncommon Ground;* Descola and Pálsson, *Nature and Society;* and Teich, Porter, and Gustafsson, *Nature and Society in Historical Context.* Essays I found particularly helpful include Cronon, "In Search of Nature"; Cronon, "Trouble with Wilderness"; and Descola and Pálsson, "Introduction." Other recent works in this area include Schama, *Landscape and Memory,* and Wrede and Adams, *Denatured Visions.* See also above, chapter 4, note 35.

fixed point. Geographical position itself conferred a meaning in one of those frames (that of the cartographic coordinate field), but others were equally important. A site with a history was preferable to one without, so the passages of previous travelers, as well as Amerindian "traditions," were recorded as they related to particular points in the landscape. Invoking aesthetic impressions highlighted the significance of the point. Natural history afforded a framework in which it was possible to assert that a given site was literally *re-markable,* that it demanded the repeated attentions of the communities of geologists or botanists or hydrographers because of its innate characteristics. A site, like a promontory, that could be clearly seen, and from which extended vision was possible (in practice or in imagination), was ideal. It could more easily be distinguished from the homogeneity of the surroundings, even as it possessed the power to turn those surroundings into an object of the gaze. The concatenation of these characteristics, and their representation in images and narrative, installed the "significant stones" of the landscape. By reproducing the positions of such sites in the coordinate field of the map, the cartographic representation could be anchored to the place itself.

These landmarks became key points in the construction of the boundary. Because they were known positions, they constituted the first points with respect to which the boundary could be drawn. In his memoir to the governor in 1839, Schomburgk invoked the boundary significance of Roraima, the point known to the Amerindians as the "mother of streams." The boundary needed to be traced from the upper Cuyuni "to that singular chain of mountains, the highest of which is named by the Indian, Roriema; here the natural boundaries of British Guiana, the Republic of Venezuela, and the Brazils unite."[27] From this known point, for which Schomburgk gave his coordinates, the boundary needed to run to another point he had established, that of the sources of the Essequibo. From there the boundary would descend the Corentyne, eventually linking itself to the Smyth and Barrow Falls, the site of Schomburgk's thwarted ascent of the Corentyne in 1836. Seen in this light, Schomburgk's boundary surveys can almost be thought of as a game of connect-the-dots. The key points had been made into landmarks on his earlier expeditions, and between 1841 and 1843 he faced the task of connecting these key points by a set of circumambulatory passages.

But boundary making was more than connecting key points. The same practice used to construct a landmark also had to be used to construct the boundary. A boundary had to be a line significant in as many frames of reference as possible: it needed to be a *historical* line, weighted with pre-

27. *Brazil Arbitration, British Case Annex,* 2:91.

vious passages and rooted in previous occupations; it needed to be a "*natural*" line, conforming to hydrographic and geographical features of the landscape; it needed to be a *visible* line, either seen from a distance or from which distance could be seen; it needed to be a *passable* line so that the surveyor could follow it and fix its position. It was rare that a single line could be made that could be weighted in so many ways. Such, in essence, was the origin of the disputes over the boundaries of the colony.

Of all of these characteristics, the one to which Schomburgk paid the greatest attention in sketching out his recommendations for the form of the boundary was the "naturalness" of the line. The great virtue of a "natural" line—either (generally) a river, or its corresponding watershed—lay in its capacity to absent the boundary surveyor from the boundary altogether, both in the past and in the future. If a natural boundary could not be found, the only option was what Schomburgk disparagingly referred to as "imaginary lines": "The adoption of imaginary lines, as per example an astronomical meridian, or a parallel, has always this difficulty, that, in future disputes its adjustment can only be procured by a repetition of the astronomical observations, while if the limits be followed which nature has prescribed . . . no disputes can arise."[28] That the boundary would depend on continual reenactment of the technical acts of observation and cartographic inscription was unacceptable. What was sought was an automatic boundary that could be found automatically. In an ideal state, boundaries could be ascertained by "the most untutored mind" and by the simplest equipment: "It requires merely to observe to which quarter of the compass the adjacent stream carries its waters."[29]

On the face of things, Schomburgk's insistence on a natural boundary was linked to his assertion of the importance of the boundary as a line circumscribing a refuge for Amerindians subject to Brazilian and Venezuelan depredations. Schomburgk repeatedly expressed his concern that unless it was abundantly clear where on the land such a line "lay," Amerindians would have no way to be sure they had settled in a protected zone. Ambiguities could be used to excuse further raids. The line had to be visible on the ground, not simply on a map. Schomburgk also expressed a good deal of confidence that it was possible to "see" such a natural line for British Guiana. He wrote the governor: "I do not entertain the slightest doubt of the practicability of marking the limits of British Guiana, with the exception of one point, entirely in conformity with natural boundaries."[30] That "one point"

28. Ibid.
29. Ibid.
30. Ibid.

was actually no *point* at all. It was the enormous *region* of the upper Cuyuni, where a British claim to the watershed of the river would reach practically to Angostura itself, in the very heart of Venezuela. A claim to one bank or the other of the river would make even less sense, since the upper regions of both banks had seen Spanish settlement since the early seventeenth century, and the same could be said for the Dutch lower down. The Cuyuni region remains the most vigorously disputed of Guyana's boundaries.

This contrapositive example seems to suggest that the remaining boundaries were indeed better fixed because of their correspondence with natural features. In fact, "natural features" were valuable as much for their ambiguity as for their clarity. Schomburgk's multiple "natural" boundaries to the southwest offer an example. In one of his earliest statements on the subject of the boundaries, Schomburgk included in a letter to Sir Thomas Buxton (at the Aboriginal Protection Society) a lengthy note on how the savannas of the southwest region might be best apportioned between Brazil and Britain. Citing Humboldt, Schomburgk pointed out that the regions had "never been in the actual possession" of the Brazilians or of the Portuguese. He suggested ambiguously, "for what is known, these plains may as well form part of British Guiana."[31] Then, dispelling the uncertainty with reference to a natural dividing line, he wrote, "in all cases the division of waters between those rivers which are tributaries to the Essequibo on the one side, and to the Amazon on the other would form the most natural boundary."[32] Schomburgk concluded: "Nature itself points out the southern boundary of British Guiana."[33] It was a watershed. Looking back at Schomburgk's earlier proposal map of 1835, as well as several remarks in his narrative of the 1836 expedition up the Essequibo, shows that this "natural" watershed boundary was a considerable extension of an earlier (but no less natural) one, that of the Rupununi River itself rather than its watershed. By 1839 when Schomburgk drafted the boundary memo to the governor, the natural southwest boundary had slipped still farther to the west, right to the banks of the Takutu (a tributary, ultimately, of the Amazon). Over four years the boundary had moved from the right bank of the Rupununi, to that river's western watershed (a line between the Rupununi and the Takutu), and then again right up to the right bank of the Takutu itself.[34] All were abundantly natural; none was a natural boundary.

These expanding, concentric boundaries have constituted the central

31. RH, MS, Schomburgk to Buxton, 25 August 1838.
32. Ibid.
33. Ibid.
34. Rivière argues that the bulk of this shift coincided with Schomburgk's increased humanitarian concern for the Amerindians in the Pirara region. See his "From Science to Imperialism."

evidence for the scheming "évolution de Schomburgk."[35] If one aimed to suggest that Schomburgk was, from the outset, intent on extending British territory in the region, it was easy to assert that the westward expansion of the boundary was cleverly contrived. For Schomburgk must have known very well, writing Buxton in 1838, that the watershed between the Rupununi and the Takutu was unlikely to be clearly defined. After all, Pirara, the isthmus and floodplain that *connected* the two river systems, lay at the northernmost end of where such a watershed would have had to run. Indeed, Schomburgk himself later objected to the watershed as a boundary exactly because it was "invisible." In essence, the odd hydrography of South America (chapter 2) meant there was no such "watershed" in the traditional sense. At the end of the century Britain would defend its claim to the Takutu in its dispute with Brazil using this very argument: "Directly a watershed becomes ill-defined it loses its value as a boundary and it is desirable to seek something else which is more definite and easily recognized, especially in a country where the native tribes are deeply interested in knowing accurately what their boundary is."[36]

Schomburgk understood perfectly well that a "natural" boundary by itself was not a boundary at all. Only histories of occupation and passage could weight one natural feature as a boundary while another remained bush. For this reason he included in his memoir to the governor a résumé of the extent of Dutch occupation, posts, and trade routes (because the extent of British territorial title hinged on Britain's subsumption, under the 1814 treaty with Holland, of Dutch territorial rights) as well as details about the extent of Hancock's travels in his formal capacity as an envoy to Mahanarva. Schomburgk went so far as to suggest that Hancock had "ordered a boundary pale to be planted at Pirara."[37] If it was planted, nothing of it remained.

Schomburgk went on to detail what he understood to be the extent of Dutch settlement and military authority in the swampy lowland areas of the northwest region of the colony, concluding that the Dutch had at one time extended their sway to the very mouth of the Orinoco (at the *embouchure* of the tributary Amacura), the point he suggested would make the strategically soundest northern corner of the British colony. Emphasizing the consilience of nature and history in delimiting the British territory, he ended his memo: "My deductions from the different circumstances to which I have attempted to draw the attention of your Excellency are that it is practicable to run and mark the limits of British Guiana on

---

35. *Brazil Arbitration, British Counter-case,* chap. 6, 113.
36. *Brazil Arbitration, British Case,* 141.
37. *Brazil Arbitration, British Case Annex,* 2:90.

the system of natural divisions, and that the limits thus defined are in perfect unison with the title of Her Britannic Majesty to the full extent of that territory."[38] The boundary was a deduction from nature and history.

The passage also emphasizes the place of the boundary surveyor himself in making this boundary visible. The boundary must be "practicable to run and mark." The formulation captures the place of the surveyor's passage, and his practice of landmarking, in constructing a good boundary. The words "running" and "marking" evoke the language and practice of the cadastral surveys of civil land surveyors mentioned in chapter 3. The circumambulation of the plot produced both inscriptions on the land—in the form of landmarks—and a corresponding inscription on paper of the routes between fixed landmarks. That inscription constituted the plot of a cadastral survey and a key element in the "transport" of the land—the documents that afforded a good title of possession.[39] Schomburgk's proposition, that he "run" the boundary of the colony, invokes the cadastral survey and the power to install territorial ownership that those surveys implied. He wrote to the governor saying he hoped that the home government would commission a survey to "determine" the limits of the colony and that the commissioner would be "empowered to plant along the extent of that line, at the most remarkable points, such monuments as are not likely to be quickly destroyed, either by the influence of weather or violence, such points, for better security's sake, being fixed with astronomical precision."[40] By invoking both the cadastral survey and the astronomically based surveys of the geographical explorer, Schomburgk was clearly indicating that he could "run the line" of the whole colony, inscribe the necessary landmarks on the land and on the cartographic field, and in doing so create the "plot" of the colony that would afford good possession. At the same time, this passage would merely follow the route of a line that already existed.

In a boundary survey, the circumambulatory passage of the surveyor had a particular function. In some regions the boundary could be said to conform to a given natural feature—a river, a mountain range—but the boundary of the whole colony was composed of an assortment of stretches so defined. In the southwest, part of the boundary was, in Schomburgk's eyes, the right bank of the Takutu to its sources. But reaching those sources left no further riverbank to follow. The next "key point" in his boundary proposition was the source(s) of the Essequibo to the east. Between these two sources Schomburgk suggested that the bound-

38. Ibid., 93.
39. For a discussion of the early use of this term in the colonial context (in India), see *OED*, s.v. "transport."
40. *Brazil Arbitration, British Case Annex*, 2:93.

ary "ought to stretch along the highlands that divide the rivers that fall into the Essequibo from those that empty themselves into the Rio Branco."[41] In total, the boundary would need to run from river to watershed, back to a river, and back to a watershed no fewer than six times in Schomburgk's natural boundary. The function of the passage of the boundary surveyor was to smooth over these transitions, and place markers at those regions, that were "remarkable" only for their poor definition rather than their obvious features.

Schomburgk's particular interest in having the commissioned boundary surveyor be "empowered" to install monuments reflects his concern with the construction and permanence of landmarks. Landmarks that had the power to "leave" the expedition on the ground enabled disclosure to be turned into enclosure and installed the features of the territory. The power to call into being landmarks weighted with the full authority of the British Crown would have added an entirely new dimension to the landmarking tools and practices of an interior geographical explorer like Schomburgk, whose capacity to fix and define points was not only a way to navigate the terrain but also, as I have now argued at length, the trademark of his work.

TRAVERSING THE BOUNDARY: THE SURPASSING PROBLEM

In the kind of interior geographic exploration that occupied Schomburgk between 1835 and 1839, fixed points were fixed points on the route of the explorer. It was a premise of the boundary survey that the explorer's route itself became the boundary. This introduced a curious tension into the work of the boundary survey. As discussed in chapter 3, the identity of a geographical explorer hung on his capacity to find some terra that could legitimately be called incognita. In this sense the aim of the geographical explorer conducting a traverse was to identify boundaries—of previous settlement or passage or knowledge—and transgress them. Noyes sums this idea up when he writes that "the explorer's constant self-reflection upon his own ability to cross boundaries into unknown territories is what constitutes his mythological mastery of space."[42] This identification of the place of the boundary in the identity of the geographical explorer resonates with the work of psychogeographers like G. Raymond Babineau and Avner Falk, who have identified "a group of disturbed persons who cross borders in a driven and repetitive way."[43] For Babineau these "rest-

---

41. Ibid., 91.
42. Noyes, *Colonial Space,* 280. One presumes the "self" here is redundant.
43. Babineau, "Compulsive Border Crosser," 281.

less and peripatetic" individuals are characterized by symptoms that include: an idealization of a new country and the denigration of the old; shifting allegiances (like changing one's nationality and citizenship); the wish to flee what are perceived to be persecutory authorities; and flights from intimacy, specifically the pursuit of environments "sufficiently foreign so that the dissimilarities in language, customs, and ethnic origins put up automatic barriers to intimacy." For these individuals, Babineau concludes, "distant reality has a way of looking more malleable than the locally intractable kind."[44]

Though neither Falk nor Babineau draws any link between the profile they characterize and the geographical explorer, their "compulsive border crosser" offers a useful framework for examining a number of nineteenth-century explorers. It accords well with the character of Schomburgk himself, as best as it can be discerned from the available sources. While Rivière suggests that Schomburgk may have been a less austere and distant man than has been thought, he has certainly always been seen as a rigid and disciplined individual who eschewed intimacies and was drawn to the exotic and alien. Not only did Schomburgk remake himself as an Englishman (even inquiring at the end of his Guiana explorations about being naturalized as a British citizen), he also repudiated his homeland because of what he described as persecution of his family by authorities, idealized British humanitarian virtues, romanticized British Guiana, heaped aspersions on the neighboring Iberian regions, and continually sought new regions. Within six months of his return to Britain, after nearly twelve years of almost continuous travel, Schomburgk penned a letter to Stanley that concluded, "I take the liberty to solicit your lordship's kind patronage, if a consulship should offer itself along the coast of China . . . or any colonial employment in Hong Kong, Ceylon or Mauritius."[45] Before he died he would live in Santo Domingo and Siam.

The idea of the explorer, and of Schomburgk in particular, as distinguished by a certain compulsive relationship with crossing boundaries may shed some light on the ambiguous boundaries of British Guiana. For perhaps the ever expanding ambit of the colony—decried by Guyana's neighbors as the "évolution de Schomburgk" from mild-mannered geographical explorer to monomaniacal British imperialist—really represents nothing more than a manifestation of the contradictions that arise when a boundary crosser is asked to delimit a boundary: each boundary must overstep the last. Assigning the task of "closing" the territory to an ex-

44. Cited in Stein and Niederland, *Maps from the Mind,* 152. Work on "compulsive border crossers" includes Falk, "Border Symbolism"; Falk, "Border Symbolism Revisited"; and Babineau, "Compulsive Border Crosser."
45. *Venezuela Arbitration, Schomburgk's Reports,* 50.

plorer whose identity (and success) had always hung on his capacity to "open" the territory yielded a boundary that meandered onto new terrain whenever Schomburgk had the occasion to take a new route back to his point of origin. Having drawn the boundary, Schomburgk had no choice but to cross it. As I suggested in chapter 1, the traverse survey embodied the expansive quality of imperial territory as much as trigonometric surveys embodied its stability.

The virtue of Rivière's account of Schomburgk and Pirara lies in its placing the agency of the geographical explorer back into the matrix of colony and empire making. This is a welcome contribution to the growing trend of literature rescuing explorers from the limbo of historical inattention to which they were relegated after the demise of Whiggish imperial history in the 1960s.[46] The danger of Rivière's account is that his somewhat tongue-in-cheek invocation of Sir John Seeley—"We [the English] seem, as it were, to have conquered and peopled half the world in a fit of absence of mind"—will be taken seriously. Correcting unwarranted emphasis on rational state behavior and nuancing overly monolithic pictures of a coordinated, nefarious imperial enterprise both advance the understanding of imperialism. But the suggestion that Europeans sleepwalked into territorial hegemony risks serving as an exculpatory shrug. A revised Seeleyism may still serve, however. My interpretation of the boundary surveys indeed suggests that territorial appropriation occurred in a fit; less a fit of absentmindedness, however, than a fit of compulsive boundary crossing. An etiology of that affliction would have to reach back to the core of European ideas about knowledge and transgression.[47] If there is truth to this explanation for the ambiguity of the boundaries of the colony, it would suggest that the imperial strategy of territorial codification may have been subverted from the start by the incompatible tactics of the traverse surveyor.

NARRATIVE PART I: NORTHWEST

On 18 April 1841 the boundary commission left Georgetown by steamer for the mouth of the Waini River in the northwest. Richard Schomburgk described how punters on the wharf made odds against the success of the boundary mission as a whole as well as of its various parts. As parts went, the northwest district presented the most difficult terrain. The region, in effect part of the delta of the Orinoco, is a swampy, low-lying area, criss-

---

46. Bridges, "Historical Role of British Explorers," 2.
47. For a discussion of exploration, enlightenment epistemology, and transgression, see Outram, "On Being Perseus."

crossed by a labyrinth of narrow waterways. It is sparsely inhabited and without any significant promontories. The area was best known as a warren for smugglers who ran contraband British goods into the neighboring settlements on Venezuelan territory.

Schomburgk's effort to bring order to this maze can be best appreciated by a look at the first map he and the artist for the expedition produced for the Colonial Office (fig. 29).[48] Its long title, "The Limits of British Guiana Surveyed under Her Majesty's Commission by Robert H. Schomburgk," suggests that Schomburgk surveyed the boundary, but the very detailed map discloses that little of Schomburgk's actual surveying was done along the line he indicated as the boundary. On several of the rivers, lines mark the bearings from which the course of the river was constructed, but no part of the boundary line shows any such bearings. Moreover, the map is keyed to the methods of its own production in a way that few of Schomburgk's maps were: each fixed point not only is marked but has been coded to show the kinds of observations that fixed it. This affords a unique opportunity to see in detail the data out of which the map was made. Points determined from observations of longitude and latitude are in short supply along the boundary itself.

Considerably easier to see than the small runes that indicate astronomically determined points, however, are the framing landmark views painted by Edward A. Goodall.[49] The map of the boundary deploys the micro/macro strategy introduced in chapter 4. The sketch prominently displayed under the title of the map depicts "Victoria Point from the Mouth of the Amacura" (fig. 30). It is a scene that Schomburgk would have wanted to highlight, representing as it did the clearest point established in the boundary survey of the northwest district. On 10 May the surveying party of the boundary commission, consisting of Schomburgk, William Leary Eichlin (ad hoc draftsman), Lieutenant Glascott (assistant surveyor), Robert King (a justice of the peace), medical school dropout Thomas Hancock (son of more celebrated John), and a crew (eleven mixed-race boatmen and seven Warrau Amerindians), set out from the

---

48. Figure 29 has been reproduced from *Venezuela Arbitration, British Case Atlas*, 44–45.

49. Goodall was not actually along on the boundary expedition's survey of the northwest region, and he must have prepared the views that decorate this map from sketches done by the expedition's interim artist, William Leary Eichlin. Eichlin, a medical man, escorted the northwest expedition after the commission's first official artist, W. L. Walton, resigned from the commission shortly after arrival, boarding the first ship on which he could secure passage back to England. His had been discomfited by Richard Schomburgk's brush with the "black vomit" of yellow fever. See Menezes, "Sketches of Amerindian Tribes," 12. Goodall, who replaced Walton on the recommendation of the secretary of the RGS, wrote in his journal of working on a large colored map for Schomburgk in October 1841. At the same time, the artist recorded doing copies of several of Eichlin's sketches. It seems likely that he was engaged in preparing this map and its images. See Goodall, "Diary July to December," 48–50.

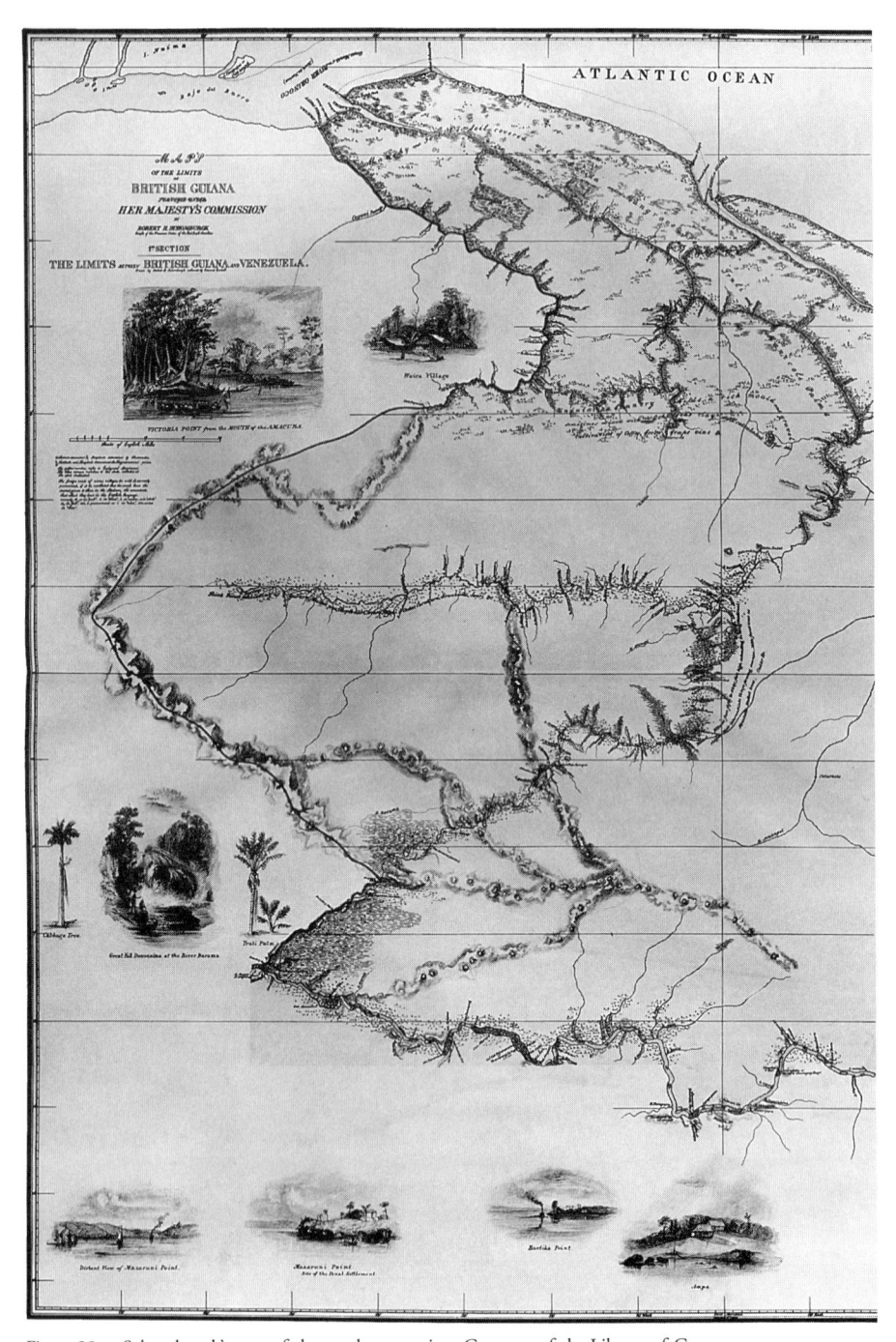

Figure 29. Schomburgk's map of the northwest region. Courtesy of the Library of Congress.

Figure 30. Detail from Schomburgk's northwest map: the point of departure, Victoria Point. Courtesy of the Library of Congress.

camp at the village of Cumaka for mouth of the Barima and Amacura Rivers.

In his report to the governor on the success of the northwest expedition, Schomburgk wrote of discovering, on arriving at the mouth of the Barima, evidence for the Dutch post that was supposed to have stood on the point and that was the source of his assertion that the British boundary began there. "The site of our camp at the mouth of the Barima," he wrote, "gave evident proof that the ground had been under cultivation, and the environs showed vestiges of trenches."[50] Admittedly shallow archaeology, but Schomburgk felt it could be buttressed: "I noted some straggling cassada [cassava] plants and a few shrubs of arnotto, which does not grow wild on grounds subject to tidal influence." Never had so much rested on an errant shrub: "These circumstances, simple as they appear, contribute to attest to the undoubted right of Her Majesty to the Barima, and all the tributary streams which flow into it."[51]

But in a region distinguished by its mazelike creeks, the Barima and all

50. *Venezuela Arbitration, Schomburgk's Reports*, 3.
51. Ibid.

of its tributaries did not constitute a sufficiently clear boundary in Schomburgk's eyes. The claim on the Barima meant a claim on the regions watered by the web of the river, and in his opinion it was impossible to distinguish this "watershed" from the "watershed" of the neighboring rivers. Hence, although he and the party commenced on Barima Point and arranged to "plant the first post with such ceremony as circumstances would permit,"[52] on the following day Schomburgk moved the party south, where they went through the same process at the mouth of the Amacura, a parallel river to the west of the Barima. As Schomburgk explained, "As in the demarcation of a territory it is of great importance to fix upon a line of boundary which is permanent and fixed in nature, and which cannot be destroyed by human hands, I thought it advisable to claim the eastern or right bank of the River Amacura."[53]

The justification for Schomburgk's first transgression of his own boundary is what the rhetorician might call "close conceit." It is subtle, obfuscatory, and dissembling, but presented as a logical argument. The watershed of the Barima could not be "destroyed by human hands" any more than the right bank of the Amacura, nor could a watershed be called artificial or unnatural. But the right bank of the Barima, in Schomburgk's opinion, would curtail the boundary of British territory too much. So after setting up a "pale" at Barima Point, "planted as an attestation of Her Majesty's undoubted right of possession to that river,"[54] Schomburgk moved south and west, to the next point, and set up an identical post with a different meaning. This one, he explained, was "a testimonial to Her Majesty's claim to its [the Amacura's] right bank as the boundary of British Guiana."[55] This more speculative point of possession, unsubstantiated by straggling arnotto plants and (owing to the uncooperative weather) impossible to fix astronomically even after seven days of camp, demanded that Schomburgk deploy his array of landmarking practices with particular assiduousness. For both points (Barima and Amacura) Schomburgk prepared an elaborate affidavit describing the posts ("branded with Her Majesty's initials") and their positions, fixed by a longitude and latitude (established by dead reckoning) as well as by compass bearings to the other spits of land visible. The document was undersigned by all the present members of the expedition, including the (illiterate) members of the crew. For Amacura Point he went further, having Eichlin do a sketch of the point, the boats, and their flags. The view (redrawn on the map by Goodall) celebrated the rechristening of the point as "Victoria

52. Ibid., 2.
53. Ibid., 3.
54. Ibid.
55. Ibid., 2.

Point," a toponym that reinforced the legitimacy of Schomburgk's first hyperbolic boundary gesture. He had marked a boundary and then transgressed it.

A nice view of the commission's admiralty-style sail-past helped clarify what Schomburgk had decided to call the true "point of departure" for the boundary survey. But it could not entirely disguise the fact that the actual surveying of the boundary was somewhere between a failure and a fiasco. The weather remained bad, no promontories could be found, and a running survey of the Barima, using sound, had led to disaster. The process—requiring the coordination of men's firing a mortar and Schomburgk's timing the flash/report interval—came to an abrupt halt when King sustained a gruesome injury by misfiring the signal cannon.[56] Schomburgk downplayed the "temporary injury" in his report to the governor and the subsequent one to the *Journal,* but King, in his enthusiasm to prepare for the second closely timed shot, had poured powder into the mortar while it was still hot, and it had gone off in his face as he stood over it (with the canister of powder in hand), blowing him thirty feet away and blinding him. An Amerindian who had been holding the swivel mount upright was also badly burned. No further reference to the wounded Amerindian exists; King recovered his sight when the swelling subsided. The blinded surveyor, however, blinded the survey, and the party returned to the base camp at Cumaka to recover.[57]

Having failed to get much of a survey for Barima Point or the Barima River, Schomburgk opted for an ascent of the Amacura, the right bank of which he had designated as the boundary. Between 20 and 27 May Schomburgk waited at Cumaka, hoping for a break in the weather and the recovery of the assistant surveyor, down with fever. When no change appeared on either horizon, Schomburgk decided he would make the Amacura ascent with the able-bodied. Given the damaged feet (ulcerated by chigoes) and fevers of the Europeans, this left only the artist, Eichlin, and sufficient crew to man one boat. In his report Schomburgk explained that the ascent was a "pioneering expedition." He brought only a handful of instruments: chronometer, sextant, artificial horizon, and prismatic compass. It was eleven days before the clouds cleared so that Schomburgk could test the chronometer, by which time he no longer trusted his ability to interpolate its rate. Since other celestial observations proved impossible, this did not matter.

The ascent of the Amacura, accompanied by the leader of a local vil-

---

56. "His whole face was blackened and trickled with blood which had formed a thick coagulum on the top of his head and on different parts of his body: his hands and arms appeared in the same condition." M. R. Schomburgk, *Travels,* 1:108.
57. *Venezuela Arbitration, Schomburgk's Reports,* 3.

lage of Arawaks and Warraus, took only two days, and for most of one day the party was unable to move because of the heavy rains. The good news was that the river's course was considerably to the west of that shown on earlier maps, extending British territory; the bad news was that navigation on the river was impeded by a twelve-foot cataract quite close to its mouth. Neither astronomical observations nor a rough local triangulation of the river could be established. If the surveying was proving disappointing, Schomburgk found among the local communities plenty of stories that bolstered his claim that the boundary mattered. Although he admitted that locals' tales of the cruelty of Venezuelan smugglers might be embellished, he reported them anyway, since they provided excellent evidence of two things: first, that the government had been right to fund the survey of the whole boundary of the colony (and not merely those regions disputed with Brazil) and, second, that the choice of the Amacura would reap rewards in the protection of these innocents. Conscripted to labor on neighboring Venezuelan plantations, those unable to finish their tasks were, he explained, "drawn up with their hands to a beam in the workhouses, and, when thus hanging above the ground, were unmercifully beaten." Schomburgk declared, "I will not relate any more of the cruelties," but added one last violation of all boundaries that occurred on the unrealized boundary itself: "It frequently occurred that Indians who travelled with their families in canoes had been overtaken by Venezuelans who, after having tied the men, had violated in their presences their wives and daughters."[58] At the village, situated on a concealed tributary on the right bank, Schomburgk left another affidavit, which he hoped might aid the inhabitants in their defense. It read in part, "I the undersigned, her Britannic Majesty's Commissioner for Surveying and Marking out the Boundaries of British Guiana, by virtue of the commission graciously granted to me by Her Majesty, and at the express desire of Her Majesty's Government 'That the native tribes within the assumed limits of British Guiana must not be molested,' hereby solemnly protest against such proceedings."[59] Schomburgk carved Her Majesty's initials on a nearby tree.

On returning to Cumaka, Schomburgk found that King had recovered his sight and had begun to perform, unbidden, his tasks as a justice of the peace. In particular, King had heard rumors from a local leader of a recent murder in a neighboring village. In his capacity as commissioned officer of the queen's peace, King felt himself obliged to investigate. Schomburgk objected to his meddling and later protested to the governor that Maicarawari, the young man King arrested, was not "amenable to the

58. This quotation and those above are from ibid., 4.
59. Ibid., 10.

Courts of Law of this colony for the deed which he has committed."[60] There was no obstructing King's sense of duty, however. It appeared that the boy, about twelve, had lost his mother and father under conditions that the Amerindians attributed to the effects of Kanaima, initiated, it was widely believed, by a feared local *piai,* or shaman. As best King could understand, the *piai* had not denied the charges but had threatened the boy. As for the outcome, little doubt could remain after King insisted on opening the shallow and recent grave of the dreaded *piai,* on whom Maicarawari had in the end taken his revenge. Richard Schomburgk recorded that it had taken considerable bribes of extra rations to persuade one of the boatmen to begin the exhumation:

> As soon as the stink reached the Indians standing at a distance, they ran off in the wildest terror, screaming with fright. The boy alone remained, as if rooted to the spot . . . after the body had been quite uncovered and the hammock unfolded there awaited us a more terrible sight, before which everybody present, except Maicerwari alone, unconsciously recoiled. The blow had crushed the whole right side of the head, and the split bone had been forced inside so that the brain lay exposed.[61]

In the event that doubts remained, Maicarawari, emboldened, offered a pantomime of the revenge killing he had perpetrated while the *piai* dozed in his hammock, drunk from feasting. Richard was disappointed that the skull was too badly damaged to take as a specimen.

King felt he had no choice but to take Maicarawari and a set of witnesses to Georgetown and hand them over to the authorities. According to Richard, Robert remonstrated with him, even contesting the legality of the arrest, given the territorial and jurisdictional uncertainty of the point where the killing took place. The prospect of a division of the expedition, the loss of King's assistance, and a delay likely angered Robert Schomburgk even more than the injustice of criminalizing the lex talionis that predominated among the Amerindians.[62] It was this that he emphasized in his report to the governor when he wrote: "But a serious question arises whether the Indian, who has no knowledge of Christian religion, and does not acknowledge our laws, can be punished for an act which civilized nations consider a capital crime, but which according to the manners and customs he has been brought up in is a meritorious deed."[63] Schomburgk emphasized, however, that the "deed was committed within the assumed

---

60. *Venezuela Arbitration, Schomburgk's Reports,* 7. I follow Menezes (*British Policy,* 138) on the spelling of Maicarawari.

61. M. R. Schomburgk, *Travels,* 1:124.

62. For a discussion of the colony's attempts to establish the character of jurisdictional authority where Amerindians were concerned, see Menezes, *British Policy,* chap. 5. She also discusses the Maicarawari case: ibid., 138–39, 143, 150–51.

63. *Venezuela Arbitration, Schomburgk's Reports,* 7.

limits of British Guiana." To underline that this was so he gave a two-part description of the locality that revealed the ambivalence of the very first boundary: the event occurred "east of the Amacura, and in a river which falls into the Barima."[64] Although this meant Maicarawari came under the jurisdiction of the colony, Schomburgk hoped the governor would throw the case out.

He did not, and Maicarawari, delivered unfettered to Georgetown, discovered he would be expected to remain there until March of the following year, when the criminal court came into session. He fled, only to be picked up and held in a small cell for nearly ten months. Richard Schomburgk explained that "this overdone zeal of Mr. King cooled my attachment to an otherwise honest man, and up to this day I cannot forget the twelvemonth which he certainly made the unhappiest of the lad's life."[65] It was not longer only because Maicarawari was acquitted.[66]

As much as the Maicarawari incident divided Schomburgk's commitments—to the integrity of the boundary (which he was loath to disavow) and also to humanitarian loyalty to the Amerindians (for whom he labored to inscribe that boundary)—it also divided the commitments of the boundary survey. The boundary commission was left with no choice but to split at Cumaka, King taking one corial and conveying Maicarawari, witnesses, depositions, and the invalid crew members back to Georgetown while Schomburgk, with the assistant surveyor, Glascott, and the artist, Eichlin, struck out via the Kaituma River to portage to the upper Barima and eventually make their way to the Cuyuni, to survey as much of the boundary as they could reach between that river and the upper Amacura. The split highlighted the double commitment of the boundary commission. Schomburgk and Glascott were surveyors charged with producing a "plot" of the territory. But just as, at the coast, the labor of the sworn land surveyor depended on the system of mobile surveillance afforded by superintendents, the boundary survey too had a function of surveillance and of enforcing British regulations.[67] King, as justice of the peace, embodied this function. Schomburgk was inscribing the territory by drawing the boundary, while King was giving substance to the assertion of British rule by enforcing the law. The two functions proved irreconcilable for the boundary commission. In the end, Schomburgk's refusal to appear at

64. Ibid.
65. M. R. Schomburgk, *Travels*, 1:123.
66. Not, it appears, on the grounds that Richard Schomburgk gave—that Robert swore Cumaka was situated in disputed territory (it is not clear that Robert ever gave such a deposition)—but, according to Menezes, for want of sworn witnesses. The issue of Amerindian oaths was a complex one. See Menezes, *British Policy*, 136–44.
67. An article on surveying and surveillance on the margins of colonial settlement in British Guiana (prepared from material in the NAG) will appear separately.

Maicarawari's trial undermined the case by the absence of a key witness, and the attempt by King and the law courts to compel Schomburgk's cooperation in the trial nearly blocked the boundary survey of the southwest. Divided, the boundary survey reflected the tensions between the desirability of the boundary and the complicated obligations the boundary entailed. The jurisdictional headache of the Maicarawari case was one of the first incidents to give pause to the Colonial Office and the home government about the wisdom of Schomburgk's boundary project. The division of the boundary commission at Cumaka was less an amicable division of labor than the first step in the unraveling of the boundary itself.

Irritated by what he perceived as King's intransigence, and likely miffed at the poor impression that would be left on the governor and the skeptical plantocracy by the return of the motley secessionist boat to Georgetown, Schomburgk and his remaining crew pushed on to extend the boundary to the south. The first step was to ascend the Barima to as close to its western sources as possible. The sources could not be reached, but Schomburgk reported that the Amerindians in the region claimed not to know a "white man ever to have penetrated so far before."[68] This assertion, included in Schomburgk's official report, provides another glimpse of the tension between Schomburgk's spatial practice as an explorer and his commission as a boundary surveyor. For while the absence of any Europeans on the river meant that Schomburgk had found some terra incognita, it also meant he had overstepped history in establishing the boundary. Overstepped it quite literally, since the Amerindians had pointed out to him the place they called "the last place of the white man," a point Schomburgk decided must have been settled by a Dutchman early in the century. Schomburgk dutifully entered the point on his map (under the full appellation) and then struck out into new territory. The name, damning to British claims to the western portions of the river in later disputes with Venezuela (where actual Dutch *occupation* was dispositive), disappeared from later British maps.

Schomburgk's intrepid boundary transgressions pushed the putative boundary as far west as he could go. He reported: "We might have stopped here, and commenced our return, the more especially since the weather was so unfavourable; but I found the course of the Barima so different from what it is laid down on maps that I considered it of importance to trace it higher up, as, by its western course on its ascent, every mile would add to the British territory."[69] He used the occasion of the river's unexpected western reach to put in a plug for the value of the sur-

---

68. *Venezuela Arbitration, Schomburgk's Reports*, 13.
69. Ibid.

vey, writing: "This course, differing so much from the Barima of theoretical geographers, will I presume, be deemed sufficient evidence of the importance of the measure which Her Majesty's government have resolved upon, namely, that an actual survey should prelude the definitive negotiations with the governments interested in the determination of these boundaries."[70] This assertion—that his activities were merely a preliminary survey for a base map on which diplomats could trace out the boundaries—was belied by Schomburgk's boundary marking: he marked three trees with the queen's insignia at Wanama junction; then, going beyond his own mark, he "armed the most effective of the crew with cutlasses and axes," and they "pathed" themselves west for two days and marked yet another tree.

The assistant surveyor, Glascott, had not held up physically, and so, preparing for the last leg of the northwest boundary—a long portage to the Barama, followed by its ascent and a further portage to the Cuyuni—Schomburgk sent him back to the coast with the chronometers, which Schomburgk did not believe would survive the trip. The trip was indeed a slog, and the second portage, across the watershed separating the tributaries of the Barama from those of the Cuyuni, followed a path Schomburgk called, simply, "tortuous." Only on reaching the top of the low ridge between a creek of the Barama (the Aunama) and a creek of the Cuyuni (the Acarabisi) did the boundary objective of this difficult trek become clear. There Schomburgk paused and looked northwest. This brief glance amounted to the "survey" of the boundary. Somewhere in that direction lay the unreached sources of the Amacura and the Barima. Schomburgk's view along the hills to the west became a claim. He wrote:

> We had now reached the most elevated spot between the Cuyuni and Barama rivers. . . . From this ridge of hills the natural configuration of the ground is sloping toward the Cuyuni southward; and I estimated the highest ridge that separated the two systems at 520 feet above the level of the sea. Heights which really deserve the name of mountains commence some 20 miles further westward. Nevertheless these ridges of hillocks are of importance in the determination of the boundary, on the principle of natural divisions.[71]

That being the case, Schomburgk "claimed them, accordingly, to form the limit from the source of the river Amacura, passing south eastward the sources of the River Barima and Barama."[72] The northern slope of the hills was British, he declared, and the southern slope Venezuelan. Schomburgk's prospect from the top of a forested hillock of 520 feet linked, in

70. Ibid.
71. Ibid., 17.
72. Ibid.

Figure 31. Detail from Schomburgk's northwest map: the range of the boundary. Courtesy of the Library of Congress.

effect, three invisible points and turned them into a line, a line that was considerably embellished on the map by the drawing of a very abrupt and strongly marked mountain range that led from the point where he stood directly to the sources of the Amacura (see the southern portion of the range in fig. 31). Because elevations were not shown, Schomburgk's 520 foot "hillock" could not be distinguished from the "mountains" to the west. Schomburgk's visual line was weighted by a mountain chain. Compiling his observations later, Schomburgk ran into the awkward fact that the Barima stretched too far to the west to make possible the configuration he believed he had seen. This obliged his ridge to bend into an elbow (see the left side of fig. 29) and meant that whether Britain possessed the northern or southern slope depended on the part of the range in question. But regardless of how they had to be shuffled in drawing the map, the hills made a visually persuasive natural boundary.

A planimetric analysis of the accuracy of Schomburgk's map is not my aim.[73] There are indeed hills that form a watershed between the Matupo

---

73. For a discussion of the methods, validity, and utility of such studies, see Blakemore and Harley, *Concepts in the History of Cartography,* 54–70, and Ballard, "Analysis in Historical Cartography Studies." For an exemplary recent effort, see Bendall, *Maps, Land, and Society,* 50–66.

and the Barama; as it happens, there really are no such hills to the north that lead to the source of the Amacura. For Schomburgk, however, the scattered hills between the low point where he stood and the mountains he could see on the horizon had to be collected into a "range" and represented as a barrier connecting the place where he stood to the last place he had reached that was truly on the boundary line: the fall on the upper Amacura. His view to the west gathered up the hillocks into a line that circumscribed a route he could not take, that enclosed all of the uppermost points he had reached on the rivers in between. This boundary line, then, overstepped all his previous marks in the region.

The boundary function of the range also terminated, as far as Schomburgk was concerned, with his person. Instead of continuing to run "naturally" along the range to the south, the boundary *crossed* the range where Schomburgk did: he carried the boundary off the range to the west. From the point on the 520 foot elevation, his route (and the boundary) crossed the watershed to the sources of the Acarabisi creek, where Schomburgk claimed the "right bank . . . as forming part of the western limit of British Guiana."[74] Trees along the way received the queen's mark. The boundary had effectively stepped off one natural line and onto another, linked by Schomburgk's eyes and feet. Where the Acarabisi flowed into the Cuyuni, Schomburgk made his last act of formal possession taking in the northwest.

Here he confronted that confusing "point" where the boundary had to transect, in midcourse, a river five hundred yards wide. This awkward moment was smoothed in Schomburgk's official report by a digression on the history of the region and the local inhabitants, all of whom affirmed that the Spaniards had never missionized this far east. Schomburgk cited several references in Humboldt and other authors to the effect that the Dutch had traded on the Cuyuni and that the last Spanish post dated to the royalist period (and had been situated about twenty miles west). As for marking the point, Schomburgk marked trees on both sides of the river, erecting, in effect, a gate on the river. The party returned down the Cuyuni, reaching the mission on the Essequibo on 27 July after a journey of one hundred days.

NARRATIVE PART 2: SOUTHWEST

Schomburgk had completed the boundary survey of the northwest, but shortly after his return to Georgetown, in the midst of his preparations for the journey to Pirara and the commencement of the survey of the boundary with Brazil, it started to become clear that the boundary survey

---

74. *Venezuela Arbitration, Schomburgk's Reports*, 17.

was raising more flags in diplomatic circles than it had managed to hoist on the western frontier. Rumors reached Georgetown that a party of Venezuelan military men had crossed the mouth of the Orinoco and cut down the posts on Barima and Victoria Points. A letter from the British consul at Caracas communicated the Venezuelans' "utmost surprise and alarm" on hearing that a "British" surveyor had staked a claim to a point in the Orinoco's mouth.[75] Light sent the message to Schomburgk, who replied by invoking his commission and Britain's "historical" right to the Barima. As for the bank of the Amacura as boundary, he pointed out again "the absolute necessity that the boundaries of British Guiana should be based upon natural divisions and not imaginary lines."[76] Schomburgk also disparaged Venezuelan geographical knowledge of the region, claiming their communication proved they did not even know where the Amacura was.[77]

Schomburgk's somewhat terse reply to Venezuelan diplomatic rancor did not suffice. Within a month, Venezuelan commissioners arrived in Georgetown to protest directly. Light again queried Schomburgk concerning the precise basis of his claim to the Barima in toto and the right bank of the Amacura as the boundary. To this request Schomburgk submitted a more weighty missive marshaling such historical evidence as he was able to find. This included the Dutch history of the colony, comments in Richard Rolt's *History of South America,* and allusions to the region in Humboldt, as well as two maps dating from the end of the eighteenth century, both of which, he claimed, predated British interest in the area and substantiated Dutch possessions ending at the Barima. On reaching the end of this evidence for the Barima as the boundary, Schomburgk acknowledged that he had not yet accounted for his overstepping the Barima in preference for the Amacura. In answering this question Schomburgk switched from a discussion of historical occupation to the story of a contemporary boundary dispute—that between the United States and Canada—in order to return to the theme of the natural boundary. This recent dispute over the correct placement of the forty-ninth parallel proved that "to prevent future misunderstandings where limits are to be determined between adjacent territories, permanent or natural boundaries ought to be selected, such as rivers, ridges of hills, etc., which ascertained with astronomical precision, leave no grounds for dispute."[78]

75. See exchanges reproduced in *Official History,* 6–19.
76. *Venezuela Arbitration, Schomburgk's Reports,* 21.
77. The Venezuelans believed the mouth of the Amacura to be five leagues up the Orinoco; it was less than five miles upstream.
78. *Venezuela Arbitration, Schomburgk's Reports,* 23.

The ground they left for dispute, of course, was the ground between one natural boundary and the next. This Schomburgk passed over, drawing the governor's attention to the bank of the Amacura, because "the savage, and the common population in general" knew which side of the river they were on, while an astronomical meridian could "only be ascertained by an astronomer."[79] The only other option was to cut a track representing the boundary, as was done to demarcate plantation and timber concession limits.[80] Schomburgk pointed out that such an alley would be nearly impossible to maintain in the swamps. Along with this letter Schomburgk sent a separate and confidential addendum, one not intended to be passed along to the Venezuelan ambassadors, which detailed the military value of Barima Point.

The Venezuelan commissioners interviewed Schomburgk, who appears to have emphasized the provisional nature of the survey and the "geographical" character of the markers erected.[81] The flag, reported by the Venezuelan commander of the post on the northern shore of the Orinoco, was said to have been raised by an overeager Amerindian (chapter 2). The Venezuelan officials were not satisfied when they departed, and the diplomatic correspondence with London continued over the next four years, culminating in Lord Aberdeen's offer of 1844, which would have given Venezuela most of the disputed region. The offer, never accepted, was retracted a decade later. At the end of January 1842, however, the Foreign Office requested of Governor Light that the boundary marks in the northwest be removed.[82]

By that time Schomburgk and the boundary expedition were well on their way to Pirara, their base of operations for the next year. There had been numerous delays in their departure, including additional disputes about funding, but most of the complications could be traced to the decision to send troops to occupy Pirara.[83] Light had advocated the idea and had thought such an occupation might be accompanied by an effort to found a colony on the site, which would both foster trade with the Amazon regions and attract Amerindians. The Colonial Office dismissed the colonization proposal but passed the issue of the occupation to the Foreign Office, where Lord Palmerston approved it. The Colonial Office fretted over costs but ultimately ratified the idea. Light selected a unit from Georgetown's West India Regiment.[84]

79. Ibid.
80. This practice is discussed in NAG, Loose reports, "Dispute over Three Friends," 1840.
81. Webber, *Centenary History,* 206.
82. *Official History,* 16.
83. Rivière, *Absent-Minded Imperialism,* 89.
84. Ibid., 83–84.

Schomburgk himself, while in favor of the occupation of Pirara, appears to have been less than entirely enthusiastic about the prospect of a military expedition, which he knew would result in delays and which he suspected would do little to "improve the morals of the natives."[85] He preferred the idea of a civilian guard, perhaps a handful of police, who could hold Pirara and assist the survey of the Cotinga and the Takutu. He ended up with some thirty-five soldiers at arms and officers, a party requiring a month of boat building just to be equipped with transportation. Schomburgk would again have his boundary survey coupled with a surveillance force. This time the military would vastly outnumber the survey party itself.

For all the delays incurred in preparing this military venture, it was the Maicarawari case that almost put a stop to the ambitious plans for the Pirara expedition. During most of the five months from late July to December 1841 that Schomburgk and the boundary commission had been in Georgetown, working up the results of the northwest surveys, Maicarawari had been languishing in jail, awaiting his March trial. Richard Schomburgk wrote of visiting him regularly and claimed that his brother had used his clout to secure Maicarawari an outdoor work detail and to have his straw mattress replaced by a hammock.[86] The preparation of the case documents and the preparations for the expedition's departure converged in late December, culminating in a court-ordered writ obliging Schomburgk to remain in Georgetown to give testimony at the forthcoming trial. The "warrant" for Schomburgk was issued the very day before his departure, and it would likely have obliged him to forgo the Pirara expedition for a full year.[87] It was a suspicious coincidence, suggesting that the affair was an effort on the part of Governor Light's enemies in the colony (of whom there were a large number) to embarrass him by obstructing the boundary surveys, which the Court of Policy had again refused to support financially.[88] If the subpoena of Schomburgk was indeed a plot to thwart the boundary commission's expedition to Pirara, it failed, but only just. According to Richard Schomburgk's account, Robert heard rumors of the proceedings on the eve of their planned departure. Gathering his effects in a rush, Schomburgk

85. Ibid., 86.
86. M. R. Schomburgk, *Travels*, 1:222.
87. The delay caused by a full trial (of such complexity) would probably have pushed his departure back enough that the expedition would have lost the better part of the season before the rains, obliging them to put the trip off until the following year.
88. Light's use of his "discretionary fund" to meet the colony's contribution to Schomburgk's expedition had outraged the governor's opponents. Light was on tender ground concerning financial matters in the colony, having precipitated in 1840 a government-collapsing crisis over executive powers of taxation. See Webber, *Centenary History*, 202, and Clementi, *Constitutional History of British Guiana*, 113–15.

slipped out of Georgetown after arranging to meet the expedition's steamer upriver the following day.[89]

The inauspicious inception of the expedition to Pirara augured ill fortune to come. Tensions between the military and civilian branches of the party ran high from the beginning, over such issues of etiquette as precisely who merited the title of officer.[90] Schomburgk's eagerness to precede the military party in order to be the first to come ashore at Pirara (perhaps to prepare the Amerindians for the arrival of the military force), and his having adopted for the occasion an elaborate military costume, led the officers of the troops taking up the rear to rule Schomburgk a cad. The regimental officers deemed pompous the message, recording toasts drunk by the boundary "officers" in honor of their military comrades, that Schomburgk left in a champagne bottle atop a bamboo pole in the midst of the Rapu rapids.[91] Given that the two branches of the expedition were engaged in a tacit race to be the first to reach Pirara, the note may have seemed like a snub. The desolate site of Pirara further disgruntled the military men, and several incidents of soldiers making attempts on Amerindian women and girls further estranged Schomburgk, the humanitarian, from his martial escort. Pirara had been deserted by the Brazilians, and by most of its Amerindian inhabitants, so that the joint forces of the expedition (the military force caught up to Schomburgk just outside Pirara) encountered neither resistance nor welcome when they made their way from the landing to the village. Nor would farewells attend the departure of the military unit when they burned their temporary fort less than six months later and returned to Georgetown, recalled by Light's order. The Colonial Office, which had had reservations about the military expedition from the beginning, seized on the first prospect of a compromise with Brazil (an agreement to leave the territory open only to the missionaries of both countries) to avoid further expense in maintaining the occupation. Declaring victory (the "liberation" of Pirara from Brazilian troops), the Colonial Office ordered Light to recall the force, which he did in August of 1842.

Though the military occupation thus proved a debacle, the surveying had gone much better. The northwest region had been a surveying nightmare—flat, overcast, labyrinthine—but the savannas around the Takutu presented a nearly ideal survey landscape. It was here that Schomburgk found his "natural heliotropes," and it was here that he was able get a bearing on the familiar landmark of Ataraipu, almost thirty miles away. The

89. M. R. Schomburgk, *Travels*, 1:223. When the warrant arrived at Richard's door he refused to touch it, so as not to be obliged (under local common law traditions) to deliver it.
90. Rivière, *Absent-Minded Imperialism*, 99.
91. M. R. Schomburgk, *Travels*, 1:295; Rivière, *Absent-Minded Imperialism*, 98.

other advantage of the region, besides the visibility, open terrain, and good weather, was the convenient correspondence of river and route that allowed Schomburgk to pass along the boundary line itself. After a brief trek from Pirara, the boundary survey reached the confluence of the Takutu and the Mahu.[92] The first step in surveying the southwest boundary was the ascent of the Takutu to its sources, conducted between March and the end of May 1842, while the troops at Pirara waited out rumors of approaching Brazilian armies and entertained themselves by shooting vultures. No sooner had the troops departed (at the beginning of September) than the boundary survey made the same short portage again, this time to descend the Takutu to the confluence of the Cotinga, which Schomburgk and crew then ascended all the way to Roraima. From there he and a light party continued north, reaching the Cuyuni on 4 January 1843. This expedition, fourteen days' march from Roraima to the Wanumu, a southern tributary of the Cuyuni, through territory still considered essentially impassable, is one of Schomburgk's most remarkable exploits.

The Takutu expedition had particular significance for Schomburgk. It was this boundary that promised to afford protection to the Macushis, whose woes at the hands of the Brazilian "recruiters" had initiated Schomburgk's boundary efforts. The four marks that inscribed the Takutu boundary clearly indicate its function: while one marked the confluence of the Takutu and the Mahu (the point of departure) and another was made at the headwaters of the river (the terminus), the two others were strategically placed to afford protection to the Amerindian villages on the right bank of the river. One mark faced Brazil at the site of Tenette, the village whose residents Schomburgk had seen delivered to Fort São Joaquim after a raid that he himself had deflected from Pirara. The other mark faced the path from the Tuarutu Mountains to the southwest. Schomburgk inscribed the boundary to defend the settlement and the means of approach to the settlement. At Tenette he spent five days "occupied from morning to night" with his trigonometric observations, "to the astonishment of the Wapishana Indians who at the commencement could not conceive for what purpose I underwent such fatigue."[93]

At the village, Schomburgk secured the assistance of an older Amerindian man who served as an onomastic aide. This man, anonymous in Schomburgk's writings, is likely the squatting figure depicted in Goodall's sketch of Schomburgk at work fixing angles on the savanna (plate

92. Also called the Ireng.
93. *JRGS* 13 (1843): 40.

18).[94] Schomburgk wrote that the man "made me acquainted with the names of the numerous groups of mountains which I could see from my principal stations."[95] This assistance, which linked Schomburgk's observations to toponyms on the map, was so helpful that he persuaded the man to accompany the survey to the sources of the Takutu. Only geographical collaboration allowed sights to become sites, and seen in this light Schomburgk's boundary map, the "Sketch Map of the River Takutu" (fig. 32), must be understood as the product of such coagency, the inscription of local knowledge onto the cartographic field.[96] It was a collaboration that was not without pitfalls. Wapishana place-names created confusion. Schomburgk fixed a particularly prominent hill from Tenette, called Kuipaiti, which he thought "promised to afford me an opportunity of verifying and extending the angles of my survey."[97] No sooner had this point been fixed and entered on the map as a boundary landmark than he discovered that the Wapishanas "call all hills which consist of solid rock and are only sparingly covered with vegetation by the general name of Kuipaiti."[98] Even more confusing, this "Kuipaiti" appeared to Schomburgk to be forested.

When Schomburgk reached the top of Kuipaiti he glimpsed a "view" that called into question the reliability of the very collaboration on which he was basing his boundary map. He wrote, "I received here proof of how fallacious it is to trust implicitly to Indian information."[99] From "Kuipaiti" Schomburgk could see that the Guidiwau River was not a tributary of the Takutu at all, as he had been led to believe, but a tributary of the Rio Branco to the west. Schomburgk had placed the river (as he had understood it to run) on an earlier map he had sent to the RGS, which had published it. In admitting that he had its orientation reversed, Schomburgk pointed out that he had drawn it on the earlier map only in the *macchie* of geographical knowledge in the making: "I . . . inserted the course in my map in dotted lines." He used the occasion of the view from a mountain (which might or might not be specified by the name Kuipaiti) to dilate on the dangers of native geographical informants.[100]

94. Plate 18 has been reproduced from BL, Goodall MS, 16936.
95. *Brazil Arbitration, British Case Annex*, 2:117.
96. Figure 32 has been reproduced from *Venezuela Arbitration, British Case Atlas*, 51.
97. *JRGS* 13 (1843): 44.
98. Ibid.
99. *Brazil Arbitration, British Case Annex*, 2:120.
100. "A traveller cannot guard himself sufficiently against false information. The wish of the Indians to be considered well acquainted with their country, or much-travelled as they express it, and occasionally misapprehension of statements made in a language which leaves much to be conjoured from the arrangement of words, or the emphasis with which they are pronounced, give rise to great mistakes in geography and natural history, especially when the traveller is under the necessity of using double interpreters." *JRGS* 13 (1843): 46.

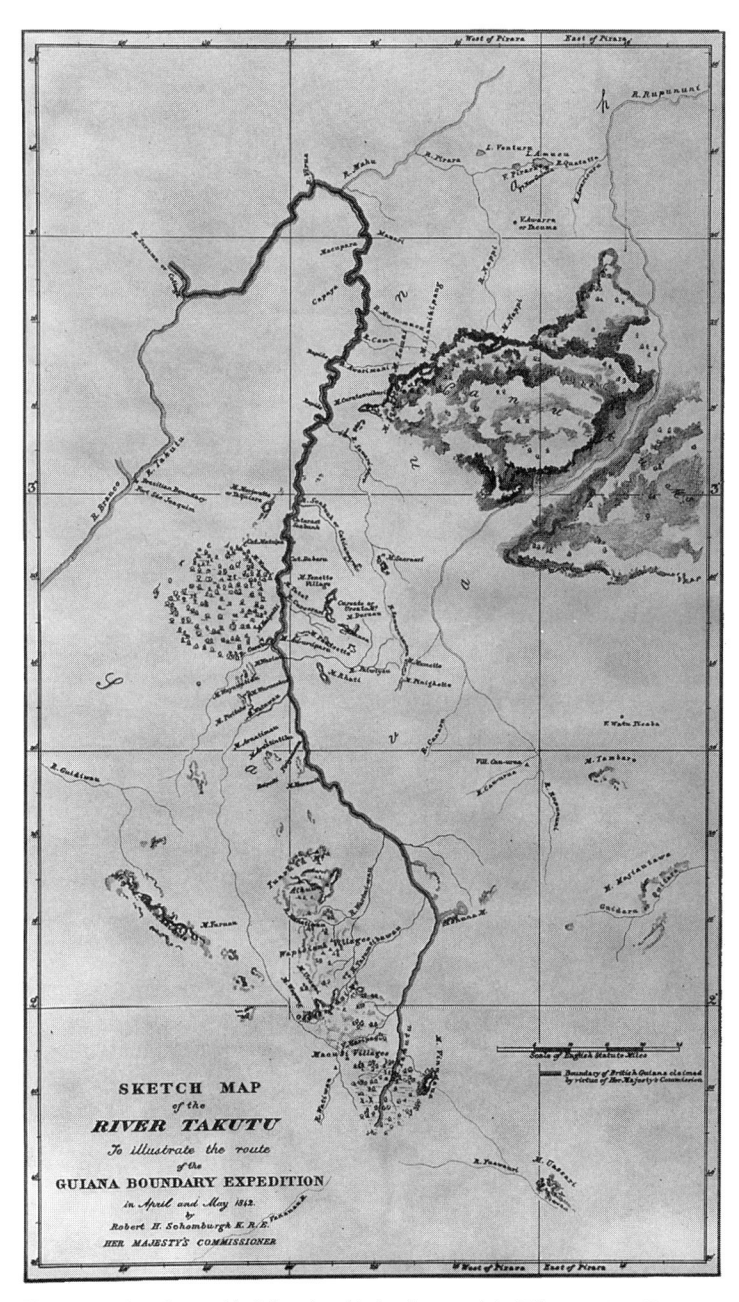

Figure 32. A region at risk: Schomburgk's sketch map of the Takutu region. Courtesy of the Library of Congress.

Richard Schomburgk would receive on the Cotinga route a still more striking example of the ambiguities implicit in exchanging goods for local knowledge. Having secured an exhaustive ornithological collection from months of trade with the local tribes, he ceased to negotiate for additional specimens. Not to be thwarted, the huntsmen appeared with skins Schomburgk described as "wondrously beautiful creatures" belonging to "the genus *Tanagra* or *Pipra*." Certain that they were novel, he paid out additional knives and combs to acquire them. They were, he discovered, neither *Tangara* nor *Pipra*, but rather both, with a clever admixture of *Fringilla* and *Euphone* thrown in for good measure: "I had secured nothing else than a monstrosity, put together with a skill so extraordinary." The Amerindians had deployed their superlative feather-weaving dexterity to create new species. "We readily forgave the bargainers," Richard wrote, "for they believed that the skins they had tampered with must have the same value for us as the other ones."[101] Similar confusions of purpose attended exchanges of geographical information. Amerindians were concerned with getting from place to place. As far as *routes* were concerned the Guidiwau *was* linked to the Takutu, by an easy portage at Curati.

A certain toponymic ambiguity lay under the surface of the whole southwest boundary, and in Schomburgk's eyes it was the fault of the Amerindians. On reaching the confluence of the Mahu and the Takutu it became clear to Schomburgk that, from the perspective of hydrography, their names were reversed. "The Takutu," he wrote, "appears more like a tributary of the Mahu than the recipient of the latter, and in reality its breadth is less."[102] A calculation of the width of the rivers at the junction showed that the upper "Takutu" was really a *tributary* of the main trunk of the river that had that name at its widest point. He repeated this later and pointed out that "a single glance at the map proves that the Mahu ought to be considered the recipient of the Takutu; its continued southwestern course after it issues from the Pacaraima mountains to its junction with the Rio Branco, and its larger mass of water, entitles it, in geographical respects, to be considered the main trunk above the junction with the Takutu."[103] If the Takutu was, in a sense, not the Takutu, what was natural about the use of the "Takutu" as the boundary with Brazil?

The way the Mahu looked disconcertingly as if it ought to be called the upper Takutu gave Schomburgk an unsettling sense that his "natural" boundary—running from the south up what was called the upper Takutu,

---

101. M. R. Schomburgk, *Travels*, 2:220.
102. *JRGS* 13 (1843): 22.
103. Ibid., 29.

then turning the corner on the Takutu to the west after the juncture with the Mahu and running to the Cotinga before continuing north—would be rejected by the Brazilians as emphatically *un*natural. In his memo to the governor he advocated that "should this point be contested by the Brazilians . . . [claiming] the principle of forming a boundary by natural divisions is violated," the government should fall back on the Mahu as the western boundary, but only if the Brazilians could not be persuaded "to accede by cession or otherwise."[104] It was indeed this region that lay at the core of the late nineteenth-century dispute between Britain and Brazil. Brazil won that portion, and the Mahu is the current boundary.

In tracing the Cotinga north Schomburgk was aware that he was overstepping a boundary that, even by his own criteria, was more "natural" than the route he chose. Using the Mahu, however, had unpleasant political implications. Not only would it remove the "security of inland trade" afforded by the "margin between the Mahu and the Cotinga," it would also "exclude Great Britain from a footing on the savanna west of the Rupununi, and deprive her of the Sierra Conocou, and a well populated district."[105] From the perspective of Schomburgk's boundary *route* the Mahu presented another difficulty: Schomburgk had only a vague sense of the location of its sources and had no reason to suspect that they adjoined Roraima, the landmark through which he had committed the boundary to pass. On the Cotinga, by contrast, he had passed close to Roraima in his expedition to Esmeralda in 1838. The Cotinga took him where he was going.

If the Cotinga as boundary represented a transgression of the Mahu as boundary, Schomburgk's expedition on the upper Takutu was not without a transgression of its own. No sooner had Schomburgk established the coordinates of the village of Tenette than he and the surveying team "found a canoe, by which we crossed to the left bank of the Takutu"— Brazilian territory by any measure, even his own.[106] But the prospect of penetrating a region that Schomburgk described as a "perfect *terra incognita* both among the Brazilians and among the Indians" was too appealing to resist.[107] Schomburgk wrote of their approach to the Tuarutu Mountains that "a number of Wapisiana arrived . . . to greet the first white man who ever visited these regions," namely, Schomburgk himself.[108] Schomburgk had crossed over onto Brazilian territory not because

---

104. *Brazil Arbitration, British Case Annex,* 2:91.
105. Ibid.
106. *JRGS* 13 (1843): 42.
107. Ibid., 49.
108. Near this point Schomburgk ascended another "Kuipaiti"—which he here explained meant "the rock *par excellence*"—and procured another set of angles back along the route. Ibid., 50.

the route was easier, but because of the appeal of crossing virgin territory, of being the first "Paranaghiri" to reach these settlements. For an explorer, boundaries were made to be crossed. As for the route on the left bank of the Takutu, Richard Schomburgk, whose boots had given out and who had been obliged to engage in the practical ethnographic experience of wearing woven *ite* palm sandals, found it brutal. They were not so much following a natural boundary as making their way across a series of such boundaries. He wrote: "In many places we had finally to employ hands as well as feet . . . to get over granite boulders that formed regular zones and barricades."[109]

Richard wrote that "it was not long before the thought of losing oneself in this rocky labyrinth gave me an uncanny shudder."[110] Indeed, as Robert Schomburgk was forced to admit in his official report to the governor, the black cook, Hamlet Clenan, did fall behind the party and, disoriented, wandered for nearly sixty hours before he was found by a search party that had doubled back to look for him. Schomburgk described finding him thus: "He was almost in an exhausted state, and fear and fatigue had operated so strongly upon him that I was at first apprehensive his reason was gone."[111] He did not, in fact, recover. Like the journey's Pip, he remained as a reminder: "His wild looks, his clothes hanging in tatters around him, and his incoherent speech, sometimes laughing, sometimes weeping showed what an impression his misfortune had made."[112]

From the perspective of the crew, Hamlet's disorientation and collapse—he was dismissed shortly thereafter—confirmed their consensus that the Takutu expedition was traveling under a bad star, a sense that had taken root among the boatmen from the first day of the trip when, immediately after the flying of the colors at the junction of the Mahu and the Takutu, a rifle misfire nearly killed one of the men accompanying the survey. His wound demanded that he be carried back to Pirara and cost the expedition the services of the assistant surveyor, who doctored him for nearly two months. The wound had not healed when the survey party returned to Pirara in May, and the man had to be sent back, with Clenan, to Georgetown.

The whole military occupying force soon followed, and when the boundary survey departed in early September to ascend the Cotinga to Roraima, lots were drawn to see who among the Europeans would be left with the lonely task of remaining behind in the deserted village to watch over the stores and instruments they would have to leave in their huts. Writing to the governor just before departing Pirara, Schomburgk ex-

109. M. R. Schomburgk, *Travels*, 2:51.
110. Ibid., 53.
111. *Brazil Arbitration, British Case Annex*, 2:124.
112. *JRGS* 13 (1843): 56.

pressed his hope that the survey would reach Roraima within a month and remain there for some weeks, to allow him to "determine this mountain with great precision."[113] Leaving Roraima, he intended to "proceed in search of the sources of the Cuyuni, which river I purpose to descend as far as its tributary the Acarabisi, at the confluence with the Cuyuni where I engraved the mark of the survey in July 1841."[114] His arrival at that mark would complete his circuit on the western frontier.

Just as the Takutu boundary survey emerged out of a negotiation with Amerindian guides, the survey of the boundary to Roraima depended, to an even greater degree, on the services of a Wapishana from the village of Nappi in the Canuku range. This man, a displaced *piai*, was deemed to look so much like Napoleon that Richard said a portrait made by Goodall elicited surprise in Paris years later.[115] Napoleon (as he was called) plotted the expedition's course, shaping in the sand an elaborate maquette of the topography. Richard wrote:

> Roraima became the subject of conversation for the rest of the day. As Napoleon had already been there once, I had the opportunity of recognizing in him a geographical genius, because he readily modeled out for us in the sand a masterly contour map of the route we had to follow, of the course of the rivers we had to cross and the lay of the mountain ranges and heights we had to climb, of the settlements we had to pass, and all of this with such precision as would later astonish us.[116]

The ascent of the mountain Maikang-Yepatori afforded the first of several opportunities to confirm the value of the "geographical gift" that Napoleon had given the survey party.[117] From the top of the hill—where Schomburgk erected a sixty-foot pole and pennant to aid in trigonometric positioning of the confluence of the Cotinga and the Zuruma—the surveyors could "follow the silver ribbon of the Cotinga . . . up to where it touched the Pacaraima Range." This view conformed to Napoleon's topographical "sketch" of "mountains and villages, according to their height and size, with varying little heaps of moist sand, the courses of the streams with broad or narrow furrows."[118] But Amerindian assistance again introduced confusion. Schomburgk had labeled a key boundary landmark "Makunaima-aute" only to discover later that this appeared to be a general term for flat-topped mountains.[119]

---

113. *Brazil Arbitration, British Case Annex*, 2:133.
114. Ibid.
115. M. R. Schomburgk, *Travels*, 2:116. Carter examines the function of such monikers for indigenous people: "Names, in short, made them white history." Carter, *Road to Botany Bay*, 332.
116. M. R. Schomburgk, *Travels*, 2:128.
117. See above, chapter 2, note 77.
118. M. R. Schomburgk, *Travels*, 2:128.
119. Ibid., 135.

Where the Cotinga entered the Pakaraima range it ceased to be navigable, and the survey party set out on a circuitous trek to cover more than fifty miles as the crow flies, arriving at a point several miles south of Roraima, where, in proximity to a friendly village willing to negotiate for food supplies, the boundary commission erected huts. They christened the spot "Our Village," and Schomburgk began a set of trigonometric observations to establish the height and configuration of Roraima and the surrounding mountains. Schomburgk had again passed considerably outside the boundary he was laying down, and it is perhaps in recognition of this that in his official report he makes no mention of possession taking after leaving the Cotinga. The land route to Roraima took Schomburgk more than twenty miles into what even he considered Brazilian territory, and "Our Village" was situated fifteen geographical miles into Brazilian territory as defined by his own boundary.

Roraima retained for the survey party the dark and mysterious force reflected in the *Twelve Views*. The ugly, quick death of a young female porter (bitten by a labaria snake) hung over the local trigonometric operations conducted around the base of the precipices. The place seemed to defy the very instruments of inscription and transcription the surveyors wielded. "In spite of the greatest care," wrote Richard, "the astronomical instruments were nevertheless covered with rust: a loaded weapon left standing for a few hours would not fire."[120] Goodall, his hands stiffened by temperatures dipping into the fifties, labored in vain to produce sketches of the landmark; the moisture in the air defeated his watercolors. Damp drying-papers robbed Richard of most of his botanical specimens.

The mists of the Kanaiba Falls and the intransigence of Roraima—which yielded its form and position to the geographical gaze only gradually and in momentary clearings—delayed the survey for weeks and left Robert Schomburgk increasingly certain that the terra incognita lying to the northwest would be the most difficult terrain the survey had yet encountered. He elected to make the trip to the Cuyuni unaccompanied by the cumbersome entourage of the surveying party. The others were dispatched back to Pirara, and Schomburgk set out light, accompanied by four mixed-race boatmen and a party of Macushis who had assisted with the work on the southwest boundary. They carried only a compass, a sextant, and a few provisions.

The trip, across sharply terraced terrain covered with thick rain forest, roughly followed the watershed separating the tributaries of the Mazaruni from those of the Caroni. In his official report to the governor, Schomburgk wrote that he did not believe, after completing the circuit,

---

120. Ibid., 213.

that the other members of the boundary commission would have been able to make the trip; there is little reason to think this an exaggeration. After a month of trekking Schomburgk and his party reached the mark he had made on the Cuyuni near the village of Haiowa a year and a half before. In reaching it Schomburgk "passed through" the gate he had erected on the river at this problematic point. He wrote to the governor of "accomplishing thus the whole line from the sources of the Takutu to point Barima on the Atlantic Ocean."[121]

The queen's marks on the trees on opposite sides of the Cuyuni represented the "finish line" for Schomburgk's survey. But his arrival—from the west to the east—betrayed yet another question about the relation of Schomburgk's course to the boundary it was intended to trace. What line constituted the boundary with Venezuela between Roraima and Haiowa on the Cuyuni? Schomburgk's route? How had he arrived at his previous "terminal" mark *from the other side?*

Schomburgk treated this subject briefly in his report to the governor when he wrote: "I consider that Her Majesty has undoubted right to any territory through which flow rivers that fall directly, or through others, into the river Essequibo."[122] This, Schomburgk recognized, the Venezuelans would never accept, since the Cuyuni's upper reaches pass deep into territory long settled by the Spanish. Schomburgk did not acknowledge directly that the claim to the entire watershed was impossible. Rather, he explained to the governor that Britain, "as a Maritime power" should not have much interest in the upper reaches of an unnavigable river. Schomburgk drew two conclusions from this observation. The first was that he could skip the upper Cuyuni in his boundary route (a good thing, since he and his party were in no condition to continue farther west).[123] The second was that the claim on the upper Cuyuni could serve as a bargaining chip for Barima Point, the "Dardanelles of the Orinocco," which was much more valuable. He wrote: "Relinquishing the claim to the territory watered by the upper Cuyuni . . . her Majesty's government acquires additional grounds to impress the claim of point Barima."[124]

In effect, then, Schomburgk had *traced* in his route the "real" boundary, which delimited the colony considerably to the east of the watershed of the Cuyuni, but he *claimed* British rights to the whole Cuyuni basin, a hyperbolic boundary that was asserted merely as a strategic ruse. Schomburgk cast a boundary outside the one he traced, even as he was transgressing the boundary he had already established. Schomburgk delimited

121. *Brazil Arbitration, British Case Annex*, 2:137.
122. Ibid.
123. Ibid.
124. Ibid.

boundaries around boundaries, nested layers of territory that could be peeled away or exchanged. He drew lines as points of departure for his routes and then cast his eye to draw boundaries beyond his passage. The boundaries of the colony were drawn to be contested.

NARRATIVE PART 3: EAST

Contested they were, and immediately. In fact, as Schomburgk set out for Roraima from Pirara, a diplomatic exchange commenced between the Brazilian minister in London and the Foreign Office concerning the legality and propriety of the boundary marks Schomburgk had engraved on the Takutu. With a tenuous arrangement of territorial neutrality in the works and rumors circulating of a possible Brazilian detachment headed for a reoccupation of Pirara, the officials in the Foreign Office agreed to have those markers too removed, even though they were declared (by the British) to be purely scientific fixtures.[125] This meant that no sooner had Schomburgk reached Georgetown, after what he thought was the triumphant completion of the western boundary, than he was informed that the Barima and Amacura marks had been disavowed by the government and that he himself was responsible for removing the marks he had erected along the Takutu.

The news doubtless had a considerable effect on Schomburgk, since it meant he would have to carve out the markings he had placed before the village of Tenette. The key inscriptions of his most complete survey would be effaced by his own hand. The irony was richer still: in his first letter to the governor on returning from the Roraima-Cotinga expedition, Schomburgk expressed his enthusiasm to return to Pirara. He needed to do so in order to fix the longitude of the village. Sour relations with the military meant that the quartermaster had refused to make space for a large reflecting telescope Schomburgk had ordered to make precise observations of the eclipses of the moons of Jupiter (to further refine his longitude calculation). Improvising, Schomburgk had calculated all his boundary points in the southwest using Pirara as the prime meridian. Collating those observations demanded a better fix of Pirara: "The expense of the instrument would not only prove useless, if by proceeding to Pirara another opportunity was not afforded to me to fill up this great desideratum, but the series of astronomical observations would remain incomplete."[126]

The necessity of returning to Pirara led Schomburgk to propose that

---

125. Rivière, *Absent-Minded Imperialism*, 145.
126. *Brazil Arbitration, British Case Annex*, 2:138.

(for a small increase in cost) he could descend the Corentyne from its sources, completing his circumambulation of the colony by surveying the eastern boundary. Given that the boundary with the Dutch had been fixed by tradition for almost forty years, the survey of the Corentyne had less urgency than that of the western boundary with Brazil and Venezuela. Since interest in even that region had waned (as the diplomatic and jurisdictional disputes and their associated costs had mushroomed), there was little ardor at the Foreign Office for the Corentyne expedition. Schomburgk assured the governor that the trek to the sources of the Corentyne (overland from Pirara), and the descent of that river, would be conducted in the same way as his most recent boundary survey: by means of a stripped-down party consisting of him and four boatmen. Back in London, correspondence between the Treasury, the Colonial Office, and the Foreign Office was leaning toward calling off the boundary survey altogether, but Governor Light's approval preempted probable cancellation.

Schomburgk's enthusiasm must have been considerably tempered by the curious assignment he received from the governor. It was a dual mandate: return to Pirara to fix its location (and hence complete all his western observations); *and,* while in Pirara, efface the very marks that embodied those observations.[127] He was being sent into the interior to complete the eastern boundary and erase the western one. He did both, but in his dramatic descent of the Corentyne he did not once inscribe the queen's mark.

The evolution of the boundary surveys over the years 1841 to 1843 amounted to an attenuation, captured best by the contrast between the elaborate party (with three boats and a crew of more than thirty, all dressed in matching red-sashed uniforms) that embarked for Barima Point on 18 April 1841 and the small group of "walking skeletons" who arrived at the mouth of the Corentyne in October 1843. It would be possible to argue that it was primarily the dwindling financial support that led to the transformation. But that was not all. Schomburgk had gotten *better* at traveling in the interior of Guiana, and as he had gotten better, routes became possible for him that would have been impossible when he began. They were also impossible for the vast majority of the other Europeans in the colony. Schomburgk's traveling abilities, his capacity to trek not merely among, but increasingly in the style of the Amerindians of the interior, effectively outstripped the accoutrements of an "expedition," elaborately supplied, attended by dozens of carriers, and unable to provision itself on the sparse hunting available and the limited surplus of Amerindian villages. Schomburgk's exploratory style had outstripped the boundary commission even as his boundaries outstripped each other.

127. Ibid.

The dreaded "évolution de Schomburgk," described by adversaries in the subsequent boundary disputes,[128] was less from mild-mannered geographer to land-hungry imperialist than from novice traveler (obliged to surround himself with cumbersome expeditionary parties of limited reach) to seasoned bush explorer, able to roam with considerable independence through unfamiliar terrain, negotiating with Amerindians, making lightweight, disposable bark canoes to pass on rivers, walking distances he would have been unable to contemplate on his first trips. This increasing freedom, combined with economic constraints, transformed Schomburgk's boundary surveys from the pomp of Victoria Point to an increasingly nomadic endeavor. The light and difficult penetration from Roraima to the Cuyuni demonstrated the advantages of this agile and itinerant style. The discovery that all of his weighty marks were to be effaced further unraveled the significance of the ceremonial fixtures that had been his concern on the first boundary surveys. The boundary survey had been stripped of its moorings and its funding. Schomburgk had left off reading the commission aloud to his assembled men and stopped carrying a long sword and wearing a plumed *chapeau bras*.[129] Experience and necessity had made him into something closer to a commissioned wanderer, an "interloper" whose route had some significance but who no longer stridently claimed territory or erected marks of possession.

The attenuation of his power to install the boundary resulted not from his having failed to establish authoritative boundaries in the western region but, in a sense, from his having succeeded too well. By installing a boundary, Schomburgk had obliged the colonial and home governments to face the implications of a territorial claim. According to Robert Sack's work on territoriality, "delimiting a geographical area" represents only the first part of the process of establishing a territory. Asserting territorial possession meant something more than a line; it meant an attempt to "control" that delimited area.[130] The authorities in London and Georgetown had seen the delimitation as a good idea. No sooner had it begun, however, than the implications of such a delimitation were thrust upon them. The Maicarawari case was the first instance. The confusing issue of responsibilities for the protection of the Amerindians followed quickly. James Stephen, undersecretary of state for the colonies and an official in the Church Missionary Society, added a minute to Schomburgk's very first report as boundary commissioner, warning of the implications of Schomburgk's work. Humanitarianism was certainly noble, but, Stephen

---

128. *Brazil Arbitration, British Counter-case*, chap. 6, 113.
129. Rivière, *Absent-Minded Imperialism*, 102.
130. Sack, *Human Territoriality*, 19; Winichakul, *Siam Mapped*, 16.

pointed out, it could not be made the basis of territorial claims without considerable awkwardness: British subjects in Australia, he noted, had recently "perpetrated enormities against the Native Tribes compared with which even the alleged crimes of the Spaniards in Guiana are insignificant."[131] He warned that if the boundary was premised on the defense of Amerindians, "the published records of the Courts of Sydney would be quoted with embarrassing effect against any British Statesman" who attempted to elaborate the point. As for the issue of Maicarawari's liability and the value of the testimony of Amerindian witnesses, "these questions are not yet ripe for discussion."[132] The turn of phrase captured the fate of the boundary. It was "unripe" in the eyes of an administration suddenly challenged to give it substance on the ground. Schomburgk drew the line, and the line drew British officials into obligations from which they sought to extricate themselves. Step one was erasing the line.

This, in effect, Schomburgk did. On reaching Pirara, he sent a communication back to the governor confirming that the assistant surveyor had erased the offending marks. The governor's wry comment to the Colonial Office on the subject would doubtless have pleased Schomburgk if he could have read it. You could chop out of the tree the engraved crest that had served as boundary mark; the tree still stood: "The trees on which these marks have been cut . . . and the presumption of territory be still in existence. Let the Brazilian Government show a counter title if it can."[133]

The Brazilian government had made, of course, a considerable effort to do so. No sooner had Schomburgk's marks on the Takutu been "erased"—leaving, as Light hinted, still deeper marks than before—than the long-awaited Brazilian Boundary Commission arrived in Pirara, led by Colonel João Henrique de Mattos. The Brazilians had arrived without trade goods with which to secure the assistance of the Amerindians, and, more crippling, without any surveying instruments.[134] Schomburgk was dismissive, particularly when the colonel penned an ingratiating letter to the British Boundary Commission that concluded: "You will be pleased to inform me which are the said points determined upon and where marks have been fixed, in order that I may with greater facility and with more certainty find them."[135] The assistant surveyor refused. According to Richard, Robert gave false information. Having been obliged to unmake his marks, Schomburgk doubtless saw no reason to afford the Brazilians any orientation to the region. They were seeking marks to con-

131. PRO, CO, 111/179: J. Stephen's minute, 3 September 1841.
132. Ibid.
133. Rivière, *Absent-Minded Imperialism*, 151.
134. Ibid., 155.
135. Ibid.

test, but the marks were no longer marks; if they wished to find the lacunae marking their erasure—the phantom marks—they could do so alone.

Phantom marks were not new to the territorial claims on Pirara. In late 1838 (when the first disputes had arisen over the presence of the missionary Thomas Youd) the commander of the Brazilian fort at São Joaquim, Captain Antonio Dos Barros Leal, arrived in Pirara to find what he claimed was a mark left by a late-eighteenth century Portuguese survey of the region. Schomburgk narrated Leal's search to the governor with scorn:

> He proceeded to the mouth of the Siparuni, where he anxiously looked for the pale which he said the Brazilian Boundary Commission had planted sometime between 1780 and 1783; having at last discovered a tree, the woody tissue of which, through some influence of the weather, had decayed, excepting the mere heartwood, in which by accident some nails had been driven, he declared it to be the remains of the former boundary mark, and although his own companions drew his attention to the circumstance that this alleged boundary post had a large and sturdy root, which could not have been the case if it had been placed there by human hands, but that on the contrary it must have been for ages in the ground, this was of no importance to the zealous Captain Leal, and a part of the heartwood was taken away as proof of his success.[136]

The later Brazilian Boundary Commission had no luck finding Leal's pale, nor did they ever find Schomburgk's un-marked trees, which would shortly become just as inscrutable as Leal's gnarled and well-rooted stump. As for the map the Brazilian Commission produced, it turned out to be a key piece of evidence in the disputes at the end of the century; not, ironically, for the Brazilians, but rather for the British. It gave itself away as a plagiarism by the very care of its copywork: the thin stream of the Guidiwau was shown as a tributary of the Takutu, that error shown only on Schomburgk's first map of the region, which had been published in Britain in 1840 in a parliamentary paper. The Brazilians had copied Schomburgk's work and had not surveyed the region themselves. In a fitting irony, it was not the accuracy of Schomburgk's surveying that later preserved the British territorial claim, but rather his error.[137]

If the prospect of his boundary marks weathering into stumps left Schomburgk ambivalent about the success of his boundary commission, the prospect over Pirara cemented a sense of failure. The missionary was dead, the Macushis had abandoned the village, Fort New Guinea was an earthworks crowned with ashes, and a fire had swept through several of the remaining buildings, driven by the wind, and brought down the

---

136. *Brazil Arbitration, British Case Annex,* 2:89. Youd echoes this account. Rivière notes that it is unclear whence they had this version of Leal's search. Rivière, *Absent-Minded Imperialism,* 42.

137. *Brazil Arbitration, British Counter-case,* 150.

church. A related explosion in the powder store nearly wrecked Goodall's drawings, and several of the Macushis who had accompanied the expedition lost everything, including their wages from the trip to Roraima.[138]

Schomburgk left Pirara in a spirit of melancholy that would pervade the Corentyne expedition. The desolation of the scene and the memories of his high hopes for the community combined to shroud the farewell with sorrow. Reflecting on his Guiana expeditions in his report, he wrote: "Had the cause of religion—had humanity been advanced during the long interval of eight years? Alas! No. The ruins of Pirara, scarcely a hut inhabited, the regular paths which traversed the village during the missionary's residence among the Macusis, overgrown with rank grass; no human visible . . .—all replied in the negative."[139] Nor was this dreary picture merely a romantic plaint. The British military expedition sent to liberate Pirara had brought smallpox from the coast, which had blazed a swath of death through the region and percolated to the distant communities of Wapishanas in the south.[140] Schomburgk reported blindness and fatalities blighting the small villages of the Canuku range. Tenette itself looked unlikely to escape. Nor had security of any sort been extended to the scattered Amerindians who remained in the region. With little optimism, Schomburgk forwarded to the governor a set of depositions concerning attempted rapes by Brazilian *vaqueros* on local women, drunken assaults leading to a knifing. When warned that their actions would be reported to the commander at Fort São Joaquim, one of the attackers answered that he would be happy to take the message about his conduct there himself, since he was "out of the reach of justice." In a statement that cast the complete failure of Schomburgk's boundary enterprise into high relief, the perpetrator stood firm in the ruins of Pirara and "coolly stated that where he stood was neutral ground."[141]

Disgusted, Schomburgk set off for the upper Corentyne on 30 April 1843. On 9 October he, Goodall, the boatmen, and the Macushis with whom he had left Pirara arrived at the coastal settlement of New Amsterdam, at the mouth of the Berbice, having completed the most harrowing of the Guiana expeditions. Death seemed to haunt the trip, and in narrating it Schomburgk returned repeatedly to the theme of Amerindian extinction. Not only had *descimientos* taken their toll, but in passing the communities he had visited in his previous ascent of the Essequibo, Schomburgk witnessed the devastating effects of measles. Worse, the smallpox of Pirara had blown ahead of the expedition and ravaged these

138. M. R. Schomburgk, *Travels*, 2:279.
139. *JRGS* 15 (1845): 2.
140. *Brazil Arbitration, British Case Annex*, 2:137. See also *JRGS* 15 (1845): 27.
141. *Brazil Arbitration, British Case Annex*, 2:140.

villages too. Its effects, Schomburgk noted, were compounded by the native ideas about Kanaima: victims' kin sought revenge for untimely deaths.

The devastation was best captured in Schomburgk's affecting description of his encounter with an aging woman, "singled out by destiny to be the sole survivor of a nation."[142] Schomburgk watched her, the last of the Amaripas, move ghostlike among the foreigners with whom she resided. He tried to transcribe a few of the words of her language but acknowledged what a small part of a people could be rescued from "utter oblivion."[143] The depopulated landscape and abandoned village led Schomburgk to acknowledge himself less a savior than a fatal harbinger of decline, a profound and unsettling transformation of his optimism. Guiana, he wrote, "since the arrival of the Europeans, has become a vast cemetery of the original races."[144] As best he could understand, some of the local tribes saw his arrival just as he did. Of the Pianoghottos he wrote: "A tradition prevailed among them that the arrival of the first white man betokened the extinction of their race."[145] It seems more likely that Schomburgk was understanding what he had come to suspect most deeply: "Reluctant as I am to despair, the conviction is forced upon me that the Indian race is doomed to extinction."[146] The macabre spirit of the expedition climaxed among them when Schomburgk and Goodall negotiated secretly to exhume and purchase two skulls of women recently deceased.

The expedition party made its way southeast overland, in order to reach a settlement of Maopitians who were to be engaged as guides to lead the party along the Caphiwuin to a point where they could portage to the upper tributary of the Corentyne.[147] Relations with the Maopitians were strained from the start, and the attempt to reach the sources of the Corentyne was hampered by the central drama of the expedition: a case of mistaken identity. Schomburgk and his nomadic survey were taken for a party of Brazilian slavers by the isolated tribes near the headwaters of the Corentyne. Since the Pianoghottos were reported recently to have killed every member of such a slaving party, this mistaken identity was more than an irony for an explorer who had spent the past five years combating Brazilian aggression; it looked like a death sentence. Deteriorating rela-

142. *JRGS* 15 (1845): 27. Thomas points out that colonial fascination with "the last of the . . ." reflected a desire to have narrative authority over vanished peoples. See Thomas, *Colonialism's Culture*, 180. This is doubtless so, but the trope here must be understood in the context of Schomburgk's circumstances.
143. *JRGS* 15 (1845): 28.
144. Ibid., 27.
145. Ibid., 87. Schomburgk's "Pianoghottos" correspond to the Farakoto people.
146. Ibid., 26.
147. Schomburgk's "Maopitians" correspond to the Mawayena people.

tions with the party of Maopitians whom Schomburgk had engaged as guides further imperiled the expedition, particularly when they took advantage of the flight of the local Pianoghottos to raid their abandoned settlement for valuables. Unable to control his guides, Schomburgk faced the wrath of the Pianoghottos, who not only feared being raided but saw that fear realized. They took to the bush to prepare a counterattack on the expedition, already embroiled in controlling its erstwhile "guides," who alone could communicate Schomburgk's intentions to the Pianoghottos.

Schomburgk had long defined himself in opposition to the mercenary Brazilian interlopers who roamed these regions. The double bind on the upper Corentyne thrust him into an unseemly proximity to them. He had, however unwittingly, led a raid on a peaceful village, and in trying to ameliorate the situation (and survive) he took six Maopitians hostage, trying to force them to broker peace with the Pianoghottos on his behalf. Schomburgk and Goodall found themselves "loading our guns before their eyes with double shot" in an effort to communicate.[148] Three of the Maopitians escaped, raising the tally of hostile forces lurking in the bush around the camp. Goodall's health gave out as days stretched to a sleepless two-week standoff, and Schomburgk had to pitch his tent across the entrance to the temporary stockade they had erected. With no way to explain that he was friendly, Schomburgk had become exactly what he had always most despised: a raider of villages and the jailer of captive natives. Signs abounded that a warning had gone out among the surrounding tribes. Schomburgk and the delirious Goodall were persuaded that their chances of survival diminished by the day.

The impasse broke when some of the Macushis were able to establish friendly relations with a temporary camp of Pianoghottos hiding in the bush. But the affair was not without costs. Relations with the Maopitians were beyond repair, and the loss of porters meant the survey had to abandon nearly all kit and collections, sparing only a few instruments and documents. Being forced to leave behind so much of the fruit of his travel caused Schomburgk considerable pain: "How frequently did I reconsider our disposable force to see whether certain objects of peculiar interest to science or to myself personally could not be stowed somewhere."[149] The implications went beyond a vague sense of loss. Forced to carry fewer provisions and items of trade, the expedition would run out of food on the descent of the Corentyne, turning the race to the coast into a race against starvation. For the last four days rations fell to three ounces of farina a day, and the expedition members passed into a fog of exhaustion and mal-

148. Ibid., 69.
149. Ibid., 75.

nourishment. Of the final leg of the journey Schomburgk wrote only, "Hunger is a very disagreeable feeling; one's ideas get confused, and there is a vague sensation which ultimately engenders headache."[150]

The boundary survey of the Corentyne, rather than seeking out a terra incognita, became an attempt to get out of, not into, unfamiliar territory. The boundary was the by-product of an escape from the interior. Like the burning of ships on arrival at a foreign shore, Schomburgk's abandonment of equipment and provisions bound his own survival (and that of his party) to the boundary line. Obstacles in the completion of the survey became obstacles to survival, hence Schomburgk's dismay on seeing the treacherous watercourse lying before them:

> I gave orders to the coxswain to halt at one of the rocky islets, and climbing to the summit, our route lay before us. The remembrance of what I felt at the sight before me, will remain as long as my memory lasts. Enormous masses of stone, black as iron, extended as far as the eye could reach, against which the river dashed its waters with the greatest fury, the white foam forming the strongest contrast with the gloomy tint of the rocky masses.[151]

The sublime and life-threatening ceased to be discrete encounters, highlighting or emphasizing points along the route, and became the route itself. The obligation to complete the boundary no longer drew Schomburgk on past difficult points; the boundary became incidental to survival. No inscriptions were made, no possession formally taken. At the commencement of the boundary surveys Schomburgk set out to follow the line. By the end, the line followed Schomburgk. He was not looking for points; he was looking for home. At the mercy of each contour of the ground, Schomburgk and his party shot the rapids of the Corentyne with their eyes on each swell and dip in the surface of the water. There were no triangulating points, no overviews of the terrain. The object was to float back to the coast, as lightly as possible, following the river, going where it was going. The route was everything; boundary lines lost all meaning.

The drama of the Corentyne expedition foregrounded Schomburgk's travel and placed the boundary in the background, where governmental indecision and backtracking had already forced it. How far the expedition represented a personal passage, and a homecoming on the part of the explorer, is captured in the progression of names Schomburgk bestowed in the course of the descent. In contrast with the toponymic grandeur of "Victoria Point," Schomburgk's names in the upper Corentyne acknowledged the difficulty of cross-cultural exchange and the absurdity of his predicament. He wrote in frustration: "Our guide, who appeared the

---

150. Ibid., 99.
151. Ibid., 98.

most stupid of all the Indians with whom our travels had made me acquainted, gave me merely a broad 'ha!' for answer when I inquired the river's name; and as I could not obtain anything further from him, I introduced the river under that name in the map."[152]

Deep in the interior, confronted with the contingencies of geographical knowledge, toponymy became an ironic commentary on the ephemerality of landmarking instead of an opportunity for paeans to metropolitan authorities. Later Schomburgk deepened the linguistic muddle of the encounter when he explained that each tribe had not only different names for each river, but different names for each other as well. Since no tribe's range extended over the whole of the territory, consistency was impossible. Each name on the map was a choice among many, most of them as arbitrary as the Ha! flowing gently through the jungle.

Disillusion and dissolution had attended Schomburgk's boundary efforts, and his optimistic vision of Pirara had proved no less a mirage than the El Dorado of the conquistadors. The beguiling illusion of Guiana was inseparable from the tale of Ralegh, and on coming to rock encrusted with sparkling mica Schomburgk recalled, somewhat bitterly, "the picture Raleigh drew of Guayana, where every rock was described as argentiferous." Less as a celebration than as a concession, Schomburgk wrote, "For want of another name, I have inserted ... Sir Walter Raleigh's Cataract in my map."[153]

As Schomburgk and the survey approached the coast, the toponymic inscriptions reflected the approaching homecoming. Whereas high on the Corentyne Schomburgk wrote in an unusually self-referential name, "Goodall's Cataract" (unique in referring to the expedition itself), farther downstream the metropolitan invocations began, progressing from most distant and abstract—the Prussian King Frederick William IV Cataract—to the more immediately relevant sources of patronage—Lord Stanley's Cataract—and culminating in an offering to the most local and immediate supporter, Governor Light, whose cataract, conveniently, lay immediately beside that of Governor Carmichael-Smyth, which Schomburgk had named on his abortive ascent of the Corentyne in 1836. Schomburgk was back on familiar ground.

Whereas the onomastics of the expedition on the lower reaches of the Corentyne told the story of a homecoming, the names on the headwaters spoke to the ambiguity and complexity of the geographical exercise. Schomburgk reflected this ambiguity in his report to the governor on his return, where he explained that the upper Corentyne, at least from the to-

---

152. Ibid., 67.
153. Ibid., 90. Schomburgk did not settle on "Ralegh" as the spelling until be began his editorial work.

ponymic point of view, appeared not to exist at all. A trident of tributaries intersected at the headwaters of the river, but none of them was called the Corentyne by the local tribes. This, of course, made for considerable boundary confusion. Where was the Corentyne for the last forty miles of its upper course? Two of the tributaries were of equal size, and water color at the confluence did not tell which had precedence. So Schomburgk fell back on a linguistic argument bolstered by a political program, writing: "The selection of the Curuwuini, which name even resembles closer the Courentyne than the Cutari, adds a few thousand square miles more to British Guiana."[154]

This territorial ambiguity was rivaled only by the ambiguity of the unique "landmark" erected by members of the boundary commission in their descent of the river. Having been obliged to erase such landmarks as Schomburgk had already established, it was particularly fitting that what Schomburgk and Goodall left as a *cache* at the headwaters of the Corentyne was not a triumphant inscription or phallic pale but rather a *crèche* of abandoned goods. Forced to relinquish their provisions, equipment, and collections, they erected a thatched lean-to in which they deposited their baggage. Atop this pile of remains they left "the least valuable man" to keep "charge of our goods and chattels."[155] This "man" was an effigy of the explorers themselves: "Clothes were at a discount, since necessity forced us to limit the quantity of our baggage; and Mr. Goodall accordingly rigged out the guardian of our baggage, a figure made of straw, with some old trousers, a jacket, a pair of boots and a hat, giving him, with artistic skill, the character resemblance of a gruffy old gentleman."[156]

It was an abundantly suggestive gesture, the care lavished on preparing this mannequin and painting his mask. Surely none of them thought the scarecrow would keep the Amerindians in the region from rifling their store. They erected a guard that could not guard, a figure of the boundary surveyor as ineffectual as the boundary survey had proved to be. The surveyor was a straw man, his mark as ephemeral as the jetsam of an imperiled passage. The genius loci of the ambiguous territory of the upper Corentyne was an ambiguous caricature of themselves: rootless, without weight, subject to the elements, easily undone by wind and rain.

Schomburgk and Goodall indeed left a monument on the upper Corentyne, but instead of a monument to metropolitan territorial authority, it was a monument to the failure of that authority. It was a monument to the failure of the boundary inscription and a monument to the

---

154. *Brazil Arbitration, British Case Annex*, 2:145.
155. *JRGS* 15 (1845): 75.
156. Ibid.

reality that the featherlight and nomadic passage of the explorer could never afford the "good occupation" on which European territorial disputes ultimately turned. The cache, soon swallowed up by the forest, was a monument to *peripateia,* a monument to what they left behind.

EPILOGUE: THE WAKE OF THE BOUNDARY

The phantom figure presided over what has remained a phantom territory. In Schomburgk's urgent dash for the coast and survival, he spent no time looking back. This meant that, his eyes on the rapids and the narrow channels amid the countless islets, he apparently missed the confluence of a giant river from the west. If he had known, the specter of the ambivalent upper Takutu would have quickly come to mind. Schomburgk *thought* he had descended the upper Corentyne, but if this river had a larger tributary, that meant he was on the *tributary,* and the other branch was the main course. So said the Dutch when Charles Barrington Brown, traveling up the Corentyne in 1871 as part of his geological surveys of the colony, discovered what he called, without imagination, the "New River." If the New River catchment was larger than that of what Schomburgk called the upper Corentyne, then the new river was the *real* upper Corentyne, and the Dutch would gain 13,000 square kilometers at the expense of neighboring British Guiana. Arguments about the relative catchment sizes remained unresolved at the outbreak of the Second World War and became the inheritance of the sovereign states of Guyana and Surinam. The "New River Triangle" remains a no-man's-land.[157]

157. Prescott, *Political Frontiers and Boundaries,* 216.

CHAPTER SEVEN

Conclusion: History, Geography, Nation

> All geographers are ready to assert, no doubt truthfully, that the history of a country cannot be understood without a knowledge of its geography; but they are not quite so ready to admit that we cannot understand the geography of a country without knowing something of its history.
> Earl Curzon of Kedleston, RGS president, commenting on a paper on British Guiana delivered to the Society in 1911[1]

> In this chapter you will see how, in trying to make a home away from home, Europeans have turned Guiana from a wilderness to a modern state.
> Vere T. Daly, *The Making of Guyana*

> Every boundary line is a myth.
> Wilson Harris, *The Guyana Quartet*

UP TO THIS POINT

This book built to an examination of colonial boundaries offered in the previous chapter. There I narrated Schomburgk's boundary surveying and emphasized the tension between boundary making (the charge of the commissioned surveyor) and boundary crossing (the primary obligation of the interior geographical explorer). I have suggested that the enduring ambiguities in the boundaries of what was once the colony of British Guiana can be traced to this tension, as well as to the conflating of the roles of *delimiter* and *demarcater* in a single individual. The techniques of the traverse survey (the subject of chapter 3) were used to fix points along a boundary, but the pursuit of terra incognita led an interior explorer to work to surpass previous historical presences. This remapping and surmounting of history (discussed in chapter 2 as geographical metalepsis) could be used to root colonial claims to possession, but it also meant that each explorer sought to overstep the last, creating increasingly tenuous territorial presumptions. The spatial practices of landmarking (the sub-

---

[1] De Villiers, "Foundation and Development of British Guiana," 8.

ject of chapter 4) served the inscription of the boundary to a point, but the contingencies of surveying practice introduced ambiguities into the making of landmarks (like those discussed in chapter 5), and these ambiguities undermined the unity and clarity of the boundary. Geographical explorers never saw the fixed and stable territory imagined by imperial administrators. Chapter 6 showed that close attention to the practice of the surveyor can unravel the apparent unity of the boundary line as it appears on maps. The lay of the land, means of entry and exit, provisions, the color of rivers, the confusions of cross-cultural dialogue, all of these played roles in establishing the line, and all of them placed it on footing as slippery as that of the surveyor portaging around steep rapids.

The aim of this analysis has been to challenge stories told about the boundaries and the man who surveyed them. In place of the power-crazed "évolution de Schomburgk," I have suggested that Schomburgk's real evolution led him gradually toward a style of travel that afforded broad nomadic range at the same time that his elaborate efforts at landmarking came to be devalued. The expansive, transgressive character of the traverse survey came into conflict with a mandate to delimit, and hence stabilize, a colonial territory. The concentric and expanding boundaries Schomburgk demarcated embodied the traverse surveyor's tendency to overstep limits. The hypertrophy of the territory can be seen as an effect of a particular spatial practice; attention to the evolving spatial practices of an interior explorer has called into question a traditional interpretation of the boundary itself. As for Schomburgk's *idée raleighienne,* my account of the boundary surveys as an ambivalent process of progressive erasures (rather than a crescendo of imperial expansion) suggests that the epithet may still fit. An *idée raleighienne* he indeed had, if this means a growing disillusion on witnessing the dissolution of a dream about Guiana.

This conclusion extends these reflections on the boundary. But before taking the Schomburgkian step (beyond my own terminus), a look back over the route is in order. I began by claiming that this text would make contributions in several general areas: most important, to the understanding of geographical knowledge in the imperial context, but also to the history of exploration and cartography as well as to the history of British Guiana and the biography of Robert H. Schomburgk. Examining in detail the history of geographical exploration in the colony in the first half of the nineteenth century has allowed me to address the last two areas: my account fills in a missing chapter in the history of the colony; my retracing of Schomburgk's involvement has suggested a different approach to the man and his disputed boundary legacy. But at the same time that this book investigated British-sponsored exploration in northern

South America, it also focused on a particular style of geographical inquiry, one practiced widely over the long nineteenth century—the traverse survey. I have made several general claims about the character of the traverse survey and the labor it demanded: that fixed points were essential to the practice of a good survey; that the fixity of those points could be enhanced by reference to a variety of traditions, including Humboldtianism, nautical navigation, and military discipline; and finally (and most significant for more general thinking about the construction of colonial territory) that fixed points had to be associated with landmarks on the ground to be useful contributions to geography. The importance of landmarks in the context of the traverse survey led me to examine the function of the landmark in British Guiana more closely and afforded an opportunity for a detailed look at the relationship between explorers and their texts. I examined how these created an image of the colony among various readers, but I also suggested that the same texts could be mined for evidence of the explorer's active engagement with the land.

My observations, particularly with respect to the character of the traverse survey and the importance of the landmark, have been supported by reference to the case of British Guiana. Comparative cases would shed additional light on these subjects. How common were the micro/macro representational strategies I identify in chapter 4? Does the aesthetic overdetermination of sites constitute a general technique for constructing landmarks? Could the idea of the metonymic appropriation of colonial territory by means of iconic landmarks be extended? Do other explorers' texts show evidence of a "repatriation of perspective" through pictorial conventions? What other sites were charged with what I have called the "Icarian anxieties" of an explorer's negotiation of alien environments? How does the contestation over indigenous monuments and the construction of European landmarks play itself out in other arenas? It would be particularly interesting to have additional examples of how place-myths and metaleptic attention to precedent passages affected exploration and territorial claims in other colonies. British Guiana's distinctive associations with Ralegh and El Dorado allowed me to suggest both the importance and the ambivalence of cycles of remapping in the context of the geographical exploration and codification of the colony. I suspect other stories operate in different but equally significant ways in other areas.

AND OVER THE LINE

As I stated in the introductory chapter, I hope this history of surveying practices, landmarking, and boundary making does more than offer insights into the character of the traverse survey and shift interpretations of Schomburgk's role in the boundary disputes. I hope it also constitutes a point of departure for a revision of attitudes toward the boundary itself and its history. The basic school textbook of Guyanese history narrates the history of the geo-body of Guyana as a story of loss.[2] Although the 1899 decision of the tribunal was mostly favorable to the British case, "we did lose," the author explains, "land in the North-West District and a great part of the Cuyuni Basin."[3] The book explains that, even so, at the end of the nineteenth century "everyone was happy over the decision of the Tribunal," and "the Venezuelans were most happy, for the tribunal had given them 5,000 square miles of territory which previously did not belong to them."[4] This story sets the stage for Venezuela's twentieth-century reopening of the dispute.[5]

These anachronistic assumptions, and the accompanying map showing Guyana's geo-body "trimmed down" from its "original" size, constitute a retrospective instantiation of national territory.[6] Such a history not only denies the contingency of this territory, it also encourages modern Guyanese to adopt what is in effect an imperial perspective on their own nation. This perspective denies the vanished structures of overlapping local authorities, the Amerindian tribes who walked through and lived in the regions. It conceals the whole history of the making of a colonial and postcolonial state out of a region that did not conform, at the outset, to European concepts of territorial authority or emerging European notions about the geo-body of a European polity.[7] Distinctive elements of the history of "Guyana" get obscured.

This misleading history and the map of ahistorical projection it implies (and moves from) not only become constitutive elements of Guyanese nationhood but become inextricable from the assertion of Guyanese patriotism as well. Here the political and social implications are still more direct. Eusi Kwayana, a respected opposition intellectual, linked the perpetual boundary crisis not only to United States meddling but also to the dictatorship of Forbes Burnham: "One reading of the present trends is

---

2. See above, chapter 6, note 23.

3. Daly, *Making of Guyana*, 184. Since 1974 this textbook has been reprinted seven times, most recently in 1990.

4. Ibid.

5. The complex geopolitical context of this move deserves attention in its own right: it seems likely that the United States pressured Venezuela to raise the issue of the boundary in order to intimidate Guyana's leftist premier, Cheddi Jagan, in the aftermath of the Cuban revolution. See Sillery, "Salvaging Democracy?"

6. Although the history is by no means the same, the result can be seen as a version of the situation diagnosed by Winichakul in Thailand. See his *Siam Mapped*.

7. For an account of the evolution of a European boundary, see Sahlins, *Boundaries*.

Figure 33. "All Guyana Is Ours to Defend": a legacy of territorial anxiety. Collection of the author; courtesy of M. N. Menezes.

that Burnham is using the border issue to help [maintain] a sympathetic hearing abroad and to blackmail Guyanese into a position which seems to advise against confrontation with the dictatorship, to brow-beat the working people into a starvation or semi-starvation level of living in the interests of defense."[8] A flier printed under the Burnham administration illustrates the propagandistic uses of the boundary in a campaign to consolidate national loyalties and silence opposition parties (fig. 33). The

8. Braveboy-Wagner, *Venezuela-Guyana Boundary Dispute,* 246; Kwayana, "U.S. Card in the Venezuela-Guyana Row," 10.

ogre of Venezuela threatens to rip the fabric of Guyana's geo-body, a geo-body that on the reverse is shown superimposed with the children of the nation. A national call to arms reads: "All Guyana Is Ours to Defend."[9]

On the strength, in part, of an anachronistic projection of putative boundaries, the state of Guyana was maintained from the late 1970s through the early 1980s in a "state of emergency." By allowing that state to become not the exception but to a significant degree the rule, Burnham generated a siege mentality and undertook policies of repression and terror that spanned the savage to the burlesque: from the brutal assassination of Walter Rodney to the funding of the strongmen of Black Zionist Chicago ex-convict Rabbi Washington as an extralegal security force.[10] As Michael Taussig has suggested, the reign of order demands its daily fix of imminent disorder.[11] The besieged boundaries of Guyana afforded Burnham a stage on which he could maintain a state of siege.

Burnham's legacy remains, and while the boundaries are quiescent, they remain a focal point of national unity and potential international violence.[12] Guyana's money is emblazoned with the geo-body of the nation, and gold bangles in the same shape are available in all the jewelry kiosks of Stabroek Market. By creating a frontier zone that lies substantially outside any jurisdiction, the boundary ambiguities are as responsible for the local violence and "outlaw" character of the Guyanese interior as they are for the imminent international conflicts manipulated by the respective governments.[13] A revisionist account of the boundary—one that emphasizes its flexibility and the arbitrariness and ambivalence that have attended its inscription from the outset—takes a small step toward defusing the "enemy function," the paranoid projections that sanction oppression and serve as the trigger of violent conflict.

I am not unaware of how small this step is: Guyana's "state of siege" is not a symbol, nor is it merely a discursive proposition. People have died over the boundary of Guyana, and people are currently willing to die over not just this boundary but literally hundreds of thousands of miles of boundaries all over the world. This is unlikely to change anytime soon. Looking at the end of the millennium, a recent survey in *The Economist* asked readers to consider the world map of 2050.[14] What will it look like?

---

9. Figure 33 has been reproduced from a collection of boundary-related documents in the possession of Sister Mary Noel Menezes, University of Guyana.

10. For an assessment of the boundary dispute as a cover for the "steady militarization of Guyanese society," see Latin American Bureau, *Guyana, Fraudulent Revolution*. For an account of violent incidents in the border region (including the "Repununi Rebellion" in 1969), see Rout, *Which Way Out?* 65–69.

11. Taussig, *Nervous System*, chap. 2, "Terror as Usual."

12. For a discussion of "invasion psychosis" in Guyana, see Rout, *Which Way Out?* 73.

13. Roopnaraine discusses the fear of the boundary region: Roopnaraine, "Behind God's Back," 51.

14. Beedham, "Road to 2050."

After taking a stab at sketching the proximate future of global geopolitics, the staid paper rejected those who herald (or decry) the demise of the nation-state. The map of 2050, we learn, will look a great deal like the present one. Such prognostications have about them just a touch of the smug solidity of realpolitik. It may well be that the map will look the same. But if the human cost of its maintenance over the next fifty years comes close to the price paid in the past fifty (not to mention the fifty before that), we might do well to consider other, even highly speculative, options. At the very least we might make an effort to remind ourselves of the ephemeral nature of the lines we draw on the ground. It would be particularly responsible to explain this to those we ask to make such large sacrifices over them.

If this book has begun such a process, there might be value in concluding with a reflection on its future. Does the history of Guyana's boundary remain to be written? Perhaps it only remains to be read. One perspective on the documents, maps, testimonies, and arguments that make up the collected volumes of British Guiana's numerous boundary disputes is that they constitute a set of sources out of which a future historian will compose a future history. But another perspective might see the collected volumes of the boundary disputes as a revisionist history of the boundary waiting to be read, not written. My attempt to undermine the weight and unity of the boundary pales in comparison with the deconstructive power of the arbitration volumes themselves. Each attempt to ground the boundary is answered by a corresponding rebuttal. Every proposition meets a reply. The hundreds of maps, the appended commentaries, the sheets of tracing paper included in the volumes to allow one map to be laid over another, all of the text, evidence, and argument— all of it fails, in the end, to add up to a boundary and instead succeeds in presenting the very *consummatio* of boundary deconstruction. As William Boelhower puts it, once the interrogation of the map's mystery begins, no amount of repetition, no number of maps superadded to maps, can "conceal the script of that which took place in its establishment."[15] Instead of a mise-en-scène of the stage space of empire, the obligation repeatedly to think of the "removed and unwritten" results in a "cartographic *mise-en-abyme.*"[16] No coherent line coalesces out of the abundance. Rather, the boundary is revealed as a game of mirrors. Maps on maps add up to maps.

Read in counterpoint, the boundary arbitration documents show that no subtle strategy for unsettling the stability of a cartographic inscrip-

---

15. Boelhower, "Inventing America," 494.
16. Ibid.

tion has been overlooked. Here readers will find elaborate arguments concerning the authority of individual maps, the authority of mapmakers, and the significance of surveys being made on the ground. But these general arguments were extended to debates about the value of a "spurious watershed" as a boundary and the possessive significance of a "tentative" entry into unfamiliar territory. If the boundary on a given map followed a range of hills found not to exist, what was the status of the line? Could it be broken up? Transferred whole or in part to another map? These sorts of questions, much disputed, amounted to questions about the status of boundary lines themselves. Were they meaningful, as Britain argued, only with "respect to the internal geography of the map" on which they were drawn? If so, what relation could they be said to have to the ground?[17]

Confronted with an overwhelming number of historical maps repudiating their claim of historical occupation, Britain was compelled to contest the Brazilian cartographic evidence by stating that "an accumulation of maps showing all the same line carries no weight whatsoever."[18] Cartographic "weight" lay only in maps buttressed by "official" seals and commissions, "and such inquiry," the argument went, "brushes away all but a few maps."[19] Why, the Brazilian disputants were curious to know, had this argument not been deemed relevant in the dispute with Venezuela just a few years before? Did the same rule apply to passages? If, as Britain claimed of a Portuguese expedition, "a passage confers no right," then how could the Dutch have "gradually acquired actual right of occupation in the district" because "they were constantly traversing it in close relations with the natives who inhabited it"?[20] Britain countered: If Portuguese officials hid geographical knowledge for fear of not holding on to their claims, how could that secret geographical knowledge be the base of their claims? Was it not in the nature of a claim to be public? Only disclosure could form the basis of enclosure. But if that was so, why had the British boundary commission refused to give the locations of its markers on the Takutu? Did erasing them amount to a withdrawal of the claim? So subtle did the casuistry become that gestures and manners became the grounds for claims to vast swaths of territory. If Charles Waterton was received at the Brazilian fort of São Joaquim when he was ill, did that mean he acknowledged Brazilian sovereignty? If the commander of the fort obliged him to

17. *Brazil Arbitration, British Argument,* chap. 3, "The Evidence of Maps," 93.
18. Ibid., 87.
19. Ibid., 88.
20. *Brazil Arbitration, British Argument,* chap. 2, "The Contention," 56.

pitch his camp on the opposite side of the Takutu, did that mean it was the limit of British territory?[21]

The tangled disputation progressively undermines the very idea that further inquiry will ever "reveal" the boundary, and it winds its way toward paradoxes more perverse than could be invented by devoted Derridians: If Portuguese surveys in the region of the Rio Branco were designed to *strengthen* the boundaries, then didn't that imply that the points the surveyors were most concerned with must by definition have been the *weakest* points of that possession?[22] What interpretation should be given to Hislop's 1803 map of the colony, where the boundary *line* did not coincide with how the map was *colored?* If the Amacura and the Barima Rivers were toponymically reversed in Bouchenroeder's 1798 map, then should the Dutch post placed on his map be assumed to be in the correct "position," or on the correctly named river? In the strategic deployment of enigmas, perhaps none of the arguments in any of the disputes surpasses that of the Brazilians where Schomburgk was concerned. They crafted a catch-22 of vivid illogic: If Schomburgk took possession of the sources of the Essequibo, didn't that imply that Britain did *not* possess those sources before he did so?[23]

The boundary arbitration documents do not add up to a boundary, and there is no missing piece of evidence that could come to light and, deus ex machina, resolve the snarled plot. The mass of the volumes fails to weight the boundary, and instead of the multiple lines' resolving themselves into some boundary *mean,* they cancel each other out, not only leaving no clear boundary but ultimately calling into question just what a clear boundary is. The language of the social scientists again proves inadequate. If an *antecedent* boundary was drawn "before the development of most of the features of the cultural landscape," doesn't that leave open the question of whose "cultural landscape" counts?[24] What boundary is ever *antecedent,* or for that matter *subsequent* (following the development of the cultural landscape)? Can such a development ever be said to have stopped? The notion that a *consequent* boundary conforms with some physical divide while a *discordant* or *superimposed* boundary does not provides a tidy boundary taxonomy, but the arbitration documents, and the previous chapter, suggest the serious problems with such terms and the premises they demand.

21. *Brazil Arbitration, British Counter-case,* chap. 5, 112–13.
22. Ibid., part 6, "Portuguese Surveys."
23. For a discussion of the question, see *Brazil Arbitration, British Counter-case,* chap. 6, 120.
24. Prescott, *Political Frontiers and Boundaries,* 14. This term and those that follow are from Hartshorne, "Suggestions on the Terminology of Political Boundaries."

Read against themselves as I am suggesting, the volumes of the boundary arbitration cease to be a hill of primary documents waiting for a historian and become a monumental literary document, the impossible novel George Perec might have written had he been born in Berbice. Read this way, the boundary documents are not so much waiting for historians as waiting for *readers,* for those able to read them not simply as a meditation on the shortcomings of a certain set of Western ideas about boundaries and territory, but as a vast subversion of imperialism itself. Such a reading could go further, for if the text of the boundary arbitration, and the arbitrary resolutions it produced, raises any question at all, it is the question of what it means to "possess" land where one does not "dwell." Every page of the tens of thousands dramatizes the intractable paradoxes and entanglements of the attempt to "divide" and "hold" land by means of *representations* of the place rather than by *participation* or *engagement* with it, by reliance on *mimetic* artifacts rather than local and dynamic *methexis.*

The volumes were designed to lie on a negotiating table, not on the ground, and Paul Carter suggests that in looking for a path along which we might move to "overcome our obsession with exclusive ownership" and diffuse the postcolonial neuroses of that obsession, "we could do worse than begin by reflecting on the mechanism of the negotiating table and the model of communication it implies." He continues: "What does this polished, horizontal surface hoisted off the ground signify? What history of violence does its pretense of smoothness, its equalization of places, conceal?"[25] The volumes meant to lie on that table reflect the perspective of those who refuse to walk the land, who believe that places can be made flat and still and therefore be possessed. This perspective means that the boundary documents present multiple sallies against a central paradox: how to *fix* territory by means of *mobile* passages; how to freeze the *peripateia* of human interactions with the land into stable boundaries and discrete regions. The dominant European anxiety about the "wandering state" is inscribed on every page.

Schomburgk wrote, mockingly, that the Brazilian boundary pale could not have roots if it had been placed there by human hands. This is true of a pale and false of a tree. Human beings can plant trees, but they take much longer to raise than a branded pike. They also have deeper roots. Orchards, in this sense, stake claims. In the end, the closest reading of the boundary documents shows that there is no way to "grasp" land separate from human interactions with it, that there is no "natural" way to divide the territory, because the very idea of "nature" defined by our absence—

---

25. Carter, *Lie of the Land,* 365.

of nature *as opposed to* culture—is a counterfactual. Nature does not divide naturally. Cultures do not live on maps.

EPIGRAM

The boundary arbitration volumes erase the boundary. But as Governor Light sagely remarked of the marks carved out of the trees on the Takutu, erasures leave marks of their own, and of necessity deeper than the marks they efface. The boundary of Guyana may be erased, but a boundary it remains; erasure remains history.

BIBLIOGRAPHY

UNPUBLISHED PRIMARY SOURCES (BY ARCHIVE)

British Library, London
  [Goodall MS] Additional MSS 16936–39
    E. A. Goodall's watercolors from the expeditions in 1842–43 (see Menezes, "Sketches of Amerindian Tribes").
  [Hilhouse MS] Additional MSS 37057
    William Hilhouse's "Account of British Guiana" (1835), with two maps and a journal of a Cuyuni expedition in 1830. The document is bound so that the recto and verso of the signature are distinct texts.
  [Schomburgk Autograph] BL call number 10470.p.p.p.1
    Robert Schomburgk's autograph copy of British Guiana, *Local Guide of British Guiana* for 1843.
  [Schomburgk MS] Additional MSS 37205
    Robert Schomburgk's journal from 1843 expedition (see *JRGS* 15 [1845]: 1–104).
[BOD] Bodleian Library, Oxford
  [MS] MS.Pigott.c.3
    Robert Schomburgk's presentation to the Antiquarian Society (n.d.) titled "Fragments of Indo American traditions and a description of the painted rocks at Warapoota."
[CRL] Caribbean Research Library, University of Guyana, Turkeyen
  [Holmes MS]
    Letters describing the colony ca. 1818.
  [Long Collection]
    A set of alphabetical scrapbooks containing assorted material. For an introduction, see Carol Collins, "Long Collection Catalogue," CRL reference Z907.G8.U5.
  [RBR] Rare Book Room
    [Alexander MS]
      Pen-and-ink sketches (colored) of British Guiana (see Alexander, *Transatlantic Sketches*).
    [Maps]
      Maps of Guyana and the surrounding regions. This collection includes accession no. 109883, the previously unidentified Schomburgk proposal map.
    [Roth Collection]
      A collection of manuscripts, notes, and translations belonging to Walter Edmund Roth (1861–1933), anthropologist and collector.
    [Sawkins and Brown MS]
      Several pencil sketches by Charles Barrington Brown and a sleeve of James G. Sawkins's watercolors.
[MQ] Private collection of Mava Quinlan, Alderbury
  [Brown MS Journal]
    Extensive journals kept by Charles Barrington Brown during his geological surveying work in British Guiana, 1867–73. Much of the material appears in revised form in *Canoe and Camp Life*. I was afforded limited access to these private papers.

[NAG] National Archive of Guyana, Georgetown
  [Loose reports]
    The National Archive, despite the labors of Sister Mary Noel Menezes and others, is in very poor condition. It has been moved, and materials have been lost. Numerous three-fold "reports" of various colonial officials are preserved in boxes, but they are no longer cataloged. The "loose reports" are just that.
  [MCP] Minutes of the Court of Policy (1803–80)
    Cited in notes by date.
  [Minutes] Minutes of the Astronomical and Meteorological Society of British Guiana
    Runs through to 1853.
[PRO] Public Record Office, Kew
  [CO] Colonial Office
    For a listing of material relevant to British Guiana, see Menezes, *British Policy,* 299–300. Menezes has also completed a considerable portion of an annotated bibliography of these documents (currently unpublished, typescript in possession of the author).
  [FO 13] Foreign Office
    General Correspondence, Brazil.
[RGS] Royal Geographical Society, London
  [Correspondence]
    These files, arranged by author (Robert Schomburgk and William Hilhouse), consist primarily of letters to the RGS secretary. Additional documents are sometimes included. In the notes I have specified author and date of letters. In referring to other documents found in these files, I have given the file and a description of the document.
  [Journal MS]
    These loose-paper files contain an assortment of manuscript reports submitted by Robert Schomburgk to the RGS. Some were marked for editing and subsequently published (in the *JRGS*). Others remained unpublished. In citing a document in these bundles I have given the file and, where possible, a folio number.
  [Map MS]
    This collection includes a number of Robert Schomburgk's original maps.
[RH] Rhodes House Library, Oxford
  [MS] MSS.Brit.Emp.S18C106/34: Robert Schomburgk's Letter to Sir Thomas Buxton, 25 August 1838.

PUBLISHED SOURCES

Adamson, Alan. *Sugar without Slaves.* New Haven: Yale University Press, 1972.
Adas, Michael. *Machines as the Measure of Men: Science, Technology and Ideologies of Western Dominance.* Ithaca: Cornell University Press, 1989.
Alexander, James Edward. *Transatlantic Sketches.* London: Bentley, 1833.
Allen, J. L. "The Indrawing Sea: Imagination and Experience in the Search for the Northwest Passage, 1497–1632." In *American Beginnings: Exploration, Culture, and Cartography in the Land of Norumbega,* edited by Emerson W. Baker et al. Lincoln: University of Nebraska Press, 1994.
———. "Lands of Myth, Waters of Wonder: The Place of the Imagination in the History of Geographical Exploration." In *Geographies of the Mind,* edited by David Lowenthal and Martyn J. Bowden. New York: Oxford University Press, 1976.
Allen, William. *Picturesque Views on the River Niger.* London: Murray; Hodgson and Graves; Ackermann, 1840.
Anthon, M. "The Kanaima." *Timehri* 36 (October 1957): 61–65.
Appun, Karl Ferdinand. *Unter den Tropen: Wanderungen durch Venezuela, am Orinoco, durch Britisch Guyana und am Amazonenstrome in den Jahren 1849–1868.* Jena: H. Costenoble, 1871.
Armitage, Christopher. *Sir Walter Ralegh: An Annotated Bibliography.* Chapel Hill: University of North Carolina Press, 1987.
Arnold, David. *The Problem of Nature: Environment, Culture and European Expansion.* Cambridge, Mass.: Blackwell, 1996.

Arnold, David, and Ramachandra Guha, eds. *Nature, Culture, Imperialism: Essays on the Environmental History of South Asia.* Delhi: Oxford University Press, 1995.
"Article XVII. [A review of] *The Lady of the Lake.*" *Quarterly Review* 3 (February–May 1810): 492–517.
Axtell, James. "The Exploration of Norumbega: Native Perspectives." In *American Beginnings: Exploration, Culture, and Cartography in the Land of Norumbega,* edited by Emerson W. Baker et al. Lincoln: University of Nebraska Press, 1994.
Baber, Zaheer. *The Science of Empire: Scientific Knowledge, Civilization, and Colonial Rule in India.* Albany: State University of New York Press, 1996.
Babineau, G. Raymond. "The Compulsive Border Crosser." *Psychiatry* 35 (1972): 281–90.
Baker, Emerson W., E. A. Churchill, R. S. D'Abate, K. L. Jones, V. A. Konrad, and H. E. L. Prins, eds. *American Beginnings: Exploration, Culture, and Cartography in the Land of Norumbega.* Lincoln: University of Nebraska Press, 1994.
Baker, William. "Local Guide for 1821." Caribbean Research Library, University of Guyana. Photocopy.
Ballard, Paul. "Analysis in Historical Cartography Studies." *Cartography* 12 (1982): 173–78.
Barnes, Trevor J., and James S. Duncan, eds. *Writing Worlds: Discourse, Text, and Metaphor in the Representation of Landscape.* New York: Routledge, 1992.
Basalla, G. "The Spread of Western Science." *Science* 156 (1967): 611–22.
Bayly, C. A. *Imperial Meridian: The British Empire and the World, 1780–1830.* London: Longman, 1989.
Beedham, Brian. "The Road to 2050: A Survey of the New Geopolitics." *Economist,* 31 July 1999, sixteen-page insert after p. 46.
Belcher, E. *A Treatise on Nautical Surveying.* London: P. Richardson, 1835.
Bell, Morag, Robin Butlin, and Michael Heffernan, eds. *Geography and Imperialism, 1820–1940.* Manchester: Manchester University Press, 1995.
Belyea, Barbara. "Amerindian Maps: The Explorer as Translator." *Journal of Historical Geography* 18, no. 3 (1992): 267–77.
———. "Captain Franklin in Search of the Picturesque." *Essays on Canadian Writing* 40 (spring 1990): 1–24.
———. "Images of Power: Derrida/Foucault/Harley." *Cartographica* 29, no. 2 (1992): 1–9.
———. "Review Article of Denis Wood's *The Power of Maps* and the Author's Reply." *Cartographica* 29, nos. 3–4 (1992): 94–99.
Bendall, A. Sarah. *Maps, Land, and Society: A History, with a Carto-bibliography of Cambridgeshire Estate Maps, c1600–1836.* New York: Cambridge University Press, 1992.
Benjamin, Anna. "A Preliminary Look at the Free Amerindians and the Dutch Plantation System in Guyana during the Seventeenth and Eighteenth Centuries." *Guyana Historical Journal* 4–5 (1992–93): 1–22.
———. "Review of Braveboy-Wagner's *The Venezuela-Guyana Boundary Dispute: Britain's Colonial Legacy in Latin America.*" *Guyana Historical Journal* 3 (1991): 53–57.
Bennett, George W. *An Illustrated History of British Guiana, Compiled from Various Authorities.* Georgetown, British Guiana: Richardson, 1866.
Bermingham, Ann. *Landscape and Ideology: The English Rustic Tradition, 1740–1860.* Berkeley: University of California Press, 1986.
Bermúdez, Juan Almécija. *La estrategia imperial británica en el Guayana Esequiba.* Caracas: Universidad Católica Andres Bello, 1987.
Bertram, Marshall. *Birth of Anglo-American Friendship: The Prime Facet of the Venezuelan Boundary Dispute, a Study of the Interreaction of Diplomacy and Public Opinion.* Lanham, Md.: University Press of America, 1992.
Bicknell, Peter. *Beauty, Horror, Immensity: Picturesque Landscape in Britain, 1750–1850.* New York: Cambridge University Press and Fitzwilliam Museum, 1981.
Blackburn, Julia. *Charles Waterton, 1782–1865: Traveller and Conservationist.* London: Bodley Head, 1989.
Blair, H. *Lectures on Rhetoric and Belles Lettres.* London: W. Strahan and T. Cadell, 1783.
Blakemore, M. J., and J. B. Harley. *Concepts in the History of Cartography: A Review and Perspective.* Edited by Edward H. Dahl. Downsview, Ont.: University of Toronto Press, 1980.

Boase, Frederic. *Modern English Biography.* London: Truro, Netherton, and Worth (for the author), 1892–1921.
Boddam-Whetham, J. W. *Roraima and British Guiana.* London: Hurst and Blackett, 1879.
Boelhower, William. "Inventing America: A Model of Cartographic Semiosis." *Word and Image* 4, no. 2 (1988): 475–97.
Bosazza, V. L., and C. G. C. Martin. "Geographical Methods of Exploration Surveys in the 19C." In *Maps and Surveys of Malawi : A History of Cartography and the Land Survey Profession,* edited by C. G. C. Martin. Rotterdam: A. A. Balkema, 1980.
"Botanical Society [Report]." *Athenaeum: Journal of English and Foreign Literature, Science, and the Fine Arts,* no. 515 (9 September 1837): 661, col. 1.
Botting, Douglas. *Humboldt and the Cosmos.* London: Michael Joseph, 1973.
Boyle, Robert. *General Heads for the Natural History of a Country Great or Small, Drawn out for the Use of Travellers and Navigators.* London: John Taylor, 1692.
Braveboy-Wagner, Jaqueline. *The Venezuela-Guyana Boundary Dispute: Britain's Colonial Legacy in Latin America.* Boulder, Colo.: Westview Press, 1984.
Bravo, Michael. "The Accuracy of Ethnoscience: A Study of Inuit Cartography and Cross-Cultural Commensurability." *Manchester Papers in Social Anthropology* 2 (1996): 1–36.
———. "Ethnological Encounters." In *Cultures of Natural History,* edited by N. Jardine, J. A. Secord, and E. C. Spary. New York: Cambridge University Press, 1996.
Brett, W. H. *The Indian Tribes of Guiana.* London: Bell and Daldy, 1868.
———. *Legends and Myths of the Aboriginal Indians of British Guiana.* London: William Wells Gardner, [1880?].
Brewington, M. V. *Navigating Instruments.* Salem, Mass.: Peabody Museum, 1963.
Bridges, Roy. "Historical Role of British Explorers in East Africa." *Terrae Incognitae* 14 (1982): 1–21.
British Guiana. *The Demerara and Essequibo Vade-Mecum.* Georgetown, British Guiana, 1825.
———. *Local Guide of British Guiana Containing Historical Sketch, Statistical Tables and the Entire Statute Law of the Colony in Force, January 1, 1843.* Demerary [Demerara], British Guiana: Royal Gazette Office, 1843.
Brock, W. H. "Humboldt and the British: A Note on the Character of British Science." *Annals of Science* 50 (1993): 356–72.
Brockway, Lucile. *Science and Colonial Expansion: The Role of the British Botanical Gardens,* New York: Academic Press, 1979.
Brown, Charles Barrington. *Canoe and Camp Life in British Guiana.* London: Stanford, 1876.
Brown, Charles Barrington, and James G. Sawkins. *Reports on the Physical, Descriptive and Economic Geology of British Guiana.* London: Her Majesty's Stationery Office, 1875.
Browne, Janet. "A Science of Empire: British Biogeography before Darwin." *Revue d'Histoire des Sciences* 45 (1992): 453–75.
Buisseret, David, ed. *From Sea Charts to Satellite Images: Interpreting North American History through Maps.* Chicago: University of Chicago Press, 1990.
Bunkse, Edmunds V. "Humboldt and an Aesthetic Tradition in Geography." *Geographical Review* 71 (1981): 127–46.
Burnett, D. G. "Science and Colonial Exploration: Interior Exploration and the Amerindians of British Guiana 1820–1845." M.Phil. diss., Cambridge University, 1994.
———. "Terra Incognita: Expedition, Narrative, and History in British Guiana." Paper presented at the Cabinet of Natural History, Cambridge University, 1 May 1995.
Burnett, D. G., and R. Craib. "Review: *The Mapping of New Spain.*" *Cartographic Perspectives* 28 (spring 1997): 40–43.
Burnham, L. F. S. *Guyana/Venezuelan Relations: Statement by the Honourable L. F. S. Burnham, Prime Minister of Guyana.* Georgetown, Guyana: Ministry of Foreign Affairs, 1968.
Burr, George Lincoln. "The Search for the Venezuela-Guiana Boundary." *American Historical Review* 6 (April 1899): 470–77.
Cain, P. J., and A. G. Hopkins. *British Imperialism: Innovation and Expansion, 1688–1914.* New York: Longman, 1993.

Cameron, Ian. *To the Farthest Ends of the Earth: 150 Years of World Exploration by the Royal Geographical Society.* New York: E. P. Dutton, 1980.

Cameron, James M. R. "Agents and Agencies in Geography and Empire: The Case of George Grey." In *Geography and Imperialism, 1820–1940,* edited by Morag Bell, Robin Butlin, and Michael Heffernan. Manchester: Manchester University Press, 1995.

Cannon, Susan Faye. *Science in Culture: The Early Victorian Period.* Kent, England: Dawson; New York: Science History Publications, 1978.

Carter, Paul. *The Lie of the Land.* London: Faber and Faber, 1996.

———. *The Road to Botany Bay: An Essay in Spatial History.* Boston: Faber and Faber, 1987.

Certeau, Michel de. *The Practice of Everyday Life.* Translated by Steven Randall. Berkeley: University of California Press, 1984.

Chapman, Walker. *The Golden Dream: Seekers of El Dorado.* New York: Bobbs-Merrill, 1967.

Clarke, G. N. G. "Taking Possession: The Cartouche as Cultural Text in Eighteenth-Century American Maps." *Word and Image* 4, no. 2 (1988): 455–74.

Clementi, Cecil. *A Constitutional History of British Guiana.* London: Macmillan, 1937.

Clementi, Marie Penelope Rose. *Through British Guiana to the Summit of Roraima.* London: T. F. Unwin, 1920.

Colson, Audrey Butt. "Routes of Knowledge: An Aspect of Regional Integration in the Circum-Roraima Area of the Guiana Highlands." *Antropológica* 63–64 (1985): 103–49.

Conrad, Joseph. *The Mirror of the Sea.* Garden City, N.Y.: Doubleday, Doran, 1932.

Cook, T. "A Reconstruction of the World: George R. Parkin's British Empire Map of 1893." *Cartographica* 21 (1984): 53–65.

Coote, Stephen. *A Play of Passion: The Life of Sir Walter Ralegh.* London: Macmillan, 1993.

Cormack, Lesley B. "'Good Fences Make Good Neighbors': Geography as Self-Definition in Early Modern England." *Isis* 82 (1991): 639–61.

———. *Charting an Empire: Geography at the English Universities, 1580–1620.* Chicago: University of Chicago Press, 1997.

Cosgrove, Denis, and Stephen Daniels, eds. *The Iconography of Landscape: Essays on the Symbolic Representation, Design, and Use of Past Environments.* New York: Cambridge University Press, 1988.

Costa, Emília Viotti da. *Crowns of Glory, Tears of Blood: The Demerara Slave Rebellion of 1823.* New York: Oxford University Press, 1994.

Cronon, William. "In Search of Nature." In *Uncommon Ground: Toward Reinventing Nature,* edited by William Cronon. New York: W. W. Norton, 1995.

———. "The Trouble with Wilderness." In *Uncommon Ground: Toward Reinventing Nature,* edited by William Cronon. New York: W. W. Norton, 1995.

———, ed. *Uncommon Ground: Toward Reinventing Nature.* New York: W. W. Norton, 1995.

Curtin, Philip D. *The Image of Africa: British Ideas and Action, 1780–1850.* Madison: University of Wisconsin Press, 1964.

D'Abate, R. "On the Meaning of a Name: 'Norumbega' and the Representation of North America." In *American Beginnings: Exploration, Culture, and Cartography in the Land of Norumbega,* edited by Emerson W. Baker et al. Lincoln: University of Nebraska Press, 1994.

Dalton, H. G. *The History of British Guiana.* 2 vols. London: Longman, Brown, Green, and Longmans, 1855.

Daly, Vere T. *The Making of Guyana.* London: Macmillan, 1974.

Daniels, Stephen. *Fields of Vision: Landscape, Imagery and National Identity in England and the United States.* Cambridge: Polity Press in association with Blackwell, 1993.

———. "Re-visioning Britain: Mapping and Landscape Painting, 1750–1820." In *Glorious Nature: British Landscape Painting, 1750–1850,* cataloged by Katharine Baetjer. London: Zwemmer, 1993.

Darnton, Robert. "Workers Revolt." In *The Great Cat Massacre.* London: Penguin, 1984.

Daston, Lorraine, and Peter Galison. "The Image of Objectivity." *Representations* 40 (fall 1992): 81–128.

Day, Archibald. *Admiralty Hydrographic Service, 1795–1919.* London: Her Majesty's Stationery Office, 1967.

Delgado, Rafael Sureda. *Venezuela y Gran Bretaña: Historia de una usurpación.* Caracas: Universidad Central de Venezuela, 1980.

Dening, Greg. *Islands and Beaches; Discourse on a Silent Land: Marquesas, 1774–1880.* Honolulu: University Press of Hawaii, 1980.

———. *Mr. Bligh's Bad Language: Passion, Power, and Theatre on the "Bounty."* New York: Cambridge University Press, 1992.

Descola, Philippe, and Gísli Pálsson. "Introduction." In *Nature and Society: Anthropological Perspectives,* edited by Philippe Descola and Gísli Pálsson. New York: Routledge, 1996.

———, eds. *Nature and Society.* New York: Routledge, 1996.

Despres, Leo A. *Cultural Pluralism and Nationalist Politics in British Guiana.* New York: Rand McNally, 1967.

Dettelbach, Michael. "Global Physics and Aesthetic Empire: Humboldt's Physical Portrait of the Tropics." In *Visions of Empire: Voyages, Botany, and Representations of Nature,* edited by David Philip Miller and Peter Hanns Reill. New York: Cambridge University Press, 1996.

———. "Humboldtian Science." In *Cultures of Natural History,* edited by N. Jardine, J. A. Secord, and E. C. Spary. New York: Cambridge University Press, 1996.

De Villiers, J. A. J. "Famous Maps in the British Museum." *Geographical Journal* 44 (August 1914): 168–88.

———. "Foundation and Development of British Guiana." *Geographical Journal* 38 (July 1911): 8–26.

De Weever, Aloysius. *A Text Book of the Geography of British Guiana, the West Indies, and South America for the Use of Schools.* Demerara, British Guiana: Argosy, 1903.

Dexter, Franklin Bowditch. *Biographical Sketches of the Graduates of Yale College with Annals of the College History.* New York: H. Holt, 1885–1912.

Doyle, Arthur Conan. *"The Lost World" and "The Poison Belt": Professor Challenger Adventures.* Introduction by William Gibson. San Francisco: Chronicle Books, 1989.

Drayton, Richard. "Imperial Science and a Scientific Empire: Kew Gardens and the Uses of Nature, 1772–1903." Ph.D. diss., Yale University, 1993. Abstract in *Dissertation Abstracts International* 54 (1993): 2283-A.

———. "Science and European Empires." *Journal of Imperial and Commonwealth History* 23, no. 3 (1995): 503–10.

Duncan, James S., and Nancy G. Duncan. "Ideology and Bliss: Roland Barthes and the Secret Histories of Landscape." In *Writing Worlds: Discourse, Text, and Metaphor in the Representation of Landscape,* edited by Trevor J. Barnes and James S. Duncan. New York: Routledge, 1992.

Dyson, Anthony. *Pictures to Print: The Nineteenth-Century Engraving Trade.* London: Ferrand Press, 1984.

E. H. B., ed. *Geographical, Commercial, and Political Essays.* London: Longman, 1812.

Edney, Matthew H. "British Military Education, Mapmaking, and Military 'Map-Mindedness' in the Later Enlightenment." *Cartographic Journal* 31 (June 1994): 14–20.

———. *Mapping an Empire: The Geographical Construction of British India, 1765–1843.* Chicago: University of Chicago Press, 1997.

———. "Mapping and the Early Modern State: The Intellectual Nexus of Late Tudor and Early Stuart Cartography—Review Article." *Cartographica* 29, nos. 3–4 (1992): 89–93.

———. "The Patronage of Science and the Creation of Imperial Space: The British Mapping of India, 1799–1843." *Cartographica* 30, no. 1 (1993): 61–67.

Edwards, W., and K. Gibbon. "An Ethnohistory of Amerindians in Guyana." *Ethnohistory* 26, no. 2 (1979): 161–75.

Einberg, Elizabeth. *The Origins of Landscape Painting in England.* London: Kenwood, 1967.

Fabian, Johannes. *Time and the Other: How Anthropology Makes Its Object.* New York: Columbia University Press, 1983.

Falk, Avner. "Border Symbolism." *Psychoanalytic Quarterly* 43 (1974): 650–60.

———. "Border Symbolism Revisited." *International Review of Psychoanalysis* 10 (1983): 215–20.

Farage, N. *As muralhas dos sertões: Os povos indígenas no rio Branco e a colonização.* São Paulo: Editora Paz e Terra, 1991.

Farley, Rawle. "The Unification of British Guiana." *Social and Economic Studies* 4, no. 2 (1955): 168–83.

Feingold, Richard. *Nature and Society: Later Eighteenth-Century Uses of the Pastoral and Georgic.* New Brunswick, N.J.: Rutgers University Press, 1978.
Fernández, Antonio de Pedro. *La historia y el derecho en la reclamación venezolana de la Guayana Esequiba.* Caracas: Editorial Mediterráneo, 1969.
Flint, Montagu. "The Ascent of Roraima." *Temple Bar* 58 (January–April 1880): 389–400.
———. "In the Wilds of Guiana." *Temple Bar* 58 (January–April 1880): 182–89.
———. "My Return from Roraima." *Temple Bar* 59 (May–August 1880): 48–67.
Ford, John. *Ackermann, 1783–1983: The Business of Art.* London: Ackermann, 1983.
Friederichsen, L. "Sir Walter Ralegh's Karte von Guayana um 1595." In *Hamburgische Festschrift zur Erinnerung an die Entdeckung Amerikas,* vol. 2. Hamburg: Mittheilungen der Geographischen, 1892.
Frome, E. C. *An Outline of the Method of Conducting a Trigonometrical Survey.* 2d ed. London: J. Weale, 1850.
Fuller, M. "Ralegh's Fugitive Gold." In *New World Encounters,* edited by Stephen Greenblatt. Berkeley: University of California Press, 1993.
Galton, Francis. *The Art of Travel.* 2d ed. London: John Murray, 1860.
Gardner, Helen, ed. *The Metaphysical Poets.* London: Penguin, 1972.
Gizycki, Rainald von. "Center and Periphery in the International Scientific Community: Germany, France, and Great Britain in the Nineteenth Century." *Minerva* 11, no. 4 (1973): 474–94.
Glacken, Clarence J. *Traces on the Rhodian Shore: Nature and Culture in Western Thought from Ancient Times to the End of the Eighteenth Century.* Berkeley: University of California Press, 1967.
Godlewska, A. "Map, Text and Image: The Mentality of Enlightened Conquerors, a New Look at the *Description de l'Égypte.*" *Transactions of the Institute of British Geographers,* n.s., 20 (1995): 5–28.
Godlewska, A., and Neil Smith, eds. *Geography and Empire.* Cambridge, Mass.: Blackwell, 1994.
Goetzman, William. *Army Exploration in the American West.* New Haven: Yale University Press, 1965.
———. *Exploration and Empire: The Explorer and the Scientist in the Winning of the American West.* New York: Norton, 1966.
Goodall, Edward Alfred. "Diary during His Sojourn in Georgetown from 28th July to 11th December 1841." *Journal of the British Guiana Museum and Zoo* 35 (1962): 39–53.
———. "Diary 23[d] December 1841–13th June 1842 (on the Essequibo and Rupununi Rivers)." *Journal of the British Guiana Museum and Zoo* 36 (1962): 47–64.
———. *Sketches of Amerindian Tribes, 1841–1843.* Edited and Introduced by M. N. Menezes. London: British Museum Publications for the National Commission for Research Materials on Guyana, 1977.
Gott, Richard. *Land without Evil: Utopian Journeys across the South American Watershed.* New York: Verso, 1993.
Gray, J. E. "Dr. Robert Schomburgk's Description of Victoria Regina." *Magazine of Zoology and Botany* (Edinburgh) 2, no. 11 (1838): 440–42.
———. "On the Names of the Victoria Water Lily." *Annals and Magazine of Natural History,* 2d ser., no. 32 (1850): 146–47.
———. "On Victoria Regia." *Annals and Magazine of Natural History,* 2d ser., no. 36 (1850): 491–94.
Greenblatt, Stephen J. *Marvelous Possessions: The Wonder of the New World.* Chicago: University of Chicago Press, 1991.
———. *Sir Walter Ralegh: The Renaissance Man and His Roles.* New Haven: Yale University Press, 1973.
———, ed. *New World Encounters.* Berkeley: University of California Press, 1993.
Grindlay, Robert Melville. *Scenery, Costumes and Architecture Chiefly on the Western Side of India.* London: Ackermann (vol. 1), 1826; Smith, Elder, and Cornhill (vol. 2), 1830.
Grove, Richard H. *Green Imperialism: Colonial Expansion, Tropical Island Edens, and the Origins of Environmentalism, 1600–1860.* New York: Cambridge University Press, 1995.
Guss, David. *To Weave and to Sing.* Berkeley: University of California Press, 1989.
Guthrie, William. *A Geographical, Historical and Commercial Grammar Exhibiting the Present State of the World.* 24th ed. London: C. and J. Rivington, 1827.
Hancock, John. "Observations on Certain Resinous and Balsamic Substances Found in Guiana." *Edinburgh Journal of Science,* n.s., 1, no. 2 (April–October 1829): 223ff.

———. *Observations on the Climate, Soil, and Production of British Guiana, and on the Advantages of Emigration to and Colonizing the Interior of That Country.* London: Fraser, 1835.

———. *Observations on the Climate, Soil, and Production of British Guiana, and on the Advantages of Emigration to and Colonizing the Interior of That Country.* 2d ed. London: C. Richards, 1840.

Harley, J. B. "Deconstructing the Map." *Cartographica* 26, no. 2 (1989): 1–20.

———. "Historical Geography and the Cartographic Illusion." *Journal of Historical Geography* 15, no. 1 (1989): 80–91.

———. "The Map and the Development of the History of Cartography." In *The History of Cartography,* vol. 1, edited by J. B. Harley and David Woodward. Chicago: University of Chicago Press, 1987.

———. "Maps, Knowledge, and Power." In *The Iconography of Landscape: Essays on the Symbolic Representation, Design, and Use of Past Environments,* edited by Denis Cosgrove and Stephen Daniels. New York: Cambridge University Press, 1988.

———. "Texts and Contexts in the Interpretation of Early Maps." In *From Sea Charts to Satellite Images: Interpreting North American History through Maps,* edited by David Buisseret. Chicago: University of Chicago Press, 1990.

———. "Victims of a Map." In *American Beginnings: Exploration, Culture, and Cartography in the Land of Norumbega,* edited by Emerson W. Baker et al. Lincoln: University of Nebraska Press, 1994.

Harlow, V. T. *Ralegh's Last Voyage.* London: Argonaut Press, 1932.

Harris, C. A., and J. A. J. De Villiers, eds. *Storm van 's Gravesande: The Rise of British Guiana.* London: Hakluyt Society, 1911.

Harris, Wilson. *The Guyana Quartet.* Boston: Faber and Faber, 1985.

———. "History, Fable and Myth in the Caribbean and the Guianas: The Amerindian Legacy, Continuity and Discontinuity." *Edgar Mittelholzer Memorial Lectures* (Georgetown, Guyana), 3d ser., February, 1970.

Hartshorne, R. "Suggestions on the Terminology of Political Boundaries." *Annals, Association of American Geographers* 26 (1936): 56–57.

Hawes, Louis. *Presences of Nature: British Landscape, 1780–1830.* New Haven: Yale Center for British Art, 1982.

Headrick, Daniel. *Tentacles of Progress: Technology Transfer in the Age of Imperialism.* Oxford: Oxford University Press, 1988.

———. *The Tools of Empire: Technology and European Imperialism in the Nineteenth Century.* Oxford: Oxford University Press, 1981.

Helgerson, Richard. *Forms of Nationhood: The Elizabethan Writing of England.* Chicago: University of Chicago Press, 1992.

Hemming, John. *The Search for El Dorado.* London: Joseph, 1978.

Henery, E. J., ed. *Almanack and Local Guide of British Guiana Containing the Laws, Ordinances and Regulations of the Colony, the Civil and Military Lists with a List of Estates from Corentyne to Pomeroon Rivers.* Demerary [Demerara], British Guiana: Royal Gazette Office, 1832. Caribbean Research Library, University of Guyana. Photocopy.

Hepple, Leslie W. "Metaphor, Geopolitical Discourse and the Military in South America." In *Writing Worlds: Discourse, Text, and Metaphor in the Representation of Landscape,* edited by Trevor J. Barnes and James S. Duncan. New York: Routledge, 1992.

Herbert, Francis. "The 'London Atlas of Universal Geography' from John Arrowsmith to Edward Stanford: Origin, Development and Dissolution of a British World Atlas from the 1830s to the 1930s." *Imago Mundi* 41 (1989): 99–123.

———. "The Royal Geographical Society's Membership, the Map Trade, and Geographical Publishing in Britain 1830 to *ca* 1930: An Introductory Essay with Listing of Some 250 Fellows in Related Professions. *Imago Mundi* 35 (1983): 67–95.

Herschel, John F. W. *A Manual of Scientific Enquiry.* London: John Murray, 1849.

Higman, B. W. *Jamaica Surveyed: Plantation Maps and Plans of the Eighteenth and Nineteenth Centuries.* San Francisco: Institute of Jamaica Publications, 1988.

Hilhouse, William. *Indian Notices, or Sketches of the Habits, Characters, Languages, Superstitions, Soil, and Climate of the Several Nations. . . . Also the Icthyology of the Fresh Waters of the Interior.* 1825.

Reprint, introduction and supplementary notes by M. N. Menezes, Georgetown, Guyana: National Commission for Research Materials on Guyana, 1978.

———. "Journal of a Voyage up the Massaroony." *JRGS* 4 (1834): 25–40.

———. "Memoir on the Warow Land of British Guiana." *JRGS* 4 (1834): 322–33.

———. "Remarks and Observations Illustrative of Hillhouse's [*sic*] General Chart of British Guiana." (Pamphlet accompanying Hillhouse's 1827 *General Chart*.) London: J. Wyld, [1827].

Holdich, Thomas. *Political Frontiers and Boundary Making*. London: Macmillan, 1916.

Hooker, W. J. *Notes on Victoria Regia, or Illustrations of the Royal Water-Lily in a Series of Figures Chiefly Made from Specimens Flowering at Syon and Kew*. London: Reeve and Benham, 1851.

Hough, Franklin Benjamin. *History of St. Lawrence and Franklin Counties, New York*. Albany: Little, 1853.

Huggan, Graham. "Decolonizing the Map." In *The Post-colonial Studies Reader,* edited by Bill Ashcroft, Gareth Griffiths, and Helen Tiffin. London: Routledge, 1995.

Hulme, Peter, and Neil L. Whitehead, eds. *Wild Majesty: Encounters with Caribs from Columbus to the Present Day*. Oxford: Clarendon Press, 1992.

Humboldt, Alexander von. "Concerning Certain Important Features of the Geography of Guiana." In *Robert Hermann Schomburgk's Travels in Guiana and on the Orinoco during the Years 1835–1839,* by Robert H. Schomburgk, edited by O. A. Schomburgk, translated by W. E. Roth. Georgetown, British Guiana: Argosy, 1931.

———. *Personal Narrative of Travels to the Equinoctial Regions of the New Continent during the Years 1799–1804*. Translated by Helen Maria Williams. London: Longman, Hurst, Rees, Orme, and Brown, 1814–29.

———. "Preface." In *Robert Hermann Schomburgk's Travels in Guiana and on the Orinoco during the Years 1835–1839,* by Robert H. Schomburgk, edited by O. A. Schomburgk, translated by W. E. Roth. Georgetown, British Guiana: Argosy, 1931.

Hutchins, Edwin. *Cognition in the Wild*. Cambridge: MIT Press, 1995.

Im Thurn, Everard Ferdinand. *Among the Indians of Guiana*. London: K. Paul, Trench, 1883.

———. "The Ascent of Mount Roraima." *Proceedings of the Royal Geographical Society* 7, no. 8 (1885): 487–521.

———. "Roraima (with Two Maps)." *Timehri* 4, no. 1 (1885): 256–67.

Ireland, Gordon. *Boundaries, Possessions, and Conflicts in Central and North America and the Caribbean*. Cambridge: Harvard University Press, 1941.

———. *Boundaries, Possessions, and Conflicts in South America*. Cambridge: Harvard University Press, 1938.

Jackson, Basil. *A Course of Military Surveying, Instructions for Sketching in the Field, Plan Drawing, Leveling, Military Reconnaissance, etc. and Embracing a Variety of Information on Other Subjects Equally Useful to the Traveller and the Soldier: and Surveying Instruments, etc*. London: W. H. Allen, 1838.

Jackson, John Brinkerhoff. *The Interpretation of Ordinary Landscapes: Geographical Essays*. Edited by D. W. Meinig et al. New York: Oxford University Press, 1979.

———. *The Necessity for Ruins*. Amherst: University of Massachusetts Press, 1980.

———. *A Sense of Place, a Sense of Time*. New Haven: Yale University Press, 1994.

Jackson, Peter. "John Tallis, 1818–1876." In *John Tallis's London Street Views, 1838–1840*. London: London Topographical Society (no. 110) in association with Nattali and Maurice, 1969.

Jacob, Christian. *L'empire des cartes: Approche théorique de la cartographie à travers l'histoire*. Paris: Albin Michel, 1992.

Jamieson, Alexander. *A Treatise on the Construction of Maps*. London: Darton, Harvey, 1814.

Jardine, N., J. A. Secord, and E. C. Spary, eds. *Cultures of Natural History*. New York: Cambridge University Press, 1996.

Jeffrey, Henry. "Bid for El Dorado: The Guyana-Venezuela Border Problem." *Guyana Historical Journal* 4–5 (1992–93): 52–67.

John Tallis and Company. *The Illustrated Atlas and Modern History of the World, Geographical, Political, Commercial, and Statistical*. London: John Tallis, 1851.

Jones, Stephen B. *Boundary-Making: a Handbook for Statesmen, Treaty Editors and Boundary Commissioners*. Washington, D.C.: Carnegie Endowment for International Peace, 1945.

Kain, Roger J. P., and Elizabeth Baigent. *The Cadastral Map in the Service of the State: A History of Property Mapping.* Chicago: University of Chicago Press, 1992.

Karsen, Sonja. "Alexander von Humboldt in South America: From the Orinoco to the Amazon." In *Studies in Eighteenth-Century Culture,* vol. 16, edited by O. M. Brack. Madison: University of Wisconsin Press, 1986.

Kinglake, A. W. *Eothen.* 1844. Reprint, Marlboro, Vt.: Marlboro Press, 1992.

Klonk, Charlotte. *Science and the Perception of Nature: British Landscape Art in the Late Eighteenth and Early Nineteenth Centuries.* New Haven: Yale University Press, 1996.

Kreuter, Gretchen. "Empire on the Orinoco: Minnesota Concession in Venezuela." *Minnesota History* 43, no. 6 (1973): 198–212.

Kuklick, Henrika, and Robert E. Kohler, eds. *Science in the Field. Osiris,* 2d ser., 11 (1996).

Kwayana, Eusi. "U.S. Card in the Venezuela-Guyana Row." *Caribbean Contact,* March 1982, 10.

Lancaster, Alan. "An Unconquered Wilderness: A Historical Analysis of the Failure to Open up the Interior of British Guiana, 1838–1919." M. A. thesis, University of Guyana, 1977.

Lanham, Richard A. *A Handlist of Rhetorical Terms: A Guide for Students of English Literature.* Berkeley: University of California Press, 1969.

Latin American Bureau. *Guyana, Fraudulent Revolution.* London: Latin American Bureau (Research and Action) and the Guyana Human Rights Association, 1984.

Latour, Bruno. *Science in Action: How to Follow Scientists and Engineers through Society.* Cambridge: Harvard University Press, 1987.

———. "Les 'vues de l'esprit': Une introduction à l'anthropologie des sciences et des techniques." *Culture Technique* 14 (1985): 4–29.

Leed, Eric. *The Mind of the Traveller.* New York: Basic Books, 1991.

———. *Shores of Discovery: How Expeditionaries Have Constructed the World.* New York: Basic Books, 1995.

Lestringant, Frank. "Fictions de l'espace brésilien à la Renaissance: L'exemple de Guanabara." In *Arts et légendes d'espaces: Figures du voyages et rhétoriques du monde,* edited by Christian Jacob and Frank Lestringant. Paris: Presses de L'École Normale Supérieure, 1981.

Lewis, Malcolm G. "La Grande Rivière et Fleuve de l'Ouest: The Realities and Reasons behind a Major Mistake in the Eighteenth-Century Geography of North America." *Cartographica* 28, no. 1 (1991): 54–87.

———. "Rhetoric of the Western Interior: Modes of Environmental Description in American Promotional Literature of the Nineteenth Century." In *The Iconography of Landscape: Essays on the Symbolic Representation, Design, and Use of Past Environments,* edited by Denis Cosgrove and Stephen Daniels. New York: Cambridge University Press, 1988.

Livingstone, David N. *The Geographical Tradition: Episodes in the History of a Contested Enterprise.* Cambridge, Mass.: Blackwell, 1993.

———. "The History of Science and the History of Geography: Interactions and Implications." *History of Science* 22 (1984): 271–302.

Lorimer, J. "The Location of Ralegh's Guiana Gold Mine." *Terrae Incognitae* 14 (1982): 77–95.

Lyell, Charles. *Elements of Geology.* 3 vols. London: John Murray, 1830–33.

MacDonald, Ian, ed. *AJS at Seventy.* Georgetown, Guyana: Autoprint, 1984.

MacInnes, Hamish. *Climb to the Lost World.* London: Hodder and Stoughton, 1974.

MacKenzie, John, ed. *Imperialism and the Natural World.* Manchester: Manchester University Press, 1990.

MacLaren, I. S. "The Aesthetic Map of the North, 1845–1859." *Arctic* 38 (1985): 89–103.

———. "The Aesthetic Mapping of Nature in the Second Franklin Expedition." *Journal of Canadian Studies* 20 (1985): 39–57.

———. "David Thompson's Imaginative Mapping of the Canadian Northwest, 1784–1812." *Ariel* 15, no. 2 (1984): 89–106.

———. "Retaining Captaincy of the Soul: Response to Nature on the First Franklin Expedition." *Essays on Canadian Writing* 28 (1984): 57–92.

———. "Samuel Hearne and the Landscapes of Discovery." *Canadian Literature* 103 (1984): 27–40.

MacLeod, Roy. "On Visiting the Moving Metropolis: Reflections on the Architecture of Imperial Science." *Historical Records of Australian Science* 5 (1982): 1–15.

———. "Passages in Imperial Science: From Empire to Commonwealth." *Journal of World History* 4, no. 1 (1993): 117–50.

MacLeod, Roy, and M. Lewis, eds. *Disease, Medicine and Empire.* London: Routledge, 1988.

Manning, Thomas. *The U.S. Coast Survey versus the Naval Hydrographic Office: A Nineteenth-Century Rivalry in Science and Politics.* Tuscaloosa: University of Alabama, 1988.

The Mapmaker's Art: Three Hundred Years of British Cartography. New Haven: Yale Center for British Art, 1989.

Markham, Violet R. *Paxton and the Bachelor Duke.* London: Hodder and Staughton, 1935.

Marshall, P. J. *The Cambridge Illustrated History of the British Empire.* New York: Cambridge University Press, 1996.

Martin, C. G. C., ed. *Maps and Surveys of Malawi: A History of Cartography and the Land Survey Profession.* Rotterdam: A. A. Balkema, 1980.

McClintock, Anne. "The Angel of Progress: Pitfalls of the Term 'Postcolonialism.'" In *Colonial Discourse, Postcolonial Theory,* edited by Francis Barker, Peter Hulme, and Margaret Iverson. New York: Manchester University Press, 1994.

———. *Imperial Leather: Race, Gender, and Sexuality in the Colonial Conquest.* New York: Routledge, 1995.

McGreevy, Patrick. "Reading the Texts of Niagara Falls: The Metaphor of Death." In *Writing Worlds: Discourse, Text, and Metaphor in the Representation of Landscape,* edited by Trevor J. Barnes and James S. Duncan. New York: Routledge, 1992.

McKeon, John. *Crystal Palace: Joseph Paxton and Charles Fox.* London: Phaidon, 1994.

Meade, Teresa, and Mark Walker. *Science, Medicine and Cultural Imperialism.* New York: St. Martin's, 1991.

Menezes, Mary Noel. "The Background to the Venezuela-Guyana Boundary Dispute." *History Gazette* (Turkeyen, Guyana) 21 (June 1990).

———. *British Policy towards the Amerindians in British Guiana, 1803–1873.* Oxford: Clarendon Press, 1977.

———. "Sketches of Amerindian Tribes." In *Sketches of Amerindian Tribes, 1841–1843,* by Edward Alfred Goodall. London: British Museum Publications for the National Commission for Research Materials on Guyana, 1977.

———, ed. *The Amerindians in Guyana, 1803–73: A Documentary History.* London: Cass; Totowa, N.J.: Bibliographic Distribution Centre, 1979.

Mentore, George. "The Relevance of Myth." *Edgar Mittelholzer Memorial Lectures* (Georgetown, Guyana), 11th ser., 1988.

Michasiw, Kim Ian. "Nine Revisionist Theses on the Picturesque." *Representations* 38 (spring 1992): 76–100.

Mignolo, Walter. *The Darker Side of the Renaissance: Literacy, Territoriality, and Colonization.* Ann Arbor: University of Michigan Press, 1995.

Miller, Angela L. *The Empire of the Eye: Landscape Representation and American Cultural Politics, 1825–1875.* Ithaca: Cornell University Press, 1993.

Miller, David Philip, and Peter Hanns Reill, eds. *Visions of Empire: Voyages, Botany, and Representations of Nature.* New York: Cambridge University Press, 1996.

Miller, J. Hillis. *Topographies.* Stanford: Stanford University Press, 1995.

Miller, Rory. *Britain and Latin America in the Nineteenth and Twentieth Centuries.* New York: Longman, 1993.

Montrose, L. "The Work of Gender in the Discourse of Discovery." In *New World Encounters,* edited by Stephen Greenblatt. Berkeley: University of California Press, 1993.

Mundy, Barbara E. *The Mapping of New Spain: Indigenous Cartography and the Maps of the Relaciones Geográficas.* Chicago: University of Chicago Press, 1996.

Naipaul, V. S. *The Loss of El Dorado: A History.* London: Deutsch, 1969.

Netscher, P. M. *History of the Colonies Essequibo, Demerary and Berbice from the Dutch Establishment to the Present Day.* The Hague: Martinus Nijhoff, 1888. Translated by W. E. Roth. Georgetown, British Guiana: Daily Chronicle, 1929.

Nicholl, Charles. *The Creature in the Map: A Journey to El Dorado.* New York: William Morrow, 1995.

Nicholson, Malcolm. "Alexander Humboldt, Humboldtian Science, and the Origin of the Study of Vegetation." *History of Science* 25 (1987): 167–94.

Noyes, John. *Colonial Space: Spatiality in the Discourse of German South West Africa, 1884–1915.* Studies in Anthropology and History 4. Chur, Switzerland: Harwood Academic, 1992.

———. "The Representation of Spatial History." *Pretexts* 4, no. 2 (1993): 120–27.

Núñez, Enrique Bernardo. *Tres momentos en la controversia de límites de Guayana.* Caracas: Ministerio de Educación, 1962.

Obeyesekere, Gananath. *The Apotheosis of Captain Cook: European Mythmaking in the Pacific.* Princeton: Princeton University Press, 1992.

Official History of the Discussion between Venezuela and Great Britain on Their Guiana Boundaries. Atlanta: Franklin Printing, 1896.

Ojer, Pablo. *Robert H. Schomburgk: Explorador de Guayana y sus líneas de frontera.* Caracas: Universidad Central de Venezuela, 1969.

Oldroyd, D. R. "Historicism and the Rise of Historical Geology, Part I." *History of Science* 17 (1979): 191–213.

Ophir, Adi, and Steven Shapin. "The Place of Knowledge: A Methodological Survey." *Science in Context* 4, no. 1 (1991): 3–21.

Osborne, Brian S. "The Iconography of Nationhood in Canadian Art." In *The Iconography of Landscape: Essays on the Symbolic Representation, Design, and Use of Past Environments,* edited by Denis Cosgrove and Stephen Daniels. New York: Cambridge University Press, 1988.

Osborne, Michael. *Nature, the Exotic, and the Science of French Colonialism.* Bloomington: Indiana University Press, 1994.

Ottley, H. A. *Biographical and Critical Dictionary of Painters and Engravers.* London: G. Bell, 1876.

Outhwaite, L. *Unrolling the Map: The Story of Exploration.* New York: John Day, 1972.

Outram, Dorinda. "On Being Perseus: New Knowledge, Dislocation and Enlightenment Exploration." In *Geography and Enlightenment,* edited by David N. Livingstone and Charles W. J. Withers. Chicago: University of Chicago Press, 1999.

Ovid. *Metamorphoses.* Edited by Frank Justus Miller. 2d ed. Revised by G. P. Goold. Cambridge: Harvard University Press, 1984.

Perkins, H. I. "Notes on a Journey to Mount Roraima, British Guiana." *Proceedings of the Royal Geographical Society* 7, no. 8 (1885): 522–34.

Petersen, Kirsten Holst, and Anna Rutherford. "Fossil and Psyche." In *The Post-colonial Studies Reader,* edited by Bill Ashcroft, Gareth Griffiths, and Helen Tiffin. London: Routledge, 1995.

Petitjean, Patrick, Catherine Jami, and Anne Marie Moulin, eds. *Science and Empires: Historical Studies about Scientific Development and European Expansion.* Boston: Kluwer, 1992.

Playfair, William. *Statistical Breviary.* London: J. Wallis, 1801.

Poovey, Mary. *A History of the Modern Fact: Problems of Knowledge in the Sciences of Wealth and Society.* Chicago: University of Chicago Press, 1998.

Porter, Dennis. *Haunted Journeys: Desire and Transgression in European Travel Writing.* Princeton: Princeton University Press, 1991.

Porter, Philip W., and Fred E. Lukermann. "The Geography of Utopia." In *Geographies of the Mind,* edited by D. Lowenthal and M. J. Bowden. New York: Oxford University Press, 1976.

Potter, Jonathan, ed. [Martin, Montgomery, "original editor."] *Antique Maps of the Nineteenth-Century World.* (Reprint of *Illustrated Atlas and Modern History of the World.* London: John Tallis, 1851.) Introduction by Jonathan Potter. London: Portland House, division of Dilithium Press, 1989.

Prakash, Gyan. "Science 'Gone Native' in Colonial India." *Representations* 40 (fall 1992): 153–78.

———. "Writing Post-Orientalist Histories of the Third World: Indian Historiography Is Good to Think." In *Colonialism and Culture,* edited by Nicholas B. Dirks. Ann Arbor: University of Michigan Press, 1992.

Pratt, Mary Louise. *Imperial Eyes: Travel Writing and Transculturation.* London: Routledge, 1992.

Prescott, J. R. V. *Political Frontiers and Boundaries.* London: Allen and Unwin, 1987.

Price, Edward T. *Dividing the Land: Early American Beginnings of Our Private Property Mosaic.* Chicago: University of Chicago Press, 1995.
Price, Richard. *Alabi's World.* Baltimore: Johns Hopkins University Press, 1990.
Price, Uvedale. *An Essay on the Picturesque.* London: J. Robson, 1794.
Priestman, Martin. *Cowper's Task: Structure and Influence.* New York: Cambridge University Press, 1983.
Procter, Adelaide, ed. *The Victoria Regia, a Volume of Original Contributions in Poetry and Prose.* London: Victoria Press, 1861.
Quelch, John Joseph. "Journey to the Summit of Roraima." *Timehri,* n.s., 9, no. 1 (1895): 107–88.
Quinn, D. B. *Raleigh and the British Empire.* London: Holder and Stoughton, 1947.
Ralegh, Walter. *The Discovery of the Large, Rich and Beautiful Empire of Guiana, with a Relation of the Great and Golden City of Manoa (Which the Spaniards Call El Dorado) . . . by Sir W. Ralegh, Knight.* Edited by R. H. Schomburgk. London: Hakluyt, 1848.
Ramos Pérez, Demetrio. *El mito del Dorado: Su génesis y proceso.* Caracas: Academia National de la Historia, 1973.
Raper, Henry. *Practice of Navigation and Nautical Astronomy.* 2d ed. London: R. B. Bate, 1842.
Redgrave, S. A. *Dictionary of Artists of the English School.* London: Longmans, Green, 1874.
Rehbock, P. "The Victorian Aquarium in Ecological and Social Perspective." In *Oceanography: The Past.* Proceedings of the Third International Congress on the History of Oceanography, edited by Mary Sears and Daniel Merriman. New York: Springer-Verlag, 1980.
Reingold, Nathan, and Marc Rothenberg, eds. *Scientific Colonialism: A Cross-Cultural Comparison.* Washington, D.C.: Smithsonian Institution Press, 1987.
Reis, Arthur Cezar Ferreira. *A Amazônia e a cobiça internacional.* Rio de Janerio: Gráfica Record Editôra, 1968.
Revel, Jacques. "Knowledge of the Territory." *Science in Context* 4, no. 1 (1991): 133–61.
Richards, Thomas. *The Imperial Archive: Knowledge and the Fantasy of Empire.* New York: Verso, 1993.
Ridley, Hugh. *Images of Imperial Rule.* London: Croom Helm, 1983.
Riffenburgh, Beau. *The Myth of the Explorer.* New York: Belhaven; London: Scott Polar Research Institute, University of Cambridge, 1993.
Ritchie, G. S. *The Admiralty Chart: British Naval Hydrography in the Nineteenth Century.* New ed. Edinburgh: Pentland Press, 1995.
Rivière, Peter. *Absent-Minded Imperialism: Britain and the Expansion of Empire in Nineteenth-Century Brazil.* New York: Tauris Academic Studies, 1995.
———. "From Science to Imperialism: Robert Schomburgk's Humanitarianism." *Archives of Natural History* 25, no. 1 (1998): 1–8.
Roberts, Emma. *Hindostan, Its Landscapes, Palaces, Temples, Tombs: The Shores of the Red Sea and the Himalaya Mountains.* London: Fisher, 1846.
Rodway, James. *Handbook of British Guiana.* Georgetown, British Guiana: Columbian Exposition Literary Committee, 1893.
———. "The Schomburgks in Guiana." *Timehri,* n.s., 3, no. 1 (1889): 1–29.
Roe, Peter G. *The Cosmic Zygote: Cosmology in the Amazon Basin.* New Brunswick, N.J.: Rutgers University Press, 1982.
———. "Of Rainbow Dragons and the Origin of Designs: Waiwai and the Shipibo Ronin Ehua." *Latin American Indian Literatures Journal* 5, no. 1 (1989): 1–67.
Roopnaraine, Terence. "Behind God's Back: Landscape and Cultural Geography." In "Freighted Fortunes," Ph.D. diss., Cambridge University, 1996.
Roth, Vincent. "Hilhouse's Book of Reconnaissances and Indian Miscellany." *Timehri,* 4th ser., no. 25 (1934): 1–52.
Roth, Walter E. "The Amerindian Kanaima or Avenger." In *Roth's Pepper Pot.* Georgetown, British Guiana: Daily Chronicle, 1958.
———. *Roth's Pepper Pot.* Georgetown, British Guiana: Daily Chronicle, 1958.
Rout, Leslie B. *Which Way Out? An Analysis of the Venezuela/Guyana Border Dispute.* East Lansing: Michigan State University Press, 1971.
"Royal Water-Lily" [Advertisement]. *Hooker's Journal of Botany and Kew Garden Miscellany* 2, no. 20 (1850): back cover, recto.

"Royal Water-Lily" [Advertisement]. *Hooker's Journal of Botany and Kew Garden Miscellany* 2, no. 21 (1850): back cover, recto.

Rozwadowski, Helen M. "Small World: Forging a Scientific Maritime Culture for Oceanography." *Isis* 87 (1996): 409–29.

Ruskin, John. *Sesame and Lilies: Two Lectures Delivered at Manchester in 1864.* New York: J. Wiley, 1866.

Ryan, Simon. *The Cartographic Eye: How Explorers Saw Australia.* Cambridge: Cambridge University Press, 1996.

Sack, Robert David. *Human Territoriality: Its Theory and History.* Cambridge: Cambridge University Press, 1986.

Sahlins, Peter. *Boundaries: The Making of France and Spain in the Pyrenees.* Berkeley: University of California Press, 1989.

Said, Edward W. *Culture and Imperialism.* London: Vintage, 1993.

———. *Orientalism.* New York: Vintage, 1978.

Salvin, O., and F. B. Godman. "Notes on Birds from British Guiana." Parts 1 and 2. *Ibis*, ser. 4, 6 (1882): 76–84; ser. 5, 1 (1883): 203–12.

Sandilands, R. W. "The History of Hydrographic Surveying in British Columbia." In *Explorations in the History of Canadian Mapping,* edited by Barbara Farrell and Aileen Desbarats. Ottawa: Association of Canadian Map Libraries and Archives, 1988.

Sangwan, Satpal. *Science, Technology and Colonisation: The Indian Experience, 1757–1857.* Delhi: Anamika Prakashan, 1991.

———. "Technology and Imperialism in the Indian Context: The Case of Steamboats, 1819–1839." In *Science, Medicine and Cultural Imperialism,* edited by Teresa Meade and Mark Walker. New York: St. Martin's, 1991.

Schaffer, Simon. "Accurate Measurement Is an English Science." In *The Values of Precision,* edited by M. Norton Wise. Princeton: Princeton University Press, 1995.

———. "Astronomers Mark Time: Discipline and the Personal Equation." *Science in Context* 2 (1988): 115–45.

———. "Contextualizing the Canon." In *The Disunity of Science,* edited by Peter Galison and David Stump. Stanford: Stanford University Press, 1996.

———. "Empires of Physics." In *Empires of Physics,* edited by J. Bennett et al. Cambridge: Whipple Museum of the History of Science, 1993.

———. "Visions of Empire: Afterword." In *Visions of Empire: Voyages, Botany, and Representations of Nature,* edited by David Philip Miller and Peter Hanns Reill. New York: Cambridge University Press, 1996.

Schama, Simon. *Landscape and Memory.* London: Harper-Collins, 1995.

Scholes, Robert. *Structuralism in Literature.* New Haven: Yale University Press, 1974.

Schomburgk, Moritz Richard. *Richard Schomburgk's Travels in British Guiana, 1840–1844.* Translated and edited by W. E. Roth. Georgetown, British Guiana: Daily Chronicle, 1922–23.

Schomburgk, Robert Herman. *A Description of British Guiana, Geographical and Statistical.* London: Simpkin, Marshall, 1840.

———. "Diary of an Ascent of the River Berbice, in British Guayana." *JRGS* 7 (1837): 301–50.

———. "Diary of an Ascent of the River Corentyn in British Guayana." *JRGS* 7 (1837): 285–301.

———. *The History of Barbados.* London: Longman, Brown, Green, and Longmans, 1848.

———. "Introduction." In *The Discovery of the Large, Rich and Beautiful Empire of Guiana, with a Relation of the Great and Golden City of Manoa (Which the Spaniards call El Dorado) . . . by Sir W. Ralegh, Knight,* by Walter Ralegh. London: Hakluyt, 1848.

———. "Journal of an Expedition from Pirara to the Upper Corentyne." *JRGS* 15 (1845): 1–104.

———. *The Natural History of the Fishes of Guiana.* 2 vols. Edinburgh: W. H. Lizar, 1841–43.

———. "On the Lake Parima, the El Dorado of Sir Walter Ralegh and the Geography of Guiana." *Report of the British Association for the Advancement of Science* 15 (1845), "Notes and Abstracts," 50–51.

———. "Remarks on Anegada." *JRGS* 2 (1832): 152–70.

———. "Remarks on the Geology of British Guiana." *Quarterly Journal of the Geological Society* 1 (1839): 298–300.

———. "Report of an Expedition into the Interior of British Guayana." *JRGS* 6 (1836): 224–84.
———. "Report of the Third Expedition into the Interior of Guayana." *JRGS* 10 (1841): 159–267.
———. *Robert Hermann Schomburgk's Travels in Guiana and on the Orinoco during the Years 1835–1839*. Edited by O. A. Schomburgk. Leipzig: Georg Wigand, 1841. Translated by W. E. Roth. Georgetown, British Guiana: Argosy, 1931.
———. *Twelve Views in the Interior of Guiana*. London: Ackermann, 1841.
———. "Visit to the Sources of the Takutu, in British Guiana." *JRGS* 13 (1843): 18–74.
Scott, Walter. *The Lady of the Lake*. Introduction by Wilkie Collins. Edinburgh: William Blackwood, 1900.
Sears, Mary, and Daniel Merriman, eds. *Oceanography: The Past*. Proceedings of the Third International Congress on the History of Oceanography. New York: Springer-Verlag, 1980.
Secord, James A. "Extraordinary Experiment: Electricity and the Creation of Life in Victorian England." In *The Uses of Experiment: Studies in the Natural Sciences*, edited by David Gooding, Trevor Pinch, and Simon Schaffer. Cambridge: Cambridge University Press, 1989.
Seddon, G. "On the Road to Botany Bay." *Westerly* 33, no. 4 (1988): 15–26.
Seed, Patricia. *Ceremonies of Possession in Europe's Conquest of the New World, 1492–1640*. Cambridge: Cambridge University Press, 1995.
Shapiro, J. "From Tupa to the Land without Evil: The Christianization of Tupi-Guarani Cosmology." *American Ethnologist* 14 (1987): 126–39.
Sheets-Pyenson, Susan. "War and Peace in Natural History Publishing: The *Naturalist's Library*, 1833–1843." *Isis* 72 (1981): 50–72.
Sillery, J. L. "Salvaging Democracy? The United States and Britain in British Guiana, 1961–1964." D.Phil. thesis, Oxford University, 1997.
Simpson, John, and Hazel Fox. *International Arbitration: Law and Practice*. New York: Praeger, 1959.
Skelton, Raleigh A. *Explorers' Maps: Chapters in the Cartographic Record of Geographical Discovery*. New York: Praeger, 1958.
———. "Map Compilation, Production and Research in Relationship to Geographical Exploration." In *The Pacific Basin: A History of Its Geographical Exploration*, edited by Herman R. Friis. New York: American Geographical Society, 1967.
———. "Raleigh as a Geographer." *Virginia Magazine of History and Biography* 71, no. 2 (1963): 131–49.
Smith, Anthony. *Explorers of the Amazon*. New York: Viking, 1990.
Smith, Bernard. *European Vision and the South Pacific, 1768–1850*. 2d ed. Melbourne: Oxford University Press, 1984.
Smith, David. *Victorian Maps of the British Isles*. London: B. T. Batsford, 1985.
Smith, John, and George Bentham. "Contributions towards a Flora of South America: Enumeration of Plants Collected by Mr. Schomburgk in British Guiana." *London Journal of Botany*, n.s., 1 (April 1842): 193–203.
Sparke, Matthew. "Between Demythologizing and Deconstructing the Map: Shawnadithit's New-found-land and the Alienation of Canada." *Cartographica* 32, no. 1 (1995): 1–21.
Sparks, C. "England and the Columbian Discoveries: The Attempt to Legitimize English Voyages to the New World." *Terrae Incognitae* 22 (1990): 1–12.
Spurr, David. *The Rhetoric of Empire: Colonial Discourse in Journalism, Travel Writing and Imperial Administration*. Durham, N.C.: Duke University Press, 1993.
Stafford, Barbara. *Voyage into Substance*. Cambridge: MIT Press, 1984.
Stafford, Robert. "Annexing the Landscapes of the Past: British Imperial Geology in the Nineteenth Century." In *Imperialism and the Natural World*, edited by John MacKenzie. Manchester: Manchester University Press, 1990.
———. "Geological Surveys, Mineral Discoveries, and British Expansion, 1835–1871." *Journal of Imperial and Commonwealth History* 12 (1983–84): 5–32.
———. *Scientist of Empire: Sir Roderick Murchison, Scientific Exploration and Victorian Imperialism*. New York: Cambridge University Press, 1989.
Stein, Howard F., and William G. Niederland. *Maps from the Mind: Readings in Psychogeography*. Norman: University of Oklahoma Press, 1989.

Stewart, Susan. *On Longing: Narratives of the Miniature, the Gigantic, the Souvenir, the Collection.* Baltimore: Johns Hopkins University Press, 1984.

Stilgoe, John R. *Alongshore.* New Haven: Yale University Press, 1994.

Stoddart, D. R. *On Geography and Its History.* Oxford: Basil Blackwell, 1986.

Stone, Jeffrey C. "Imperialism, Colonialism and Cartography." *Transactions of the Institute of British Geographers,* n.s., 13 (1988): 57–64.

Swijtink, Zeno. "Alexander von Humboldt's Instrumental Life." History of Cartography Project, University of Wisconsin, Madison. Photocopy.

Tamayo, Isbelia Sequera, et al. *Guayana Esequiba: Espacio geopolítico.* Caracas: Academia Nacional de Ciencias Económicas, 1992.

Tate, Thomas. *A Treatise on Practical Geometry, Mensuration, Conics, etc. and Land Surveying.* Dublin, 1834.

Taussig, Michael T. *The Nervous System.* New York: Routledge, 1992.

Taylor, E. G. R. *The Haven-Finding Art: A History of Navigation from Odysseus to Captain Cook.* London: Hollis and Carter, 1956.

Taylor, Iain C. "Official Geography and the Creation of 'Canada.'" *Cartographica* 31, no. 4 (1994): 1–15.

Teich, Mikuláš, Roy Porter, and Bo Gustafsson. *Nature and Society in Historical Context.* Cambridge: Cambridge University Press, 1997.

Thomas, Nicholas. *Colonialism's Culture: Anthropology, Travel and Government.* Princeton: Princeton University Press, 1994.

Tillotson, John. *Beauties of English Scenery.* London: T. J. Allman, [1860].

Tinne, J. E. "Opening up the Country." *Timehri,* n.s., 3, no. 1 (1889): 31–34.

Tooley, R. V. *English Books with Coloured Plates, 1790–1860.* London: Dawsons, 1973.

Tuan, Yi-Fu. *Space and Place: The Perspective of Experience.* Minneapolis: University of Minnesota Press, 1977.

——— . *Topophilia: A Study of Environmental Perception, Attitudes, and Values.* Englewood Cliffs, N.J.: Prentice-Hall, 1974.

Tufte, Edward R. *Envisioning Information.* Cheshire, Conn.: Graphics Press, 1990.

Twyman, Michael. *Lithography, 1800–1850.* London: Oxford University Press, 1970.

Van Heuvel, J. A. *El Dorado.* New York: J. Winchester, 1844.

Veness, W. T. *El Dorado or British Guiana as a Field for Colonisation.* London: Cassell, Petter, and Galpin, 1866.

Von Hagen, Victor Wolfgang. *Frederick Catherwood, Archt.* New York: Oxford University Press, 1950.

——— . *Search for the Maya.* Westmead, England: Saxon House, 1973.

——— . *South America Called Them; Explorations of the Great Naturalists: La Condamine, Humboldt, Darwin, Spruce.* New York: Alfred A. Knopf, 1945.

Walcott, Derek. *Collected Poems, 1948–1984.* Boston: Faber and Faber, 1986.

Wallis, Helen, and Arthur H. Robinson, eds. *Cartographical Innovations: An International Handbook of Mapping Terms to 1900.* London: Map Collector Publications, 1982.

Walter, E. V. *Placeways: A Theory of the Human Environment.* Chapel Hill: University of North Carolina Press, 1988.

Warren, Adrian, ed. "Report of the 1971 British Expedition to Mount Roraima in Guyana, SA." University of Guyana. Photocopy.

Waterton, Charles. *Wanderings in South America, the North-west of the United States and the Antilles in the Years 1812, 1816, 1820 and 1824.* New ed. Edited and introduced by J. G. Wood. London: Macmillan, 1885.

Webber, A. R. F. *Centenary History and Handbook of British Guiana.* Georgetown, British Guiana: Argosy, 1931.

Webber, Edward John. *British Guiana, the Essequibo and Potaro Rivers with an Account of a Visit to the Recently-Discovered Kaieteur Fall.* London: Stanford, 1873.

Whitehead, Neil L. "Carib Ethnic Soldiering in Venezuela, the Guianas, and the Antilles, 1492–1820." *Ethnohistory* 37, no. 4 (1990): 357–85.

——— . "El Dorado, Cannibalism, and the Amazons: European Myth and Amerindian Praxis in the

Conquest of South America." In *Beeld en verbeelding van Amerika,* edited by W. Pansters and J. Weerdenburg. Utrecht: Bureau Studium Generale, 1992.
———. "Ethnic Transformation and Historical Discontinuity in Native Amazonia and Guayana, 1500–1900." *L'Homme* 126–28, no. 33, 2–4 (1993): 285–305.
———. "The Historical Anthropology of Text: The Interpretation of Ralegh's *Discoverie of Guiana.*" *Current Anthropology* 36, no. 1 (1995): 53–74.
———. *Lords of the Tiger Spirit: A History of the Caribs in Colonial Venezuela and Guyana, 1498–1820.* Dordrecht: Foris, 1988.
———. "The Mazaruni Pectoral: A Golden Artifact Discovered in Guyana and the Historical Sources concerning Native Metallurgy in the Caribbean, Orinoco and Northern Amazonia." *Archaeology and Anthropology: Journal of the Walter Roth Museum of Anthropology* (Guyana) 7 (1990): 19–38.
———. "Tribes Make States and States Make Tribes: Warfare and the Creation of Colonial Tribes and States in Northeastern South America." In *War in the Tribal Zone: Expanding States and Indigenous Warfare,* edited by R. Brian Ferguson and Neil L. Whitehead. Santa Fe, N.Mex.: School of American Research Press, 1992.
Whitely, H. "Explorations in the Neighborhood of Mount Roraima and Kukenam, in British Guiana." *Proceedings of the Royal Geographical Society* 6, no. 8 (1884): 452–63.
Wilford, John Noble. *The Mapmakers.* New York: Alfred A. Knopf, 1981.
Williams, Brackette. "Nationalism, Traditionalism, and the Problem of Cultural Inauthenticity." In *Nationalist Ideologies and the Production of National Cultures,* edited by Richard G. Fox. Washington, D.C.: American Anthropological Association, 1990.
Williams, Dennis. "The Forms of the Shamanic Sign in the Prehistoric Guianas." *Archaeology and Anthropology: Journal of the Walter Roth Museum of Anthropology* (Guyana) 9 (1993): 3–21.
Wilme, B. P. *A Handbook for Mapping, Engineering and Architectural Drawing.* London: John Weale (for the author), 1846.
Winichakul, Thongchai. *Siam Mapped: A History of the Geo-body of a Nation.* Honolulu: University of Hawaii Press, 1994.
Wolfe, Patrick. "History and Imperialism: A Century of Theory, from Marx to Postcolonialism." *American Historical Review* 102, no. 2 (1997): 388–420.
Wood, Denis. "How Maps Work." *Cartographica* 29, nos. 3–4 (1992): 66–74.
———. *The Power of Maps.* New York: Guilford Press, 1992.
Woods, M., and A. Warren. *Glass Houses: A History of Greenhouses, Orangeries and Conservatories.* London: Rizzoli, 1988.
Wrede, Stuart, and William Howard Adams, eds. *Denatured Visions: Landscape and Culture in the Twentieth Century.* New York: Museum of Modern Art (distributed by Harry N. Abrams), 1991.
Wright, John Kirtland. *Human Nature in Geography.* Cambridge: Harvard University Press, 1966.
———. "Terrae Incognitae: The Place of the Imagination in Geography." In *Human Nature in Geography.* Cambridge: Harvard University Press, 1966.
Zahl, Paul A. *To the Lost World.* New York: Alfred A. Knopf, 1939.

PRINTED DOCUMENTS RELATED TO THE BOUNDARY

These documents were prepared in limited numbers by the disputants, and I have found the collections of various libraries to be incomplete. (The U.S. Library of Congress, for instance, has lost the atlas that accompanied the British case against Brazil.) The full *Brazil Arbitration* is listed at the British Library as Maps.21.a.(5–25), but the atlases are listed separately: Brazil's as Maps.15.e.4 and Britain's as Maps.145.e.33. Volumes of the *Venezuela Arbitration* are listed at the British Library under the call number BS.14/393. The atlas prepared by Great Britain in that dispute is also cataloged separately (as Maps 145.e.32). I list below only the volumes referred to in the text. For more complete listings, see Núñez, *Tres momentos;* Menezes, *British Policy,* 303–4; and Rivière, *Absent-Minded Imperialism,* 183–84.

Great Britain. British Guiana Boundary Arbitration with the United States of Brazil. London, 1903.
  [*Brazil Arbitration, British Case*]: *The Case on Behalf of the Government of His Britannic Majesty.*

[*Brazil Arbitration, British Case Annex*]: *Annex to the Case on Behalf of the Government of His Britannic Majesty.*
[*Brazil Arbitration, British Counter-case*]: *The Counter-case on Behalf of the Government of His Britannic Majesty.*
[*Brazil Arbitration, British Counter-case Notes*]: *Notes to the Counter-case on Behalf of the Government of His Britannic Majesty.*
[*Brazil Arbitration, British Argument*]: *The Argument on Behalf of the Government of His Britannic Majesty.*

Great Britain. British Guiana Boundary Arbitration with the United States of Venezuela.

[*Venezuela Arbitration, Schomburgk's Reports*]: *Question of Boundary between British Guiana and Venezuela.* Number 5. London, 1896.
[*Venezuela Arbitration, British Case Appendix*]: *Appendix to the Case on Behalf of the Government of His Britannic Majesty.* London, 1898.

Venezuela. Venezuela–British Guiana Boundary Arbitration. New York, 1898.

[*Venezuela Arbitration, Venezuelan Case*]: *The Case of the United States of Venezuela before the Tribunal of Arbitration.*
[*Venezuela Arbitration, Venezuelan Counter-case*]: *The Counter-case of the United States of Venezuela before the Tribunal of Arbitration.*

INDEX

Italic numerals refer to page numbers of illustrations

abandonment, in interior, 176–77, 238, 249
Aberdeen, Lord, 52, 53, 230
Aboriginal Protection Society, 21, 211
absentee proprietorship, 171
Acarabisi (creek of Cuyuni), 226, 228, 239
Ackerman, Rudolph, 123
Ackerman and Company, 121–23
Adams, William Howard, 208n
Adamson, Alan, 126n
Additional MSS 17940a, 62, *63;* history of, 65–66. *See also* Ralegh, Ralegh map
Adelaide, queen dowager of Great Britain, 125
administrators vs. explorers, relationship to colonial territory, 10, 67–83, 114–17, 167–69
Admiralty Manual for Scientific Enquiry, 84, 101
aesthetics and colonial exploration: appropriation by means of aesthetic convention, 131–65; "aesthetic graticule," 131; aesthetic overdetermination, 147, 257; aesthetics and circuits, 145–48
affidavit, use in claiming territory, 222
Akawoi Amerindians, 35. *See also* Amerindians
Albújar, Juan Martínez de, 27
Alexander, Captain James Edward (later Sir), 34, 105–6, 126n; map for RGS, 74, *75*
Allen, J. L., 25n
Allen, John, 94
Allen, William, 122–23
allocation, 204
altazimuth theodolite, 91n. *See also* instruments
Amacura Point, 220
Amacura River, 217, 220, 242, 263; bank of, as boundary, 227–30; Victoria Point from mouth of, *219*
Amaripa Amerindians, last of, 248
Amazon River, 28, 30, 211
American Revolution, 19
Amerindians: control over routes and territory, 186, 186n; cosmology of, 181–89; crimes of, 222–24; and geographical knowledge, 233–37,
239, 252; extinction of (foreseen), 248; and oaths, 224n; as "professors of Demonology," 183; *See also* Akowoi; Amaripa; Arawak; Carib; Macushi; Maiongkong (Yekuana); Maopitian (Mawayena); Patamona; Pianoghotto (Farakoto); Shipibo; Waiwai; Wapishana; Warrau
Amiens, Treaty of, 20
Amucu, Lake, 30–31, 36, 41
Anegada, island of, 21, 104, 150n
Anglo-Dutch War, 18
Angostura, Venezuela, 211
Annai: lake at, 185; village of, 181–82
Anthon, M., 108n
anthropologists, 4
anti-conquest, 11n
anti–El Dorado, 54–60
apprenticeship, 80–81
Appun, Karl Ferdinand, 184
aquariums, 151
Arawak Amerindians, 18. *See also* Amerindians
archaeology: function of, in exploration, 178–81; and possession, 219
Armitage, Christopher, 25n
Arrowsmith, John, 84, 164, 171; map of Guiana, *85*
Articles of Capitulation, 20, 25
Articles of War, 109
artificial horizon, 86, 91n. *See also* instruments
artistic ability, importance of in exploration, 110
asceticism, of explorer, 106
Astronomical and Meteorological Society of British Guiana, 126n
astronomical observations, 15, 36, 86, 106, 176; and boundaries, 210; in traverse survey, 111–17
Atakaleet, 161n
Ataraipu ("Devil's Rock"), 130, 143, 160, 165, 175, 179, 181, 195, 232; disturbing power of 185; as symbol for British Guiana, 155–56; in *Twelve Views,* 136–37, plate 6

Aunama (creek of Barama), 226
Axtell, James, 26n
Ayangike Mountain, 175

Baber, Zaheer, 5n, 7n
Babineau, G. Raymond, 214–15
Back, George, 195n
"backdam," 59
Baigent, Elizabeth, 112n
Baker, Emerson, et al., 26n
Ballard, Paul, 227n
balloons, and exploration, 158, 160
Barama River, 226, 228
Barbados, 19, 20
Barima Point, 47, 200, 220, 229–30, 241; Ralegh's presence at invoked by Schomburgk, 50–53. *See also* "Dardanelles of the Orinoco"
Barima River, 220–21, 225–27, 229, 242, 263
barometer, Bunten's, 99. *See also* instruments
Barrow, Sir John, 49, 67, 114
Barthes, Roland, 7, 130
Basalla, G., 5n
Batavian Republic, 20
batons, as gifts to Amerindians, 18
Bayly, C. A., 8n, 26n
Beechey, Vice Admiral Frederick William, 84, 101–2
Beedham, Brian, 260n
Belcher, Commander Edward, 102–3, 174
Bell, Morag, 5n
Bell, Thomas, 125
Belyea, Barbara, 7, 41n, 132n, 139n, 198n, 206n; on subversions of picturesque conventions, 194–96
Bendall, A. Sarah, 227n
Benjamin, Anna, 18n, 201n
Bennett, George, 156
Bentham, George, 97, 97n, 159n
Bentham, Jeremy, and Benthamite panopticons, 6
Bentley, Charles, 159–61, 165, 176, 179, 198; depiction of *Victoria regia*, 153–56; picturesque conventions in work of, 135, 194–96; work on *Twelve Views*, 133–36, 138–48
Berbice (region), 150n; map of, 154, *155*
Berbice expedition, 82
Berbice River, 21, 68, 71, 82, 128, 148, 189, 191
Berbice slave rebellion, 19
Bermingham, Ann, 131
Bermúdez, Juan Almécija, 200n
Bernard, George, 196
Bertram, Marshall, 199n, 200n
bibliographical essay, treating historiography of cartography, science, exploration, imperialism, 5–13
Bicknell, Peter, 127n
bird's-eye view, 171
black vomit (yellow fever), 217n
Blackburn, Julia, 33n
Blair, H., 39n
Blakemore, M. J., 227n
Boase, Frederic, 134n
Boca de Navíos, 52
Boddam-Whetham, John, 156, 158
Boelhower, William, 167, 261
Bonpland, Aimé, 29, 149
Bosazza, V. L., 86n
botany, 97, 120, 205, 240
Botting, Douglas, 29n, 35n
Bouchenroeder's map (1798), 263
boundaries, 16, 199–253; allocation, delimitation, and demarcation of, 204; antecedent, 263; arbitrations over, 200; and astronomy, 213, 229–30; with Brazil, 46–50, 228–42; as connected landmarks, 209; consequent, 263; diplomatic confrontations over, 229–30, 242; discordant (superimposed), 263; disputes concerning, 1, 16, 19, 46, 199–202, 229, 257–63; documentation of, 261; with Dutch territory, 242–53; false uniformity of lines representing, 206; at a glance, 226; and history, 25, 53, 212, 229; how given significance, 205, 209; and humanitarianism, 244–45; hyperbolic, 241; as "imaginary line," 210; and landmarks, 209, 213–14; marks indicating, 233, 264–65; and mountain ranges, 226–28; natural, 208–14, 229; between nature and culture, 208; and political opportunism, 259; postcolonial legacy of, 16, 22–23, 255–65; as route, 210; and routes (of explorer), 228, 237; as scientific, 242; and "state of siege," 260; strategic considerations concerning, 241; subsequent, 263; to be run and marked, 213; transgression of, 221, 225; with Venezuela, 46–66, 216–28; visibility of, 210. *See also* watershed
boundary survey in British Guiana, 16; ambiguity of mission, 205; east region, 242–53; northwest region, 216–28; to smooth discontinuities, 214; southwest region, 228–42
bounty hunting, 76
Branco (river), 41. *See also* Rio Branco
Braveboy-Wagner, Jacqueline, 1n, 200n, 201n, 259n
Bravo, Michael, 41n
Brazil, 21, 141; Brazilian Boundary Commission, 245; Brazilian slave raids, slavers, 140–41, 248

Brett, Reverend William H., 29n, 33n, 156, 193n; sketches, *157, 158*
Brewington, M. N., 86n
Bridges, Roy, 8n
Briggs, John, 53
British Association for the Advancement of Science, 46
British Empire, origin of, 17
British Guiana: boundaries of, 199–253; conventions of spelling, 3n; and El Dorado 25–66; history to 1831, 13, 17–23; later geographical representations, 119–65; mistaken for Caribbean island, 25, 49; seal of, 19, *20;* status of geological knowledge in mid-1830s, 67–92; *See also* Guyana
British Guiana–Venezuela Boundary Commission, 1, 64
Brock, W. H., 94n, 115n
Brown, Charles Barrington, 39, 40n, 53n, 156, 160, 175, 178, 184n, 191, 193, 253
Browne, Janet, 5n
Buisseret, David, 127
Bunske, Edmunds V., 181
Bunten's siphon barometer, 99
Burke, Edmund, 132, 141, 194
Burkhardt, J. L., 120n
Burnham, Forbes, 258–59; administration flyer, *259*
Buro-buro River, 169
Burr, George Lincoln, 2
bush, tools for interpreting, 77
"Bush-Negroes," 18. *See also* "Maroons"
Butlin, Robin, 5n
Buxton, Sir Thomas, 21, 122, 211–12

caches, 177–78; left by explorers, 114, 190, 252–53. *See also* landmarks
cadastral survey, 213
Cain, P. J., 5n
"calenture," 49
Cameron, James M. R., 50n, 67n
cannibalism, 19, 182
Cannon, Susan Faye, 86n, 92–93, 94n, 96–97
canon shots, used in conducting itinerant surveys, 102
Canuku range, 247
Caphiwuin River, 248
Carib Amerindians, 40. *See also* Amerindians
Caribbean island, Guiana mistaken for, 25
Carmichael-Smyth, Sir James (Governor), 78, 81, 251
Caroni River, 30, 240
Carter, Paul, 8, 91, 131, 140, 168n, 170–72, 174, 195n, 196, 197n, 208n, 239n, 264; *Road to Botany Bay,* 8n, 10–12
cartography, 31, 67–117; and ambiguity, 44, 263; and anachronism, 13; and blood, 122; and concealment, 167–73; cultural history of, 8; and death, 189–94; and "deconstruction," 206–8; epistemological dilemma of, 88; and erasure, 169, 174, 243, 245; and its ideal, 88; and "leaving the ground," 171, 208; and obscurity, 173; and omission, 178; as panopticon, 11; and plagiarism, 246; postcolonial legacy of, 171, 10–13, 16, 22–23, 255–65; professionalization in, 90n; as proof, 66; and representation, 6; representational styles in, 43, 90; role in distorting history, 207; role in shaping explorer's vision, 69–71; and secrecy, 61–66; and spatial history, 6; symbolism in, 172–74, 43–44, 217, 69–71, 90–91; as weapon, 206; weight of, 262
cartouches, interpretation of, 77–78
Caruma Mountain, 186–87
Casiquiare Canal, 29, 119, 146, 155
Cassini, Giovanni Domenico, 170n
catastrophe, geological, 36n
Catherwood, Frederick, 145n
Caura River, 30
Cavendish, William Spencer, duke of Devonshire, 125, 150, 152, 154
Cayenne, 31
celestial observations. *See* astronomical observations
Certeau, Michel de, 10n, 12n
chain (surveying tool), 112
Chapman, Walker, 25n
charts (and maps): computational character of, 100. *See also* cartography; maps
Chatsworth, 150–52, *151*
chigoes (parasite), 221
Chimborazo, 145n
Cholula, pyramid of, 145n
chorography, 129
Christmas Cataracts, 165, 178, 189–91, 193–94, plate 15, plate 17 (detail); genius loci of, 191; Reiss's death at, 191
chronometer, 86, 221, 226. *See also* instruments
Church Missionary Society, 244
circumambulation, and possession, 213
Clementi, Cecil, 231n
Clementi, Marie Penelope Rose, 159n
Clenan, Hamlet, 238
Cleveland, Grover, 1, 1n, 64
closure, in surveying, 113
cocoa, 18

coffee, 18
collation of observations, 90, 242
Collins, Wilkie, 137n
Colonial Office, 50–51, 232, 243, 245
color-coded map, at beginning of *Twelve Views,* 128, plate 4
"Committee on the Guayana Expeditions," 167
compass, 174; and boundary, 210; Schmalcalder, 86. *See also* instruments
compulsive boundary crossing, 214–15
computational character, of charts and maps, 100–103. *See also* cartography; maps
Comuti (or Taquiari) Rock, 130, 135–36, 143, 145, 147, *157,* 178–79, 195–96, plate 5
Conan Doyle, Sir Arthur, 159
Conrad, Joseph, 100n
Constable, John, 127n
consummatio, 133, 194, 208, 261
contract, as used by explorers, 107
Cook, Captain James, 39
Cook, T., 6n
Coote, Stephen, 25n, 66n
Corentyne River, 21, 68, 71, 80–81, 84, 106, 168, 169, 178, 187, 193, 204, 209, 243; boundary survey of, 247–53
Cormack, Lesley B., 10n
Cosgrove, Denis, 127
Costa, Emília Viotti da, 3n, 34n
Cotinga River, 231, 233; Schomburgk's route on, 236–40
cotton, 18
Court of Policy, 55
Cowper, William, 141
Cronon, William, 208n
cross-cultural encounter, 3. *See also* Amerindians
Crown commission, won by Schomburgk, 21
Crystal Palace, 152
cultural geography, 4
cultural history of cartography, 8
Culture and Imperialism (Said), 3, 3n
Cumaka, village of, 219, 221–22, 224–25
Curtin, Philip D., 53n, 122, 124n, 126
Curuwuini River, 252
Curzon, earl of Kedleston, 255
Cutari River, 252
Cuyuni River, 34, 42, 68, 71, 209, 211, 224, 226, 228, 233, 239–41, 244; gate on, 228, 241

D'Abate, R., 26n
D'Anville, Jean-Baptiste, 31
D'Orbigny, Alcide, 149
Dalton, H. G., 17n, 28n
Daly, Vere T., 17n, 255, 258n

Daniels, Stephen, 127
"Dardanelles of the Orinocco," 51, 241. *See also* Barima Point
Darnton, Robert, 2n
Daston, Lorraine, 99–100
Day, Archibald, 86, 101n
dead reckoning, 86
De la Rochette, L. S., 171
Delgado, Rafael Sureda, 200n
delimitation, 204, 255
demarcation, 204, 255
Demerara East Coast Railway, 145n
Demerara River, 18, 40, 68, 69, 71
Dening, Greg, 23, 59n, 103n, 109n, 170n, 183n
Derrida, Jacques, 7, 206, 207n
descimientos, 47, 247
Descola, Philippe, 208n
Description de l'Égypte, 110, 123
A Description of British Guiana, Geographical and Statistical (Schomburgk), 124
Despress, Leo A., 208n
Dettelbach, Michael, 49n, 116, 126n, 137n
De Villiers, J. A. J., 1n, 18n, 62n, 255n
Devil's Rock. *See* Ataraipu
Devonshire, duke of. *See* Cavendish, William Spencer
De Weever, Aloysius, 161n
De Wolff, N., 82n
diplomatic controversies. *See* boundaries
discipline, of surveyors, 15, 99–109
Discovery of Guiana (Ralegh), 14, 17, 26, 53. *See also* El Dorado; Ralegh
disorientation, 175–78
disputed territory, 1–2, 25–53, 199–265. *See also* boundaries
Donne, John, 49n
dotted lines (in cartography), 169, 172–74
Downing, William, map of Berbice, 154, *155*
Drayton, Richard, 5n, 50n, 53n, 97n
Duida, Mount, 145–46, 195, plate 9
Duncan, James S., 130n, 207n
Duncan, Nancy G., 130n, 207n
Dutch occupation in Guianas, 33, 199, 212, 219, 228; laws during, 25
Dutch West India Company, 18
Dyson, Anthony, 121

Eclectic and Verulam Philosophical Society of London, 69n
Edney, Matthew H., 87n, 127; *Mapping an Empire,* 3n, 7n, 9–10, 86, 88, 115, 117n, 170n, 198n
Edwards, W., 17n
Eichlin, William Leary, 217, 220, 221, 224

Einberg, Elizabeth, 127n
El Dorado, 4, 13–14, 17, 25–66, 139, 196, 251, 257; and British colonial projects, 28; as "classical soil," 36; and colonial territory, 29; dual meaning of, 60–61; as encouragement to colonization, 28; finding site of, 25–33; imperial function of (introduced), 25; and metalepsis, 39; reaching site of, 33–36; and Satan, 139; as wandering site, 29–46. *See also* anti–El Dorado; Manoa; Parima; Ralegh
El Dorado (Van Heuvel), 40–46
emancipation, 80–81
empires: and visibility, 126; rooted in history and myth, 37
enchantment, 139, 195; colonial function of, 135; Guiana as enchanted land, 135
enclosure (principle of surveying), 171
Enclosure Acts, 131
"enemy function," 260
engagement with place (vs. representation), 169. *See also* methexis
Equinoctal Plants (Humboldt), 93
erasure, and history, 265
Esmeralda: mission town of, 29, 94–95, 97, 98, 119, 144–46, 179–80, 195, 237, plate 9; mythology, 181
Essequibo Falls ("Great Falls of the Essequibo"), 191. *See also* King William IV Cataract
Essequibo River, 2, 18, 21, 30, 34, 68–71, 78–79, 113, 136, 179, 180, 182, 209, 211, 213, 241, 263
ethnography, 31, 181–89
ethnohistorians, 4
Et in Arcadia, as trope, 176
etymology, 31
expedition: as mobile punitive force, 107; as pageantry, 81
exploration, 8, 67–117, 167–253; and aesthetics, 145–48; and collapse, 191; control over, 83; correct scale, 167–68; and death, 189–94, 240, 247–48; and etiquette, 78, 232; and fantasy, 139, 142, 184, 195 (*see also* enchantment); financing, 78–79, 106; and history, 180–83; and inscriptions, 177; and madness, 184; "mere scientific," 47; and missionizing, 182; and nomadism, 244; and photography, 160; and relation with Amerindians, 82, 106–8; and rituals, 190; and spiritual apotheosis, 95; and void, 181. *See also* explorers
explorers, 8, 67–117, 167–253; active engagement with land, 11–12, 88; ambivalence toward Amerindian place mythology, 185; and Amerindians, 181–89, 105–9; authority of, 37–46; bodies of, as bond of territory, 187; as boundary crossers, 214; confidence of, 185; in conflict with colonial administrators, 10; control over, 80; erased, 16, 197; and history, 14; identity of, 177; idiosyncrasies of, 194; perspective of repatriated, 196–97; role of in unseating genius loci, 187; shift in perspective of, 71; and surmounting history (*see also* metalepsis), 37–40; and waterfalls, 189–94

Fabian, Johannes, 41n, 159
Falk, Avner, 214–15
fantasy and exploration, 139, 142, 184, 195. *See also* enchantment
Farage, N., 201n
Farakoto Amerindians, 248n. *See also* Amerindians; Pianoghotto
farina (farine, food), 249
Farley, Rawle, 17n
Fels, John, 7
Fernández, Antonio de Pedro, 200n
Fitch, Walter, 148
fixed points, 14, 91, 114; and boundary, 217; and Humboldt, 93; importance of, 90; and landmarks, 15; not adequate for geographical representation, 15, 111; overall numbers on globe, 91n; techniques for establishing, 91n; as ultimate fruit of geographical exploration, 91. *See also* astronomical observations; landmarks; latitude and longitude; precise coordinates
fixing territory, 264
Flint, Montagu, 159n
flooding, in interior, 41. *See also* isthmus; Parima
Ford, John, 121, 122n, 123n, 153n
Foreign Office, 21, 51, 242, 243
"forest ogres," 188n
Foucault, Michel, 7, 206
foundational moments, ambivalence of, 46
Fox, Desrey, 183n
Fox, Hazel, 1n
Franklin, Sir John, 28n
French Revolution, 19
Frome, E. C., 87, 87n
Fuller, M., 25n

Galison, Peter, 99–100
Gallagher, John, 5
Galton, Francis, 104n, 105, 178
gardens and aquariums, 151
gates (symbolic): and boundary, 228; in interior, 179, 241; on Cuyuni, 228, 241
Gauci, P., 196
Geertz, Clifford, 7n
General Chart of British Guiana (Hilhouse), 69–71, *70,* 76–78, *76* (detail), *77* (detail), 128

genius loci, 140, 142, 183, 187, 188, 252; of Christmas Cataracts, 191
"geo-body," 207–8, 258
geographical construction of colonial territory: and boundaries, 199–253; and geographical exploration, 67–117; and representational practices, 119–98; role of maps and history in, 25–66
geography: establishing geographical authority, 179; geographical exchange, 41; geographical exploration, 1–23, 57–118; geographical explorer, proper view of, 168; and history, 255; and nationhood, 255–65; role of image and narrative in, 111
geology, 137, 205–6; geological catastrophe, 36n; and loss, 258
Georgetown, 20, 126, 186, 204, 224, 238, 242. *See also* New Town; Stabroek
Gibbon, K., 17n
Gilpin, William, 132, 136n, 144
Girtin, Thomas, 138
Gizycki, Rainald von, 5n
glacial lakes in Alps, 37n
Glacken, Clarence J., 127n
Glascott, Lieutenant, 217, 224, 226
glimpses, how mapped, 174
Godlewska, A., 8n, 110n, 128n, 198n
Godman, F. B., 159n
Goetzmann, William, 8, 93
"going native," implications, 31
gold, 41. *See also* El Dorado
Gold Medal, of Royal Geographical Society, 84
Goodall, Edward Alfred, 108n, 109–10, 194n, 217, 220, 233, 239, 247–49, 252; "Goodall's Cataract," 251; sketch of Schomburgk fixing angles on savanna, 233–34, plate 18
graticule, 15n
Gravesande, Larens Storm van 's, 18
Gray, John Edward, 149n
"Great Cataract," 80
Great Exhibition, 83, 152
Great Trigonometric Survey, 9
Greenblatt, Stephen J., 25n, 170n
Greenwich, 96
Grindlay, Robert Melville, 134n
Grove, Richard H., 5n
Guatavita, Lake, 35n
Guayana, 3n
Guiana. *See* British Guiana; Guyana
Guidiwau River, 234, 236, 246
Gullifer (ill-fated traveler in Guayana), 182–83, 185–86. *See also* Smith
Guss, David, 183n, 188n

Gustafsson, Bo, 208n
Gutherie, William, 164
Guyana, 3, 16, 255–65; conventions of spelling, 3n; disputed boundaries, 199–201; Guyanese nationality, 208; National Archives of, 55. *See also* British Guiana
Guyana (Walcott, poem), 1

Ha! River, 251
Hades, 172
Haenke, Tadeás, 148
Haiowa, village of, 241
Hakluyt Society, 26, 37, 52, 61
Hamilton, William, 98
Hancock, John, 25n, 34, 34n, 40n, 69, 88, 92, 136, 179, 212; father of Thomas, 217; on Mahanarva, 55–57; sketch map of, 88–91, *89*
Hancock, Thomas, 217
Hanhart, M. and N., press, 156
Harley, J. Brian, 6, 7, 8n, 170n, 206, 227n
Harlow, V. T., 66n
Harris, C. A., 18n
Harris, Wilson, 133n, 255; on Mahanarva, 56–59
Hartshorne, R., 263n
hatching (in cartography), 169
Hawes, Louis, 138n
Headrick, Daniel, 5n
Heffernan, Michael, 5n, 8n
heliotropes, natural, 175
Hemming, John, 25n
Hepple, Leslie W., 207n
Herbert, Francis, 6n, 123n
Herschel, John F. W., 102n
Higman, B. W., 112n
Hilhouse, William, 34, 67–69, 74, 80, 92, 110, 173, 185: and El Dorado, 34–6; map of Mazaruni, 173, *173*; *General Chart of British Guiana,* 69–71, *70,* 76–78, *76* (detail), *77* (detail), 128
Hilhouse MS, 25n, 29n
Hislop's map, 263
historical maps, as evidence, 2. *See also* cartography
historical "shallowness" of Americas, 143–45
history: and boundary, 225; and geography, 255; as intervention, 22; and territorial myths, 14
Holdich, Thomas, 49n
Hondius, Jodocus, 42; map of Guiana, *43,* 62n
Hooker, Sir Joseph, 158n
Hooker, Sir William, 97, 125, 148, 149, 150n, 151, 152n
Hopkins, A. G., 5n

Horstman, Nicholas, 30–33, 55, 69, 176; as interloper, 31; sketch map of, *32*
hostages, 249
Huggan, Graham, 6n, 206n
Hulme, Peter, 41n
humanitarianism, and boundary, 244
Humboldt, Alexander von, 15, 20, 21, 45n, 92–99, 145n, 173, 179–80, 185, 203, 228; British ambivalence toward, 115n; cultural significance of his instrumentalism, 93; and hagiography, 95; Humboldtianism, conceptual character, 97; Humboldtian science and traverse survey, 92–99; and isthmus theory, 29–33; and literary style, 94; *Personal Narrative,* 29–31, 34; *Views of the Cordilleras,* 98
hummingbirds, 120
Hutchins, Edwin, 100n, 101n, 174
hydrography, 29–31, 49n, 180, 206, 212; and ambiguity, 253; and natural boundaries, 236. *See also* isthmus
hyperbolic boundary, 241

Icarian perspective: attendant anxieties, 189, 198, 257; consequences, 176; as ideal 171–72
ichthyology, 68n, 120
Iconuri mine, 17
Illustrated Atlas and Modern History of the World (Tallis), 163–65, plate 14, *162* (detail)
immutable mobile, 170n
imperial space, 10
Im Thurn, Everard, 40n, 156, 158–59, 184n, 192
"incident at Karinambo," 185–86
informal empire, 5
instrumentalism, 96
instruments, 240, 242, 94; artificial horizon, 86, 91; Bunten's siphon barometer, 99; chain, 112; chronometers, 86, 226; compass, 86, 174, 210; failures of, 96; metonymic character of, 99; protection of, 105; reflecting telescope, 242; sextant, 86, 91; theodolite, 87, 91n
interior: and Amerindian presences, 181–89; and explorers' experience, 169, 181, 189–97; explorers' perspective on, 67–84; as featureless, 100; as sensory deprivation, 175
"interloper," 31, 244
Ireland, Gordon, 199n
Ireng River, 233n. *See also* Mahu River
island, British Guiana as, 25, 49
isoline map, 97
isthmus: Humboldt's theory of, 30; Pirara as, 212. *See also* El Dorado; Humboldt
itabbos, 29, 34, 35

ite palm, 238
itinerant survey, using canons, 102, 221

Jackson, John Brinkerhoff, 127n, 171n
Jackson, Major Basil, 86n, 87n, 104n, 167, 169, 170
Jacob, Christian, 7n
Jagan, Cheddi, 258n
Jami, Catherine, 5n
Jamieson, Alexander, 69n
Jansson, Jan, 42; map of Guiana, *44*
Jardin des Plantes, 149
Jardine, William, 120
Jeffrey, Henry, 201n
Jerrold, Douglas, 152
Joaquim, 92. *See also* São Joaquim
Jones, Stephen B., 204n
jubliee stamps, 160, plate 12
jungle, as prison, 175

Kaieteur Falls, 192, plate 13; as symbol of interior of Guiana, 160–61. *See also* "Old Kai"
Kaieteur valley, 184n
Kain, Roger J. P., 112n
Kaituma River, 224
Kamaiba (cataracts), 142–43, 240
Kanaima (Amerindian concept of revenge murder), 88n, 108, 223, 248
Kantian aesthetics, 132
Karinambo, incident at, 185–86
Karsen, Sonja, 29n
Kew Gardens, 150, 152
Keymis, Lawrence, 17, 26, 37
King, Robert, 217, 221–25; injured, 221
King Fredrick William IV Cataract, 251
Kinglake, Alexander William, 37n
King William IV Cataract 114, *115,* 128
Klonk, Charlotte, 117n, 137n
Knight, Richard Payne, 139n
knowledge, and transgression, 216
Koch-Grünberg, Theodore, 184
Kohl, Johann G., 62n
Kohler, Robert E., 5n
Kreuter, Gretchen, 199n
Kuipaiti, 234, 237
Kuklick, Henrika, 5n
Kundanama Junction, 196, plate 16
Kwayana, Eusi, 258, 259n
Kykoveral, 17

labaria snake, 240
labyrinth, 238
La Condamine, Charles-Marie de, 31

Lake Parima. *See* El Dorado; Manoa; Parima
Lancaster, Alan, 17n
Lander, John, 122
Lander, Richard, 122
landmarks, 4, 111–98, 257; aesthetics of depiction, 131–33, 148, 133–45; and Amerindians, 181–89; and appropriation of colonial territory, 161; and British reading public, 148–65; central to understanding nineteenth-century geography, 112; and chains of representation, 155–61; and curses, 181–89; and death, 181–94; explorer's work in creating, 15; function in geographical exploration and representation, 111–17; and geographical construction, 16; and geology, 137–39; and graves, 190, 191n; as icons 12, 130, 148–61; how given significance, 15; as means of seeing passage, 174–81; and natural theology, 137–38; as nodal points, 129, 161; pictorial conventions and, 131–33, 148; and practice of exploration, 16, 167–98; role in shaping colonial space, 130; role of narrative, astronomy, and image in creating, 113–16; and ruins, 143–45; as superposition of structure and direction, 174; and time, 137; use of picturesque and sublime in constructing, 134–48. *See also* cartography; fixed points; marks; queen's mark
landscape: and enchantment, 184; imagery, 4, 119–98; and spirit presences, 181–89
land surveying: and closure, 113; and land tenure, 112; relation to traverse, 87, 112–14. *See also* cartography
language, linguistic confusion and boundary, 236, 251–52
Lanham, Richard A., 39n
latitude and longitude, 15, 86, 90. *See also* fixed points
Latour, Bruno, 170; notion of immutable mobile, 6
Leal, Antonio Dos Barros, 246
Leed, Eric, 28n, 37n, 41n
"The Legend of Kaieteur" (Seymour), 161n
Lestringant, Frank, 6n
Lewis, M., 5n
Lewis, Malcolm G., 26n
lex talionis, 223
Light, Henry, 50, 202, 229–32, 243, 251, 265; and taxation, 231n
Lindley, John, 97, 149, 150, 154
Livingstone, David (nineteenth-century explorer), 28n, 191n
Livingstone, David N. (historian of geography), 53n, 79n, 116, 124n
Lizar, William, 119

Logan, John, 137n
London Convention of 1814, 20
longitude and latitude. *See* latitude and longitude
Look-out Island, 188
Lord Stanley's Cataract, 251
Lorimer, J., 27
Lyell, Sir Charles, 37n

maada, 188n
macchie, 172–74, 195, 199, 234; *macchiare*, 172
MacInnes, Hamish, 159n
MacKenzie, John, 5n
MacLaren, Ian S., 131, 141n, 146, 176n, 191, 194, 196
MacLeod, Roy, 5n
Macushi Amerindians, 139, 141, 176, 246, 247, 249. *See also* Amerindians
Madness: and interior, 238; in bush, 177, 191
Mahanarva, 40, 88, 114, 212; myth of, 54–60; as shamanistic trickster, 56
Mahu River, 22, 233, 236–38. *See also* Ireng River
Maicarawari, trial of, 222–25, 231–32, 244, 245
Maikang-Yepatori (mountain), 239
Maiongkong Amerindians, 188n. *See also* Amerindians; Yekuana
Makunaima, 138
"Makuniama-aute," 239
Malcolm, Sir Charles, 53
Mallet-Prevost, Severo, 64–5
mannequin, left in interior, 252
Manning, Thomas, 101n
Manoa, 27, 50. *See also* El Dorado; Parima
Maopitian Amerindians, 248, 249. *See also* Amerindians; Mawayena
"Maple White Land," 159
Mapping an Empire (Edney), 3n, 7n, 9, 10, 86, 88, 115, 117n, 170n, 198n
maps: and charts, computational quality, 103; as evidence, 64–66; function in authenticating expedition, 62; as "memorandum," 90–91; and metalepsis, 37–38; and particular views of explorer, 90; reciprocity with passage, 100–103; and territorial possession, 1–53, 67–117, 255–65; as tools for exploration, 69; as weapons, 6–7. *See also* cartography; land surveying
"mar de aquas blancas" (Parima), 61. *See also* El Dorado
Maranika-mama ("mother of salamanders"), 187; cavern, *187*
Markham, Violet R., 152n
marking territory, practice of, 226, 228. *See also* landmarks

marks: and erasure, 242; and possession, 233, 241, 245–46. *See also* landmarks
"Maroons," 18, 19
Marshall, P. J., 19
Martin, C. G. C., 86n
Martin, Montgomery, 164
"masters of all they surveyed," literal significance of, 188
Mattos, João Henrique de, 245
Matupo River, 227
Mawayena Amerindians, 248n. *See also* Amerindians; Maopitian
Mazaruni River, 34, 35, 40, 68, 71, 169, 240; Hilhouse's map of, 173, *173*
McClintock, Anne, 4, 126n
McGreevey, Patrick, 192, 193n
McKean, John, 152n
Mea Kaumen, 37n
measles, 247
Menezes, Mary Noel, 17n, 18n, 34n, 40n, 54, 68n, 109n, 183n, 201n, 217n, 224n, 259, 260n
Mentore, George, 58, 183n
metalepsis, 37–39, 47, 95–96, 178–79; and claims to colonial territory, 46–54; equivocal role in establishing territory, 61–66
methexis, 170–71, 264
metonymy, in context of geographical exploration, 196; metonymic disasters, 192
Michasiw, Kim, 132
micro/macro strategy in geographical representation, 161, 165, 169, 217, 257
Mignolo, Walter, 7n
Miller, Angela L., 127, 170n
Miller, Rory, 67n
mimesis, 170–72, 264
miniatures, and control, 170
mining concessions, 199
mise-en-abyme, 261
mission work, 141, 189n
mollusks, on trees, 27
Monroe Doctrine, 199
Montrose, L., 25n
Morrison, John, 134, 156n, 196
Moulin, Anne Marie, 5n
mountains, significance in establishing boundaries, 226–28
Mount Roraima. *See* Roraima
Mundy, Barbara E., 7n, 129n
Murchison, Sir Roderick Impey, 52, 53, 98
murder, in northwest region, 222–24. *See also* Maicarawari
Murray, John, 122, 123n
myth: ambivalence of, 60–61; as intervention, 60; and margins, 37. *See also* El Dorado

Naipaul, V. S., 25n
National Archives (Guyana), 55
Naturalist's Library, 21, 119–21
nature: and culture, 265; as possession, 16
nautical almanac, 104
navigation: in absence of markers, 100; disciplined protocols, 102; duties of, 101–3; and maintaining positioned consciousness, 103; naval, 15, 100–105; in relation to traverse survey, 100–102; role of astronomy in, 101
negotiating table, symbolic significance of, 264
Netscher, P. M., 17n
neutral ground, 247
New Amsterdam, 247
New Guinea, Fort, 61
New River, 253; "New River Triangle," 253
New Town, 20. *See also* Georgetown; Stabroek
New World, historical shallowness of, 180–81
Niagara Falls, 192
Nicholl, Charles, 25n, 62n
Nicholson, Malcolm, 5n
Nickerie, Dutch settlement of, 81
Niederland, William G., 215
Niger expedition, 122–23
nodes, 129, 161; in geographical construction of Guyana, 15
nomadism, 16, 244
Northumberland, duke of, 150
Noyes, John, 10n, 11n, 22, 130n, 194n, 214

Obeyesekere, Gananath, 25n
objectivity, moral character of, 99
obscurity, and topographic views, 195
occupatio, 63
Ojer, Pablo, 20n, 201n
"Old Kai," 193. *See also* Kaieteur Falls
Ophir, Adi, 130n
Oreala, 188
Orinoco River, 2, 29, 30, 51, 97, 199, 216, 229–30; sources, 180
ornithology, 120, 159n; and exchange with Amerindians, 236
Orpheus, 172
Osborne, Brian S., 130n
Osborne, Michael, 5n
Ottley, H. A., 134n
"Our Village," 240
Ovid, 172

Pacaraima Mountains, 30, 142, 175, 176, 183n, 196, 236, 239, 240
Palmerston, Lord, 51, 230; on question of boundary surveys, 202–4
Pálsson, Gísli, 208n

panopticon, 11
paradoxes, of possession taking, 46
Paragua River, 30
"Paranaghiri" (Europeans), 188, 238
Parima (lake), 27–28, 30, 31, 33, 42, 46, 139; size and shape, 29, 30n; Van Heuvel's map of, *42, 45* (detail). *See also* El Dorado; Manoa
Parima River, 41, 192
Paris Observatory, 170n
Park, Mungo, 28n
Patamona Amerindians, 161n. *See also* Amerindians
paths, representation of, 172. *See also* routes
Paxton, Joseph, 152
peripateia, 171–72, 174, 253, 264
Perkins, H. I., 159n
Personal Narrative (Humboldt), 29–31, 34, 93
Peru, 31
Petersen, Kirsten Holst, 133n
Petitjean, Patrick, 5n
petroglyphs, 188
phantom marks, 245–46, 253, 265. *See also* landmarks
philately, 160
photography, 160
piai (shaman), 223, 239
Pianoghotto Amerindians, 248, 249. *See also* Amerindians; Farakoto
Pickles, John, 7
pictorial conventions: and misprision, 196; subversion of, 194–96. *See also* picturesque; sublime
picturesque, 131–48; and traveler 132, 194–97
pinpricks (cartographic transfer technique), 86
Pirara, 47–49, 108, 109, 138–42, 144, 156, 160, 175, 178, 179, 186, 201, 202, 204–5, 228, 230–31, 238, 240, 242, 243, 245, 246, 247, plate 1; abandoned chapel at, *203;* incident at, 48, 201; isthmus, 212; landing at, 61; military expedition to, 231–33; mythic stature of (as El Dorado site), 48–49; village of, 36, 47, plate 1. *See also* El Dorado
place: and erasure, 167, 265; mythology of, 181–89
planisphère terrestre, 170n
plantation, 77
plantocracy, 225; attitude toward boundary surveys, 203n; attitude toward interior, 74
Poovey, Mary, 124n
Pöppig, Dr. Eduard, 149
Porter, Dennis, 132
Porter, Roy, 208n
possession: and secrecy, 262; symbols of, 18; taking of, 217

postcolonial legacy of colonial boundaries, 16, 22–23, 255–65
postcolonial states, 22
posts (colonial government's upriver establishments), 83; established by Dutch in Guiana, 18
Potaro River, 160, 192
Potter, Jonathan, 164n
Prakash, Gyan, 26n, 58n
Pratt, Mary Louise, 8, 11n, 130n, 170n
precise coordinates, 205. *See also* fixed points
Prescott, J. R. V., 199n, 204n, 253n
presents, 55, 57; to Amerindians, 18
Prester John (Buchan), 59n
Price, Edward T., 112n
Price, Uvedale, 139n
Procter, Adelaide, 153n
Prussia, queen of, 125
Puré-Piapa, 137–38, 143, 147, 196, 202, plate 7
Purumama Falls, 192, 196

"Quartermaster General to the Indians," colonial position, 35. *See also* Hilhouse
queen's mark, 228, 243. *See also* landmarks; marks
Queen Victoria, 83, 126, 150
Quelch, John Joseph, 159n
Quinn, D. B., 25n

Ralegh (also Raleigh), Sir Walter, 13–14, 17, 179, 251, 257; as colonial "father," 25, 27; *Discovery of Guiana,* 17, 26; Ralegh map, 62–66; Raleigh Club, 53; "Raleigh's Peak," 35, *173. See also* El Dorado
Ramos Pérez, Demetrio, 25n
rape: and boundary survey, 222; and territorial possession, 247
Raper, Henry, 101
Rapkin, John, 163, 169, 174; Tallis map, plate 14
Rapu rapids, 232
recalcitrance, of environment in interior, 15. *See also* interior
reciprocity of map and passage, 15, 103–5, 109
reconnoiter, 86
Rehbock, P., 151n
Reingold, Nathan, 5n
Reis, Arthur Cezar Ferreira, 201n
Reiss, Charles, 189–91, 193, 198, plate 11
repatriation, of perspective, 16, 194–97, 257
revisionist histories of imperialism, 8
RGS. *See* Royal Geographical Society
Richards, Thomas, 6n, 170n
Richmond, Virginia, 20
Ridley, Hugh, 147n

Riffenburgh, Beau, 194n
Rio Branco, 30, 73, 146, 234, 263
Rio Negro, 28, 29, 146, 182
Rio Pardo, 149
Ritchie, G. S., 101n
ritual blinding, 188, 188n
rituals of possession, 180, 220
rivers: charts, 84; names, confusion about, 236; sources, 180
Rivière, Peter, 4, 47–48, 50n, 79n, 104n, 201, 202n, 205, 215–16, 230n, 232n, 242n, 244n, 246n
Road to Botany Bay (Carter), 8n, 10–12
Roberts, Emma, 134n
Robinson, Arthur H., 15n
Robinson, Ronald, 5
rock crystal, 31
rocks, and Amerindian mythology, 183
Rodie, Hugh, 150
Rodney, Walter, 260
Rodway, James, 18n, 19n, 20n, 78n
Roe, Peter G., 183n
Rolt, Richard, 229
Roopnaraine, Terence, 59n, 175n, 183n, 260n
Roraima (also Roriema), Mount, 22, 130, 142–44, 147, 155–56, *158*, 158–60, 176, 179, 195–96, 202, 209, 233, 237–42, 244, plate 8; as "forgotten world," 159; as intersection of Brazil, Venezuela, Guyana, 209; as "mother of streams," 209; summited, 158
Rosa, Salvator, 136
Roth, Walter E., 39, 108n, 160n, 161n, 175n
Rothenberg, Marc, 5n
Rotscher, C. G., 123n
Rout, Leslie B, 1n, 260
routes: and boundaries, 199; selecting of, 68–82
Royal Geographical Society (RGS), 14, 34, 42, 49, 78–80, 96, 110, 125, 185; and conflicts with local colonial governments, 14, 80; geographical desiderata of, 170n; Gold Medal of, 84
Rozwadowski, Helen M., 103n
ruins, and landmarks, 143
Rupununi River, 41, 69, 79, 113, 211–12
Rutherford, Anna, 133n
Ryan, Simon, 8, 10n, 127, 131n, 170n

Sack, Robert, 244
Sago, Nurse, 183n
Sahlins, Peter, 204n, 258n
Said, Edward, 6, 170n; *Culture and Imperialism,* 3
Salvin, O., 159n
Sandilands, R. W., 100n, 101n
Sangwan, Satpal, 5n

Santo Domingo, 215
Santo Thomé, 17
São Gabriel, Fort, 146, 195, plate 10
São Joaquim, Fort, 69, 80, 146, 233, 246, 247, 262
Satan, 139; and El Dorado, 36
Sawkins, James, 53n, 156, 160
scalar shifts, in perspective of explorer, 71–75
Schaffer, Simon, 5n, 39n, 99n, 102n, 103
Schama, Simon, 25n, 41n, 127n, 136n, 196n, 208n
Schiller, Friedrich von, 137n
Schmalcalder compass, 86. *See also* instruments
Schomburgk, M. Richard (brother of Robert), 46, 176–77, 184, 216, 223, 231, 236, 238; retelling of Mahanarva story, 54
Schomburgk, Sir Robert H., 4, 14; appointed magistrate, 107; assessment by contemporaries, 98, 109; as autodidact, 98; awarded RGS Gold Medal, 84; and boundary surveys, 16, 204; and control over expedition, 110; *Description of British Guiana, Geographical and Statistical,* 124; deserted by Amerindians, 108; development as explorer, 243–44; discipline, 100; dismissal of Van Heuvel, 45; early surveying, 104; editing of Ralegh, 26–28; "évolution de," 212, 215, 244, 256; expedition proposal map to governor, 71, *73,* 74; expedition proposal map to RGS, 71–73, *72;* first expedition, 79; first survey map, 217, *218, 219* (detail); as humanitarian, 201–2, 247, 222; as imperial megalomaniac, 201; instrumentalism of, 96; as interloper, 244; interpretation of character and motivations, 256; introduced, 4, 20–22; knighted, 21, 99; map of Ralegh's voyage, *38;* as merchant, 20; northwest map, 227, *227* (detail); portrait of, *121;* preparation to explore, 67; publications, 119; punitive measures used by, 108; and quest to complete work of Humboldt, 95; and Ralegh, negotiation of precedent, 38; reaching El Dorado, 28; relation to Humboldt, 92–99; reports edited, 114; "Schomburgk Line," 200; sketch fixing angles on savanna, 233–34, plate 18. See also *Twelve Views*
science and imperialism, 5–6
Scott, Sir Walter, 137
Secord, James, 149n
Seddon, G., 10n
Seeley, Sir John, 216
sermocinatio, 141
sextant, 86, 91. *See also* instruments
Seymour, A. J., "The Legend of Kaiteur," 161n
Shanks, James, 81n

Shapin, Steven, 130n
Shapiro, J., 183n
Sheets-Pyenson, Susan, 119n, 120
Shipibo Amerindians, 188n. *See also* Amerindians
Siam, 215
Siam Mapped (Winichakul), 8, 12–13, 207–8, 244n, 258n
Sierra Accarai, 68, 81n
Sillery, J. L., 258n
Simpson, John, 1n
Siparimer, 175
Siparuni River, 169
Sir Walter Raleigh's Cataract, 251
Skelton, R. A., 8, 62n, 122n
slaves and slavery, 18, 19, 54–55, 59, 74, 76, 80–81, 82n
smallpox, 247
Smith (ill-fated traveler in Guayana), 181–82, 185–86. *See also* Gullifer
Smith, Anthony, 29n
Smith, Bernard, 127
Smith, Coke, 196
Smith, David, 164n
Smith, John (botanist), 159n
Smith, John (missionary), slave rebellion of, 34
Smith, John "Warwick" (watercolorist), 144
Smith, Neil, 8n
smugglers, 217
Smyth and Barrow Falls, 82, 209
Sparke, Matthew, 41n, 207n
Sparks, C., 26n
spatial history, 7–8
spatial perspectives, diverse, 10
Stabroek, 20; Stabroek Market, 260. *See also* Georgetown; New Town
Stafford, Barbara, 116–17, 126
Stafford, Robert, 5n, 49n, 52n, 53, 53n, 160n
stage space, 11
Stanley, Sir Henry Morton, 28n
Stanley, Lord, 21, 125, 126
starvation, 249
Stein, Howard F., 215
Stephen, James, 244, 245n
Stephens, Lloyd, 145n
Stilgoe, John R., 127n
Stoddart, D. R., 121n
sublime, 131, 134–48, 194. *See also* Burke
sugar, 18, 77, 155
superintendents of rivers and creeks, 107
Surinam, 200, 253
survey: and navigation, bodily discipline in, 103; of boundary, 199–253; and cross-cultural exchange, 233–37; by sound, 221; stripped down, 240, 243; symbolic value, 9–10. *See also* instruments; traverse survey; trigonometric survey
Sussex, duke of, 125
Syon, 150, 151

Takutu River, 22, 41, 80, 181, 184, 211–13, 231, 232, 236–38, 241, 242, 245, 253, 262, 265; sketch map of, 234, *235;* survey expedition, 233–34, 239
Tallis, John, 163–65; *Illustrated Atlas and Modern History of the World,* 163, plate 14, *162* (detail)
Tallis map, 198–99
Tamayo, Isbelia Sequera, et al., 200n
Taussig, Michael T., 186n, 260
taxonomic monstrosities, 184, 236
Taylor, E. G. R., 101n
Taylor, Iain C., 3n
Teboco Falls, 35, 36
Teich, Mikuláš, 208n
telescope, reflecting, 242. *See also* instruments
Tenette, village of, 233, 237, 242, 247
terra incognita, 10, 67–84, 178; and Amerindians, 183; meaning of, 14; political character of, 83; and politics, 80; transformed into territory, 3, 189
terra nullius, 194n
terrestrial magnetism, 97
territorial claims: and Amerindians, 59; and cycles of exploration, 26; origin and meanings, 3; and pain, 194n; and picturesque, 131–48
territorial practices, legacy of, 22, 255–65
territory: European notions of, 58; made legitimate, 244; and manners, 262; strategies, 39; strategies vs. tactics, 11
terror, and landmark, 143
theodolite, 87, 91n. *See also* instruments
Thomas, Nicholas, 8n, 59n, 248n
"Three Rivers" region, 18. *See also* British Guiana, history to 1831
Tillotson, John, 134n
time and travel, 40–41
Tooley, R. V., 121
topographic views: concealment in, 195; formal character of, 174; linked to fixed points, 128; role in geographical representation, 127; situated in cartographic field, 161; as "typical" of place, 127
toponyms: and confusion, 234, 236; geographical significance of, 31, 43, 114, 221, 251; and irony, 251
Tortola, 21, 97, 104
"transport" of land, 213
travel writing, 4, 7, 115
traverse survey, 9–10, 84–117; and boundaries,

214–16; discipline in, 99–109; and expansive quality of colonial territory, 216; hardships of, 105; idiosyncrasies, 88; importance of fixed points in, 91; importance of instrumentation, 92; instrumentalism of, 87, 112; military character of, 86–87; naval character of, 87, 99–106; role of land surveying in, 112; significance of, 257; tensions inherent to, 116–17, 214–16; tools of, 86; and travel writing, 115; ubiquity of, 4, 86; weaknesses of, 87
Treasury, London, 243
Treaty of Amiens, 20
Treaty of Versailles, 19
triangulation, 187
trigonometric observations, 240
trigonometric survey, 87; rhetoric vs. practice, 9
Trinidad, 27
Tuan, Yi-Fu, 127n
Tuarutu Mountains, 233, 237
Tufte, Edward, 128
Turner, J. M. W., 127n
Twelve Views in the Interior of Guiana (Schomburgk), 36n, 50, 98, 121–26, 128–30, 133–48, 153–56, 159, 161, 164–65, 168–71, 173, 174, 176, 178–79, 182, 185, 189–98, 199, 202, 240; plate 1, plates 3–11, plates 15–17; key map of, plate 4, plate 11 (detail); posing question of territoriality, 202; preparation, significance, and function of, 133–45; as solution to problem of explorer's view, 167; subscribers, 125
Twyman, Michael, 121n, 123n

uniformitarianism, 137
United States and Canada, boundary, 229
United States involvement in Caribbean, 1–3, 199
Ursato Mountains, 140

Van Heuvel, Jacob Adrian: *El Dorado,* 40–46; map of Parima (from *El Dorado*), *42*
Veness, Rev. W. T., 60
Venezuela, 1, 141, 199, 211; British negotiations with, 51–53, 64–66, 228–30
Versailles, treaty of, 19
Victoria amazonica, 148n. See also *Victoria regia*
Victoria Point, 51, 217, 219, 229, 244, 250; from mouth of Amacura River, *219*
Victoria regia, 82–83, 135n, plate 2; on British Guiana stamps, 160, plate 12; as centerpiece of Great Exhibition, 152; as frontispiece of Schomburgk's *Twelve Views,* 128, 148, 153–55, plate 3; as hydrographic evidence, 155; as landmark, 154; location of discovery, 148, plate 11; on show at Chatsworth, 152, *153*; as symbol of tropical fertility, 154, 148–55
view, linked to overview, 169
Views of the Cordilleras and Monuments of the Indigenous People of America (Humboldt), 98
Virgin Islands, 68
visibility: colonial myths concerning, 159, 188; of colonial possessions, 126
vision and power, in exploration, 188
Vittorio Emanuele, king of Italy, 22
Vogel, F. R., 123n
Von Hagen, Victor Wolfgang, 8n, 145n
vulcanism, 137

Waini River, 216
Waiwai Amerindians, 188n. See also Amerindians
Walcott, Derek, 1
Wallis, Helen, 15n
Walter, E. V., 127n
Walton, W. L., 217n
Wanama junction, 226
Wanderings in South America (Waterton), 33, 33n, 34, 54, 110, 135
Wanumu River, 186n, 233
Wapishana Amerindians, 233–34, 237, 247. See also Amerindians
Waraputa, 178
Wardian cases, 150
Warrau Amerindians, 217. See also Amerindians
Warren, A., 150n, 151n, 159n
Washington, John, 123n
Washington, Rabbi, 260
waterfalls: as landmarks, 189–94; symbolic structure of, 192; as vortexes, 189
"Water-mama" spirit, 182
watershed, 31, 241, 262; as boundary, 211–12; doctrine of, in determining boundaries, 220
Waterton, Charles, 69, 171–72, 179, 184, 262; *Wanderings in South America,* 33, 33n, 34, 54, 110, 135
Watu-Ticaba, settlement of, 108
Watuwau River, 175, 181
Webber, A. R. F., 52n, 81n, 82n, 107n, 231n
Webber, Edward John, 192
Whiggish imperial histories, 7, 216
Whitehead, Neil L., 3n, 18n, 26n, 40n, 41n, 56n, 58n, 182n, 183n
Whitely, H., 159n
Wilford, John Noble, 101n
Williams, Amai, 183n
Williams, Brackette, 208n
Williams, Dennis, 183n
Wilme, B. P., 172n

winged gaze, 170
Winichakul, Thongchai, 8, 12–13, 207–8, 244n, 258n
Wolfe, Patrick, 5n, 6n
Wolseley, Garnet, 103
Wonotobo Falls, 191, 193–94, 196, *197*
Wood, Dennis, 6, 7, 206
Woods, M., 150n, 151n
Woodward, David, 7
Wrede, Stuart, 208n
Wright, John Kirtland, 2n, 25n

Wurucokua, 181
Wyld, James, 69n, 164

Yekuana Amerindians, 188n. *See also* Amerindians; Maiongkong Amerindians
Youd, Thomas, 201, 246
Yucuribi Falls, 178

Zahl, Paul A., 159n
Zuruma River, 176, 239